Cross-Disciplinary Approaches to Action Research and Action Learning

Tome Awshar Mapotse
University of South Africa, South Africa

A volume in the Advances in Knowledge Acquisition, Transfer, and Management (AKATM) Book Series

IGI Global
DISSEMINATOR OF KNOWLEDGE

Published in the United States of America by
IGI Global
Information Science Reference (an imprint of IGI Global)
701 E. Chocolate Avenue
Hershey PA, USA 17033
Tel: 717-533-8845
Fax: 717-533-8661
E-mail: cust@igi-global.com
Web site: http://www.igi-global.com

Library of Congress Cataloging-in-Publication Data

Names: Mapotse, Tome Awshar.
Title: Cross-disciplinary approaches to action research and action learning / Tome Awshar Mapotse, editor.
Description: Hershey PA : Information Science Reference, [2018]
Identifiers: LCCN 2017008483| ISBN 9781522526421 (hardcover) | ISBN 9781522526438 (ebook)
Subjects: LCSH: Action research in education. | Active learning.
Classification: LCC LB1028.24 .C76 2018 | DDC 370.72--dc23
LC record available at https://lccn.loc.gov/2017008483

This book is published in the IGI Global book series Advances in Knowledge Acquisition, Transfer, and Management (AKATM) (ISSN: 2326-7607; eISSN: 2326-7615)

British Cataloguing in Publication Data
A Cataloguing in Publication record for this book is available from the British Library.

Advances in Knowledge Acquisition, Transfer, and Management (AKATM) Book Series

Murray E. Jennex
San Diego State University, USA

ISSN:2326-7607
EISSN:2326-7615

Mission

Organizations and businesses continue to utilize knowledge management practices in order to streamline processes and procedures. The emergence of web technologies has provided new methods of information usage and knowledge sharing.

The **Advances in Knowledge Acquisition, Transfer, and Management (AKATM) Book Series** brings together research on emerging technologies and their effect on information systems as well as the knowledge society.**AKATM** will provide researchers, students, practitioners, and industry leaders with research highlights surrounding the knowledge management discipline, including technology support issues and knowledge representation.

Coverage

- Cognitive Theories
- Cultural Impacts
- Information and Communication Systems
- Knowledge acquisition and transfer processes
- Knowledge management strategy
- Knowledge Sharing
- Organizational learning
- Organizational Memory
- Small and Medium Enterprises
- Virtual communities

IGI Global is currently accepting manuscripts for publication within this series. To submit a proposal for a volume in this series, please contact our Acquisition Editors at Acquisitions@igi-global.com or visit: http://www.igi-global.com/publish/.

The Advances in Knowledge Acquisition, Transfer, and Management (AKATM) Book Series (ISSN 2326-7607) is published by IGI Global, 701 E. Chocolate Avenue, Hershey, PA 17033-1240, USA, www.igi-global.com. This series is composed of titles available for purchase individually; each title is edited to be contextually exclusive from any other title within the series. For pricing and ordering information please visit http://www.igi-global.com/book-series/advances-knowledge-acquisition-transfer-management/37159. Postmaster: Send all address changes to above address.

Titles in this Series

For a list of additional titles in this series, please visit: www.igi-global.com/book-series

Techniques for Coding Imagery and Multimedia Emerging Research and Opportunities
Shalin Hai-Jew (Kansas State University, USA)
Information Science Reference • copyright 2018 • 483pp • H/C (ISBN: 9781522526797) • US $235.00 (our price)

Cross-Disciplinary Approaches to Action Research and Action Learning
Tome Awshar Mapotse (University of South Africa, South Africa)
Information Science Reference • copyright 2018 • 380pp • H/C (ISBN: 9781522526421) • US $215.00 (our price)

Cross-Cultural Analysis of Image-Based Assessments Emerging Research and Opportunities
Lisa Keller (University of Massachusetts Amherst, USA) Robert Keller (Measured Progress, USA) and Michael Nering (Measured Progress, USA)
Information Science Reference • copyright 2018 • 166pp • H/C (ISBN: 9781522526919) • US $130.00 (our price)

Social Media for Knowledge Management Applications in Modern Organizations
Francesca Di Virgilio (University of Molise, Italy)
Business Science Reference • copyright 2018 • 384pp • H/C (ISBN: 9781522528975) • US $215.00 (our price)

Philosophical Perceptions on Logic and Order
Jeremy Horne (International Institute of Informatics and Systemics, USA)
Information Science Reference • copyright 2018 • 402pp • H/C (ISBN: 9781522524434) • US $225.00 (our price)

Handbook of Research on Tacit Knowledge Management for Organizational Success
Dhouha Jaziri-Bouagina (University of Sousse, Tunisia) and George Leal Jamil (Informações em Rede Consultoria e Treinamento Ltda, Brazil)
Information Science Reference • copyright 2017 • 542pp • H/C (ISBN: 9781522523949) • US $275.00 (our price)

Bio-Inspired Computing for Information Retrieval Applications
D.P. Acharjya (School of Computing Science and Engineering, VIT University, India) and Anirban Mitra (Vignan Institute of Technology and Management, India)
Information Science Reference • copyright 2017 • 388pp • H/C (ISBN: 9781522523758) • US $205.00 (our price)

Managing Knowledge Resources and Records in Modern Organizations
Priti Jain (University of Botswana, Botswana) and Nathan Mnjama (University of Botswana, Botswana)
Business Science Reference • copyright 2017 • 280pp • H/C (ISBN: 9781522519652) • US $185.00 (our price)

IGI Global
DISSEMINATOR OF KNOWLEDGE

701 East Chocolate Avenue, Hershey, PA 17033, USA
Tel: 717-533-8845 x100 • Fax: 717-533-8661
E-Mail: cust@igi-global.com • www.igi-global.com

Editorial Advisory Board

List of Reviewers

Monkie Moseki, *University of South Africa, South Africa*
Isabel Payne-van Staden, *North-West University, South Africa*
Karie Peralta, *University of Toledo, USA*
Vishal Pradhan, *Symbiosis International University, India*
Moshe Shachak, *Dead Sea and Arava Science Centre, Israel*
Dan Tlale, *University of South Africa, South Africa*
Marius Venter, *University of Johannesburg, South Africa*

Table of Contents

Detailed Table of Contents

It is the supreme art of an action research practitioner to awaken the joy of tapping into professional development review of Technology Education (TE) knowledge with the teachers as co-researchers. TE has been introduced as a new subject nationally and globally just few decades ago. Teachers and learners are still experiencing hurdles in implementing TE. Most teachers are poorly grounded in pedagogy and content knowledge of Technology Education. This AR study does not blame the limited teacher training in TE, as its intention was to empower such. The study was underpinned by critical theory and following the AR strategies and Technology Education Cascading Theory was envisaged to be incorporated. Focus group (interviews) was used as the method to engage these TE teachers. From the findings of the study, it has been proven that the AR approach study can be used in didactic and pedagogic situation to emancipate unqualified and under qualified Technology teachers.

In today's world, 'global' problems increasingly require global solutions. In order to realize these solutions, innovation processes are 'opening up'. Both of these developments adds a level of complexity to the Action Research process that is traditionally, and perhaps inherently local in nature. However, it also offers opportunities. This chapter explores the case of an Action Research innovation that reached a global scale beyond the initial Action Research process that started it, with the help of Open Innovation strategies. From this case we learn that Open Innovation has a significant potential for sustaining Action Research 'action' beyond its initiation and make them transferable across contexts. At the same time, such 'open' innovations can grow very complex, and therefore so can 'open' Action Research solutions – especially when they concern (free and open source) information systems. The concept of 'Action Research Systems' is introduced as one way in which Action Research can help ground Open Innovation processes in dealing with this complexity.

Action learning coaching is advocated as an aspirational means to develop novices and emergent leaders that builds personal capacity as well as ability to meet new challenges. In this chapter, action learning coaching will be considered in three different ways. The first marks a broad overview of action learning and coaching, along with its integrated practice. The second way presents a practical model of action learning coaching as an innovative method to fortify novices and emergent leaders in the workplace. Finally, insights are drawn from an action learning action research (ALAR) case study and an object lesson to show as to how five novices in human resource development (HRD) in roles such as instructional design and training and development gained heightened capacity and efficacy in performing their roles through action learning coaching.

Design process is a dominant method (backbone) for teaching the Technology Education subject in schools which demands critical and creative thinking skills. Moreover, the design process is research orientated, which could jell well with action research. This chapter explores the synergy between Technology Education and action research, and ultimately contributes a theoretical pedagogical framework which could help in teaching Technology by consciously pulling in action research principles into the design process, thus helping also to approach design process as a cyclical rather than a process.

The South African government has mandated national universities to emancipate and capacitate the teachers of the selected schools in science and technology through an Open Distance Learning (ODL) mode of which the University of South Africa form part of such as Higher Educational Institutions cohort. This chapter reports on the observations, field notes and interviews that were conducted with a group of four teachers and eight learners as a case study at Lovemore Primary School. This Community Engagement chapter focuses on teaching science and technology. In this chapter, researchers argue that focus group participants were able to give enough ideas, thoughts and points to enable the team to design an intervention programme through an Action Research approach. A developmental theory was used to underpin the study while the researcher and participants were engaged in a Community of Practice paradigm. More sessions will be planned to jointly structure the way forward and cultivate the nature of the intervention strategies.

This chapter will highlight how practitioners use trans-disciplinary approaches to conduct Action Research (AR) in the context of e-schools, Community Engagement (CE) and Information and Communication Technologies for Development (ICT4D). The objectives provide details on how: AR is used to align e-education with life-enhancing values and in situations dealing with partnerships towards CE and/or ICT4D matters at local, national and international levels; to promote e-learning and development among those previously excluded from formal education; AR is contributing to transformation and equity in the context of e-schools, CE and ICT4D; AR CE is contributing to improving the quality of life for all; changes and/or growth in the way in which AR is utilized and interpreted can be explored when such questions are raised; AR can stay relevant in light of the ever-increasing speed of change in terms of technological innovations; and examples of ground-breaking AR work can thus be achieved.

The objective of this chapter is to narrate the initiations of sustainability lessons for inclusion within an MBA-IT Business Management Curriculum. This process is elucidated in a narrative style. Initiation of AL began in the year 2013 and improvised in 2014. 'Trial and Error' experimentation in the year 2013, with 'out of the box' lessons, achieved initial success with highly positive student feedback. AL was improvised in 2014 with industry visits, reflecting on corporate social responsibility and collaborations with civil society organizations. For easy comprehension of the thought process of AL iterations, it is recommended to understand Figure 1. Each AL activity is marked with year-wise sub-headings and AL reflections. For the stakeholders interested in modifying in higher education with AL, this institutional choice of sustainability and emphasis on sustainability AL is useful. Unlearning of previous social responsibility action patterns, usage of multiple intelligence, and experiential learning of social issues by IT Business students are the highlights of this chapter.

The objective of this chapter is to narrate the realizations while developing sustainability lessons by Action Learning (AL) for inclusion within an MBA-IT Business Management curriculum. This process is elucidated in a narrative style. The premise was set by ALs of earlier cohorts' trials, errors, improvisations, and reflections. In the year 2015, the Institutional Social Responsibility (ISR) team attempted scaling up 'Learning by Doing.' In 2016, a 'Consolidate and Optimize' approach was adopted. Each activity used a reflexive approach during AL evaluations. Together, AL practitioner and learners developed incrementally maturing sustainability lessons for the subsequent cohort. The chapter aims to share experiential learning with all stakeholders in higher education. An overall shift from apathy to empathy by virtue of AL to implement sustainability lessons is the highlight of this chapter.

Action Learning has been and is regularly implemented within Australian schools as a platform for teacher professional development and professional learning. The following chapter reports on a decade-long association between one Australian government primary school, an Academic Partner and the process of Action Learning. Initially, Action Learning was implemented in 2005 in the form of a small-scale, more traditional Action Learning project; In 2009, Action Learning was then modified and stretched to involve the whole school's teaching, support and executive staff; In 2013, the principles of Action Learning were extended as the school executive and teaching staff began to take ownership over their own learning. This case study demonstrates that Action Learning can be implemented on a school-wide basis for the effective professional development of the whole school's teaching and executive staff. Recommendations are made for similar schools who are considering extending Action Learning across the whole school.

This chapter examines the expansion of volunteer participation in a participatory research project, which explored possibilities to enhance volunteer involvement in an academic camp operated by a non-governmental organization in the Dominican Republic. Although there are numerous studies that identify how organizations should involve volunteers, these studies typically represent an outsider perspective. Therefore, the concerns, ideas, and agency of volunteers—all of which may facilitate improvements in their participation—are often overlooked. For the purpose of this chapter, volunteers are elevated as key actors in the development of the organization. Specifically, community-based philosophy's concept of participation is applied to identify key moments in the project when the broadening of volunteer involvement was important for achieving a greater understanding of the issue of interest. The objective is to advance an appreciation of how projects with volunteer participation, albeit in fluctuating levels, can result in the co-creation and elevation of local knowledge.

A case study of a Soil Science teaching intervention for second year university students is presented here. The purpose was to address post-first year student underpreparedness for higher education. The curriculum was redesigned to iteratively develop students' academic writing skills: with particular regard to environmental report writing, whilst imparting Soil Science content. To this end, students had to integrate practical and theoretical learnings into a soil erosion modelling report. 'Micro action research

Participatory action learning and action research in sustainable local economic development strategies is not a common practice. In this chapter, the author firstly gives an introduction of the concept of sustainable local economic development; the importance of local economic development is outlined, as well as the aims. The author proceeds by discussing participatory action learning and action research as integrated empowering concepts. Lastly, the author describes how participatory action learning and action research were used by the Overstrand Local Economic Development Agency in partnership with the Hawston community to assist them to develop a sustainable neighbourhood development strategy. The steps of the participatory action learning and action research cycle are discussed coupled with the learning and actions taken, that emanated from the reflections during the research.

The aim of this chapter is to provide delegates with a fresh perspective on the littering challenges facing schools and homes in developing countries today, as well as the established skills and awareness intervention strategies necessary to overcome these challenges hence sustaining the teaching of Environmental Education (EE). The extend of Africa's ignorance of EE subject amongst teachers has intensified and reinforced that action research (AR) be regarded as a tool for learner emancipation in the teaching of EE. EE is still seen as a foreign concept by many teachers and a notion that can be down played in school curriculum and home territory. In this chapter EE has find its way into school environment successfully and effectively through engaging learners as co-researchers with action research approach. The study was designed from educational living paradigm and is underpinned by collaborative learning theory. An intervention in the form of action research spiral cycle activities with the learners has managed to make them conscious about littering.

Diversity, inclusion and equity are complex, interdependent and interlinked concepts and embedded in social justice. Diversity summarises numerous differences that exist among people. It means not only people of differing race and culture but also differing languages, religions, values, abilities, socioeconomic status, gender and any other aspect that makes people different from one another. It, therefore, recognises that each person is different even if they belong to the same ethnic group, community or family. Diversity is an evolving concept in inclusive education. While it once referred to children from different ethnic backgrounds, it now encapsulates every kind of individual difference that exists for children, families and professionals. Diversity is the celebration of difference and the acknowledgement that difference is complex and that it exists in some way for every child. Diversity refers to attributes that contribute to each person's uniqueness, such as social class, gender, occupational status, income, sexual orientation, ability, disability, religion, and education.

Foreword

Cross-Disciplinary Approaches to Action Research and Action Learning (cda-ARAL) is a deliberate move driven through a global symmetrical research approach in knowledge revolution to simplify research processes for knowledge creation. Yet relatively few research practitioners are close to respond to the old Marxist question: how can we change the world to become a better place for human beings? This book provides practical knowledge, innovative, creative for progressive transformative and sustainable learning environments towards answering the old global question. Action Research and Action Learning (ARAL) is a constructive, innovative and creative qualitative way to solve problems and to develop professionally while recognising human beings irrespective of their status in society, forming relationships and regular reflections on the forma and the later concepts. ARAL is a process that alternates between action and reflection and learning continuously. ARAL enables the production of research outcomes that are rigorous and valid as well as community relevant, practical, and collaborative – therefore transformative and sustainable. ARAL involves participants as co-researchers throughout the extensive literature review and curriculum mapping process, which involves reviewing research practice across different disciplines. This book provides knowledge on Action Research and Action Learning, which is more-inclusive, power-sensitive, reflective and often focus on people who are marginal and who possess the least power for enacting change, such as the disenfranchised.

In this book, the first four chapters demonstrated that partnerships were established in cross disciplinary approaches by community members who were concerned about what action research is about, how action research can be localised for communal innovation and benefit. How coaching in action research has empowered novice researchers as co-researchers - using their own voices and experiences to jointly draw a plan that will ensure capacity and efficacy in leadership development a way that guaranteed a holistic leadership concluding reflective cycles for technology educational advancement.

In the ensuing four chapters (Chapter 5-8), the authors developed a solid relationship which they demonstrated the manner in which how the co-researchers in their study respected one another's experience, and understood one another's role and responsibility. These chapters provide critical reflections relating to community engagement on teaching science and technology with emphasis on group participation. A further emphasis on community engagements and information communication technology for development reflect heavily on e-Education policy for e-schooling raised in the light of ever-increasing speed of unprecedented technological changes. Chapters 7 and 8 demonstrate sustainability to the readerships on how participants disseminates knowledge based on apathy to empathy relating to IT business students.

Chapter 9 used social constructionism as its epistemological stance because participants jointly perceived the challenges posed by action learning on the whole school's teaching, support and executive staff to take ownership in the school. Chapter 10 brings a practical aspect relating to volunteering as key elements in organisational renewal. Chapter 11 crystallises Action Research and Action Learning in cross disciplinary areas of soil science and relied on research participants' construction of knowledge and solution to their problems they encountered as they saw and experienced them. In Chapter 12, meaningful learning occurs when there are real-world-related authentic tasks and by means of interaction and collaboration between experts and peers. Chapters 13 to 16 show the afford taken by participants with an opportunity to describe their experiences in depth and detail in order to find out what their lives, experiences and interactions mean to them in their own expressions, and to instil collaborative approach in finding solutions to their problems, without imposing pre-existing expectations, in the context of HIV/Aids infection and affection, teaching of environmental education and diversity, equity and inclusivity in education.

This book is long overdue. In *Cross Disciplinary Approach to Action Research and Action Learning* the reader is taken through transformative research methodologies, and research for the instant empowerment of participants; it means enchanting a viewpoint to create emancipatory endeavours. The book can serve as a guide to ARAL practitioners on how to conduct ARAL in one's diverse disciplines across specific field. *Cross-Disciplinary Approaches to Action Research and Action Learning* differentiates the concepts Action Research and Action Learning in a convincing manner.

Molaodi Tshelane
University of the Free State, South Africa

Preface

The book intent to inspire, motivate and encourage Action Learning/Action Research practitioners the world over, whether working in resource rich or more socio-economically challenged contexts, to explain how these practitioners are contributing to the emancipation of conducting an Action Research/Action Learning (AR/AL) in their specific field. The book has created a space for sharing different cases from diverse fields in conducting AR/AL. In this case, the scholars/academics/readers will be empowered to conduct AR related investigations in their field of specialization. As the book editor-in-chief, I am an AR practitioner and my field of specialization, which is Technology Education, and I am grounded in Electrical Engineering, it was interesting to hear how chapter authors from other fields conduct their AR/AL within their communities. It should be noted that there is a thin line between Action Research and Action Learning concepts even though some scholars use these terms interchangeably. The author believes that AR has to take place first between the AR practitioner and co-researchers and once the emancipated co-researchers rollout a plan to empower their other colleagues or peers then AL will take place. This alluded argument is driven by the purpose of AR and the goal of AL.

Edge (2001), who draws on Altrichter, Posch, and Somekh (1993), points out that the purpose of Action Research "is not simply to describe, interpret, analyze and theorize—the stuff of traditional research—but to act in and on a situation to make things better than they were" (p. 4).

The goals of Action Learning can be summarized as follows from Quirke (n.d.):

- To encourage best practices and enhance quality outcomes within the working environment as lead by the emancipated co-researchers;
- To develop effective Professional and Personal Development programs within the working environment by encouraging all stakeholders to get involved in Action Learning projects;
- To encourage all stakeholders involved to define the various issues and problems that they encounter in the workplace and to address them through a process of Action Research lead by the empowered AL practitioners;
- To encourage and support emancipated co-researchers with their future proposal as a way forward to put their colleagues or peers on par with them.

The National Staff Development Committee (1996) outline the difference between AL and AR in the following manner:

Action Learning is the coordinated focus on professional development by the participants themselves. The participants will form groups to research a similar problem, concern or interest. The problems, around which these collaborative learning groups are formed, are taken directly from the participants' daily working lives; in our case, that is the classroom. These action learning groups then look at a variety of ways in which that problem can be solved and try out, actively, each approach. Finally, the conclusions reached or solutions found are presented to a wider audience. The difference to Action Research is noticeable in that there are no rigid criteria for the cycle; and the phases, such as problem identification and investigation, are not necessarily required to follow strict research principles. (p. 17)

AR ATTACHED FOUNDER

Kurt Lewin (1958) then a professor at MIT first coined the term "action research" in about 1944. In his 1946 paper "Action Research and Minority Problems," he described action research as "a comparative research on the conditions and effects of various forms of social action and research leading to social action" that uses "a spiral of steps, each of which is composed of a circle of planning, action, and fact-finding about the result of the action."

If one idea can be said to summarize OD's underlying philosophy, it would be action research as it was conceptualized by Kurt Lewin and later elaborated and expanded on by other behavioral scientists (Kemmis & McTaggart, 1988; Reason & Bradbury, 2001; Altrichter, Posch, & Somekh, 1993; Carr & Kemmis, 1986).

Concerned with social change and, more particularly, with effective, permanent social change, Lewin believed that the motivation to change was strongly related to action: If people are active in decisions affecting them, they are more likely to adopt new ways.

Action Research (AR) can be defined as a form of collective self-reflective enquiry undertaken by participants in social situation in order to improve rationality and justice of their own social or educational practices, as well as their understanding of these practices and the situation in which these practices are carried out. Groups of participants can be teachers, students, principals, parents and other community members – any group with a shared concern (Kemmis & McTaggart, 1988).

Action research is an interactive inquiry process that balances problem-solving actions implemented in a collaborative context with data-driven collaborative analysis or research to understand underlying causes enabling future predictions about personal and organizational change (Reason & Bradbury, 2001). Carr and Kemmis have another view of what AR is all about.

Carr and Kemmis (1986) describe action research as being about:

- The improvement of practice;
- The improvement of the understanding of practice;
- The improvement of the situation in which the practice takes place.

The notion of improvement can be problematic when viewed from the outside. One person's improvement can be another person's deterioration. It depends on the beliefs and values underpinning the individual's perspective. Paradoxically, however, this uncertainty is perhaps the one truth of professional practice.

AL ATTACHED FOUNDER

The author wishes to acknowledge the origins of Action Learning as being in the work of Professor Reg Revans in the early 1980s. The description of Action Learning below has been intentional as pertaining to Revans' original description. Further referencing has been used later on in describing how Action Learning has flourished since its origins.

Action Learning is a process of learning that enables organizations to adapt to change (Revans, 2011). The purpose of Action Learning is learning: a type of learning, which enables organizations to adapt during times of change. In *The ABC of Action Learning,* Revans (2011) highlights how important it is for organizations to be able to adapt in times of rapid change. An adaptation, he emphasizes, which can only occur through learning: a type of learning, he describes as:

Being able to do tomorrow that which might have been unnecessary today, or to be able to do today what was unnecessary last week. (p. 1)

Ewing et al. (2010) states that, "An Action Learning approach to professional development is based on the belief that professionals working together have the ability to ask meaningful questions and searching questions about their own practice and formulate answers to those questions" (p. 5).

Action Learning (AL) is a highly structured, facilitated process, which unlocks the expertise and knowledge in a person, group or organization to explore and then determine actions to solve problems and move forward to success.

Action Learning programs are by their very nature flexible and adjustable (Brook, Pedler, & Burgoyne, 2012; Revans, 2011). This is because the design of the program responds to the individual needs of organizations.

THE CHALLENGES

Mapotse (2015) cited the aim of Action Learning and Action Research is to "generate collaborative learning, research and action" (p. 1) to attain local and global change in pursuit of a fairer world – fairer meaning more socially just and also more attractive in terms of an improved quality of life for all. The complexity of life in the twenty-first century makes such learning essential for all, in both well-resourced and socio-economically challenged contexts. Traditional approaches to research tend to determine what people need, and to address those needs by recommending the allocation of resources – more money, more infrastructure, more services, more programs. However, due to the top-down nature of the recommendations, those who most need the resources often struggle to access and use them to their benefit, increasing the gap between rich and poor, the advantaged and the disadvantaged. Unless as practitioners we aim to develop the capacity of people to identify and use resources, the AL/AR participants are unlikely to profit from those resources. Action Learning and Action Research purports to increase the ability of people to learn how to adapt to rapidly changing social, economic and political imperatives. However, are the AR/AL practitioners really living up to these claims?

Challenges in the book writing were about chapter review and some authors not fully citing their sources adequately. The delays in review ended up extending the publication target date. Around 35 chapter proposals were sent over and 10 were rejected on the basis that they were not speaking for the theme of the book. From the 25 that were left, only 16 were approved to be published in this book, as they were scholarly written, that is minimal similarity index was observed. The challenge with the remaining nine chapters were myriad ranging from reviewers' chapter rejection to IGI Global system chapter checking against plagiarism and resulting in chapter withdrawal. Only a handful of contributors, due to some other personal challenges, could not deliver their updated reviewed chapter on time for inclusion in the book.

WHO WILL BENEFIT FROM THIS BOOK

The book might be used by academics, researchers, scholars, managers, teachers, engineers, artisans, health workers, social workers, government officials, IKS practitioners, psychologists, etc. at universities, schools, firms, companies, industries, different governmental departments, and other institutes of learning intending to conduct AR/AL for organizational development, co-researchers empowerment/ emancipation or personal growth.

Different governmental departments may include but not limited to, for example, Agriculture, Forestry and Fisheries, Art and Culture, Communications, Cooperative Governance, Correctional Services, Defense, Education, Energy, Environmental Affairs, Government Communications and Information System, Health, Home Affairs, Human Settlement, Independent Complaint Directorate, International Relations and Cooperation (Foreign Affairs), Justice and Constitutional Development, Labour, Mineral Resources, National Intelligence Agency, National Treasury, Police, Police Service, Public Administration Leadership and Management Academy, Public Enterprises, Public Service and Administration, Public Service Commission, Public Works, Revenue Service, Rural Development and Land Reform, Science and Technology, Secret Service, Social Development, Sport and Recreation, Statistics, The Presidency, Tourism, Trade and Industry, Traditional Affairs, Transport, Water Affairs, and Women, Children, and People with Disabilities can be identified as potential contexts in which this book will be used to keep up with the discourse over the latest trend in their specialized fields.

THE BOOK IS COVERING THE FOLLOWING TOPICS

The book has five areas of specialization that are covered within Action Research/Action Learning. These areas include education, organizational development in a workplace, community engagement/ outreach, and conversation of environment. The book has 16 chapters. The book sections are grouped into the following broad areas of interest:

- **Education:** This strand covers Action Learning/Action Research work being done on local/global scales in both formal and informal education contexts. How has AL/AR been used to align education with life-enhancing values? How has our research promoted learning and development among those previously excluded from formal education? How has AL/AR contributed to transformation, equity and social justice?

- **Organizational Development in a Workplace:** This strand covers Action Learning/Action Research in organizational design, organizational development, organizational change, personnel development, systems thinking, soft systems methodology, evaluation, and similar areas of interest - how has AL/AR been used, and to what effect; what could be done differently to gain greater acceptance, use and outcomes; and where to from here?
- **Community Engagement/Outreach:** Action Learning/Action Research has been used in many situations dealing with partnerships, health, and welfare, advocacy, and gender and community development matters, but are they being acknowledged and lauded at local, national and international levels? Are there examples of this work that have been groundbreaking, and should be or have been exposed to the public? How is our AL/AR community engagement/outreach contributing to improving the quality of life for all?
- **Innovations:** This strand considers changes or growth in the way in which AL/AR is utilized and interpreted, technological innovations for AL/AR, and use of AL/AR in a global economy, and questions how AL/AR will stay relevant in the ever-increasing speed of change and need for immediate knowledge.
- **Conserving the Environment: Sustainability:** Many local and community Action Learning/Action Research projects have been about or involved the environment and local resources, but what has been the impact of AL/AR on the debate and actions on conserving the environment and resources of nations and the world; what should AL/AR practitioners be doing to ensure the environment for our grandchildren's grandchildren?

THE BOOK IS ORGANIZED AS FOLLOWS

The book is organized into 16 chapters. The book sections are grouped into the five broad areas of interest. The chapter will be outlined based on those areas of interest. This is the main objective of the book, to prove that interest practitioners in all occupations can conduct Action Research and Action Learning across diverse fields. A brief description of each of the chapter within an area of specialization is outlined as follows:

Education Strand Has Four Chapters

Chapter 1 identifies the existing challenges faces by the teachers of the new developed subject called Technology Education (TE). The chapter sets the scene for discussions the how of emancipating these TE teachers to teach this subject with confidence and every chance of successes. In particular, the chapter identifies the strategies to use together with community engagement activities to take this un-or under qualified TE teachers through emancipation process. Technology Education Cascading Theory underpinned the study that employed focus group (interview) on sampled teachers.

Chapter 4 establishes the need to explore the synergy between Technology Education and action research, and ultimately contributes a theoretical pedagogical framework that could help in teaching Technology by consciously pulling in action research principles into the design process.

Chapter 9 examines how an Action Learning within a school can start on a small-scale intervention until there is a buy-in of all stakeholders within the school organization. The overall aim of the chapter is to demonstrate that Action Learning can be implemented on a school-wide basis for the effective professional development of the whole school's teaching and executive staff.

Chapter 16 paints a clear picture of diversity as attributes that contribute to each person's uniqueness, such as social class, gender, occupational status, income, sexual orientation, ability, disability, religion and education. The author recognises that each person is different even if they belong to the same ethnic group, community or family. With all differences, the author encourages inclusivity.

Organizational Development in a Workplace Has Four Chapters

Chapter 3 advocated Action Learning coaching as an aspirational means to develop novices and emergent leaders that builds personal capacity as well as ability to meet new challenges. The author alluded to the fact that Action Learning coaching can be considered in three different ways to emancipate novices in their roles such as instructional design, training and development with the results of heightened capacity and efficacy in performing their roles. The author uses Human Resource Development novices as a case in this chapter.

Chapter 10 examines the expansion of volunteer participation in a participatory research project, which explored possibilities to enhance volunteer involvement in an academic camp operated by a non-governmental organization. The authors' objective in this chapter is to advance an appreciation of how projects with volunteer participation, albeit in fluctuating levels, can result in the co-creation and elevation of local knowledge.

Chapter 11 addresses a case study of Soil Science teaching intervention issues of post-first year student who are under preparedness for higher education curriculum. Based on recommendations, the authors suggest that a deep approach to learning requires a less dense curriculum; tutors need to be more formally capacitated; and administrative hurdles should not be underestimated whilst imparting Soil Science content. The authors applied 'micro action research dialogues' to drive the development and delivery of this scaffolded curriculum.

Chapter 14 is where the authors discuss participatory action learning and action research as integrated empowering concepts in sustainable local economic development strategies. The author describes how participatory action learning and action research were used by the Overstrand Local Economic Development Agency in partnership with the Hawston community to assist in the development of a sustainable neighborhood strategy.

Community Engagement/Outreach Has Three Chapters

Chapter 5 reports on the observations, field notes and interviews that were conducted with a group of four teachers and eight learners as a case study. In this chapter, the authors argue that focus group participants were able to give enough ideas, thoughts and points to enable the team to design a Community Engagement intervention programme through an Action Research approach to address Science and Technology subjects' challenges.

Chapter 6 provides details on how Action Research (AR) approaches were incorporated to align e-education with life-enhancing values and in situations dealing with partnerships towards Community Engagement (CE). The author highlight how practitioners used trans-disciplinary approaches to conduct AR in the context of e-schools, Community CE and Information & Communication Technologies for Development.

Chapter 13 offers several examples of how researchers and practitioners are using participatory action research to select effective practices and support strategies to generate knowledge in order to initiate change. The author bring this to the readership that although teachers generally have the will and inclination to help children, they often feel they lack the necessary skills to effectively deal with challenges such as HIV and Aids in their classrooms.

Innovations Has Only One Chapter

Chapter 2 explores the case of an Action Research innovation that reached a global scale beyond the initial Action Research process that started it, with the help of Open Innovation strategies. From this case, the authors learn that Open Innovation has a significant potential for sustaining Action Research 'action' beyond its initiation and make them transferable across contexts.

Conserving the Environment: Sustainability Has Four Chapters

Chapter 7 reflects how Action Learning (AL) was phased in for two years (2013 and 2014) as a means to replicate on corporate social responsibility and collaborations with civil society. The first year was marked by 'trial and error' experimentation and 'out of the box' learning was evident. AL was found to be a useful approach that can be used to bring about changes and sustainability in higher education within an MBA-IT Business Management Curriculum. This chapter just report about the initial phase of the project with the students.

Chapter 8 reports the continuation of Action Learning (AL) activities that took place in two consecutive years (2014 and 2015) as a follow-up of Chapter 7 narratives. During this period, the learner cohort together with the A/L practitioner developed incrementally maturing sustainability lessons with evident that lead to a transition of the Institutional Social Responsibility team attempted scaling up 'Learning by Doing' to adopt Consolidate and Optimize' approach.

Chapter 12 give an account of a team of scholars from Negev Highlands of Israel who uses some approaches informed by both transdisciplinary research and participatory action research to identify trajectories of human influence on the ecosystem. The approach identifies several cascades of effects for allowing diverse stakeholders to build capacity for sustainable management.

Chapter 15 outlines how teachers can intergrade environmental education in teaching so as to make learners' conscious of their cleanliness and littering in home and school environment. Action Research can be employed as a vehicle to emancipate learners who serve as co-researchers to develop enmity against littering if the AR strategies can be designed from the educational living paradigm and use collaborative learning theory as the lenses to guide the practitioner.

The above five areas of specialization within this book hang on two factors outline below by Quirke as central to AR in all studies conducted.

CENTRAL TO THE CONCEPT OF ACTION RESEARCH ARE TWO KEY FACTORS

Firstly, the practitioners themselves undertake the research rather than a third party. This was one of the bases of criticism of Action Research in the 1960s when it was suggested that any conclusions that emerged from such research would be invalid because of the subjectivity involved and the necessarily small scale. This was, however, to miss the point: the idea is not to emerge from the research with some scientifically valid generalization, but to benefit the participants directly. For these participants, Action Research provides both a way of knowing and a means of empowerment.

Secondly, Action Research is essentially a problem-solving strategy. It encourages practitioners to question what they are doing, to identify their concerns, to do some research and then to take action. This is not research for the sake of research but research in search of improvements.

Summary of Action Research by Waters-Adams (2006)

- Action research is a practical way for individuals to explore the nature of their practice and to improve it.
- Action research encourages practitioners to become knowledge-makers, rather than merely knowledge-users.
- Action research uses action as a means of research; planned change is implemented, monitored and analyzed.
- Action research proceeds in an action-reflection cycle or spiral.
- The process can be messy; as research proceeds, wider links are likely to be identified.
- Action research is carried out by individuals, but these individuals may work collaboratively.
- Action researchers may use a variety of research methods, both qualitative and quantitative.
- Action researchers must ensure triangulation in their methods.

Tomé Awshar Mapotse
University of South Africa, South Africa

REFERENCES

Altrichter, H., Posch, P., & Somekh, B. (1993). *Teachers investigate their work: An introduction to the methods of action research*. London: Routledge.

Brook, C., Pedler, M., & Burgoyne, J. (2012). Some debates and challenges in the literature on action learning: The state of the art since Revans. *Human Resource Development International*, *15*(3), 269–282. doi:10.1080/13678868.2012.687926

Edge, J. (Ed.). (2001). *Action research: Case studies in TESOL practice*. Alexandria, VA: TESOL Inc.

Ewing, R., Groundwater-Smith, S., Mockler, N., Loughland, T., Simpson, A., Smith, D., & Books, D. et al. (2010). *Meta-analysis of quality teaching action learning project*. Sydney, Australia: University of Sydney.

Kemmis, S., & McTaggart, R. (1988). *The action research planner* (3rd ed.). Victoria, Australia: Deakin University.

Lewin, K. (1958). *Group decision and social change*. New York: Holt, Rinehart and Winston.

Mapotse, T. A. (2015). *From etic action research practitioner to emic action learning facilitator: A community engagement case*. Paper Presented at ALARA World Congress: 9th Action Learning Action Research and 13th Participatory Action Research, Centurion, South Africa.

National Staff Development Committee. (1996). *Action learning in vocational education and training*. Chadstone, Australia: NSDC Press.

Quirke, P. (n.d.). *Action learning & research*. Retrieved from http://philseflsupport.com/actionlearning.htm

Reason, P., & Bradbury, H. (Eds.). (2001). *The SAGE handbook of action research: Participative inquiry and practice* (1st ed.). London: SAGE.

Revans, R. W. (2011). *ABC of action learning*. Farnham, UK: Gower.

Sagor, R. (2000). *Guiding school improvement with action research*. Alexandria, VA: Association for Supervision and Curriculum Development.

Waters-Adams, S. (2006). *Action research in education*. Faculty of Education, University of Plymouth.

Chapter 1
Technology Education Teachers' Professional Development Review Through Community Engagement:
An Action Research Inquest

Tome' Awshar Mapotse
University of South Africa, South Africa

ABSTRACT

It is the supreme art of an action research practitioner to awaken the joy of tapping into professional development review of Technology Education (TE) knowledge with the teachers as co-researchers. TE has been introduced as a new subject nationally and globally just few decades ago. Teachers and learners are still experiencing hurdles in implementing TE. Most teachers are poorly grounded in pedagogy and content knowledge of Technology Education. This AR study does not blame the limited teacher training in TE, as its intention was to empower such. The study was underpinned by critical theory and following the AR strategies and Technology Education Cascading Theory was envisaged to be incorporated. Focus group (interviews) was used as the method to engage these TE teachers. From the findings of the study, it has been proven that the AR approach study can be used in didactic and pedagogic situation to emancipate unqualified and under qualified Technology teachers.

INTRODUCTION

Development in Technology Education and the acquisition of new knowledge stimulate questions about the nature of human condition, the nature of a good life and how humans ought to live (Stevens, 2017). In the same breath, Hannaway (2016:122) stressed that, "Technology in this age is ubiquitous and is changing the way that individuals live, work and play". The conundrum within Technology Education (TE) is caused by the fact that it was introduced just two decades ago in school curriculum both nation-

DOI: 10.4018/978-1-5225-2642-1.ch001

ally and internationally. TE is one of the subject where teachers have to be grounded with both content and theory because of its hands-on and mind-on nature of the subject. As teaching cannot be done in a vacuum, some content and theory are thus necessary to act as a medium through which the teaching of TE can be practiced (Pudi, 2007). Most of the TE teachers are unqualified or under qualified and they have been compelled by their seniors to accept the responsibility of teaching this Technology subject (Mapotse, 2015a). In this case, it is incumbent of TE teachers to be developed professionally and that has to be continually and continuously. As an action research (AR) practitioner one has embarked on a community engagement (CE) project as a way to intervene with teachers development at Bathabile Secondary School first (school pseudonym was used to conceal its true identity) so as to bring the change within learners' learning and teachers' teaching of Technology subject. Learners therefore need the 21st Century skills from a competent teacher to be able to comprehend development and rapid changes in this technological era.

In the Integrated Strategic Planning Framework for Teacher Education and Development (referred to as Action Plan 2014 from henceforth) in South Africa – 2011 to 2025, the stakeholders (2011: 1) report that:

The challenges facing Teacher Education and Development (TED) in South Africa are considerable. They include a lack of access to quality TED opportunities for prospective and practicing teachers; mismatch between the provision of and demand for teachers of particular types; the failure of the system to achieve dramatic improvement in the quality of teaching and learning in schools; a fragmented and uncoordinated approach to TED; etc.

The primary outcome of Action Plan 2014 is to improve the quality of teacher education and development in order to improve the quality of teachers and teaching. The plan places teachers firmly at the centre of all efforts to improve teachers' development, and enables teachers to take substantial responsibility for their own development, hence when the TE teachers from the three schools were invite to be co-researchers in action research they accept the opportunity with open arms. The invite was for emancipation purposes and to let their TE teaching light shine so as to fulfil the Plan outcome. One of the great African Head of State, Dr Nelson Mandela, once said, 'and as we let our own light shine, we unconsciously give other people permission to do the same', (Mapotse, 2017b). In undertaking and exposing TE teachers to CE project, one was trying to let their own light shine so that the Plan can have smooth and successful implementation. So many researches were undertaken and few focuses on action research within community engagement.

Technology Education scholars have for some years disseminated their research findings around TE teachers' professional development. These scholars in the Technology field did target diversity of features of TE, for example Gunther and Chatoney (2016), Hallström, Hultén and Lövheim (2014), Love (2015), Potgieter (2004), Pudi (2005), Pule (2017), Stevens (2006), Williams and Gumbo (2011), Yager (2013). These TE scholars mentioned above hail from both African community of scholars and TE researchers from the world over. The above listed scholars have used similar instruments and/or common methods in their data collection and so far little has been done in undertaking action research (AR) approach within CE stages to empower TE teachers. With this study, as an AR practitioner attempt to fill that gap by sharing on how action research can contribute to rapid changes of Technology Education landscape in South Africa through involving teachers in a CE project. The AR practitioner will be using a newly developed TE cascading theory to guide the study and also underpinning this study through critical theory as well.

STUDY'S THEORETICAL FRAMEWORK

Critical Theory Used to Underpin the Study

In his 1937 essay, *Traditional and Critical Theory,* Max Horkheimer (Buchanan, 2010) of the Frankfurt School of Sociology initially unpacked critical theory (CT). Max alluded to the fact that CT is a social theory. CT was deemed suitable to flagship this AR study in tackling the issues of socially emancipating TE teachers with the challenges they face in delivering their lessons. Technology subject is not only a minds-on enterprise but also a hands-on therefore critical theory will be imperative to underpin the study. This fact is supported by (Tooley, 2000), who stressed that an essential dialectical phenomena should be considered in Technology Education, referring to the indivisible relationship between theory and practice. Critical theory emphasises personal and social transformation (Giroux, 2003). Critical theory addresses real world challenges hence (Gibson, 1983) concluded that it has an emancipatory intent in the sense that in this study the focus is on teachers' difficulties in presenting TE core themes as prescribed by the National Ministry of Education. Technology Education is fundamentally a hands-on enterprise, as alluded earlier on, therefore teachers are expected to have Technology content knowledge, pedagogy and didactic of this subject before they can offer TE subject. Critical theory can modify things as recalled by Allison (2008), because of its emphasis on connection between theory and practice in teacher education.

It should be noted that hands-on and minds-on in Technology Education denote to learning by following five design/technological process/stages. Those stages involve learning through experiences, that is, through practical engagement in IDMEC which stand for 1) Investigating; 2) Designing; 3) Making; 4) Evaluating; and 5) Communicating ideas and plans (Department of Education [DoE], 2003). This aligns well with my understanding that approaching Technology theoretically is unfathomable. It is through critical theory that teachers can serve as agents of change (Fullan, 1993) and transformative intellectuals (Giroux, 2003). The AR practitioner share this sentiments, "in most qualitative studies, theory comes at the beginning and provides a lens that shapes what is looked at and the questions asked especially in a transformative research" (Creswell, 2014). Critical theory has been earmarked as the ontological stance in which TE teachers perceive their challenges in TE classroom practice. Critical theory helped the TE teachers who went through inferior separated training during apartheid to now realise that teaching TE need one to contextualise it for learners to comprehend it, whether the school is placed in rural, urban or semi-urban setting.

The Newly Developed Technology Education Cascading Theory: 'Each One, Teach One'

The chapter author who is an AR practitioner developed a theory being used to guide the study which is labelled Technology Education Cascading Theory (Mapotse, 2017a). The theory declares that after co-researchers are emancipated they can sustain the project with or without the intervention of the AR practitioner. The theory is ushered forth in phases; initial phase; developmental phase and sustainability phase. During the project inception, the AR practitioner first engage TE teachers from Bathabile Secondary School with an AR emancipation strategies of observation, planning, action and reflection per term covering all the TE core themes within an academic year. The strategies were applies in circular and spiral cycles during AR contact sessions. The researcher serve as an AR practitioner and TE teachers serve as participants or co-researchers. As the researcher adopt a school per year for emancipation purposes, the TE teachers from the adopted school serves as participants, the teachers who were emancipated in the previous year are elevated to be the practitioners and the researcher become the facilitator. In short, each year the AR team change their roles as a new cohort of teachers from another school is added into the emancipation programme. The three secondary schools used in this study are as follows, Bathabile the 1st; Lamola the 2[nd]and Kholofelo the 3rd. The pseudo names are used to conceal the true identity of the schools. Each school was added chronologically every year.

CONCEPTUALIZATION

To avoid miscommunication, definition and explanation of terms are necessary, and to create a common frame of understanding, which increases the effectiveness of communication (Pudi, 2007). The following three concepts will be defined or explained in the background of this study: Action Research, Action Learning, Technology Education and Community Engagement.

Action Research

The book theme is 'Cross-Disciplinary Approaches to Action Research' and the AR can be defined based on the specific field of specialization. This chapter unpack the TE teachers' professional development through community engagement using an AR approach. In this case the study is anchored by De Villiers (2005: 146) definition of AR which seems to be more relevant to the chapter's title as it emphasise that, "action research is a participative, practitioner-researcher approach which lends itself to the domain of educational research, where an evolving intervention or product is investigated over several cycles". A central idea of AR is an inclusion of practitioners from the real world (Cohen et al 2007), in this case, TE teachers. The classroom can be considered as an ideal AR milieu to be used by a teacher in attempting to find the best ways of teaching their learners (Pilkington & Gelderblom, 2010).

AR as the name suggest it is an approach which has dual aims of action and research as succinctly pointed out by Dick (2010): action to bring about change in some community or organisation or programme and research to increase understanding on the researcher or the clients or both. The study therefore intends to bring about change in TE didactics and increase understanding in Technology implementation. AR was used as a resource for deep-seated interpositions and it was implemented in the three participating schools. The AR cycles and spirals activities of observing, planning, acting and reflecting manage to professionally develop TE teachers from low self-esteem of teaching TE to a remarkable increased TE didactic and pedagogic knowledge levels. This weekly AR activities were conducted bi-monthly during CE emancipation sessions with the TE teachers. After unpacking AR, the next section will discuss the concept action learning, more details of this concept are avers at the book preface.

Action Learning

Most scholars' uses action research and action learning (AL) interchangeably, but there is a thin line between these two concepts. First the participants can be taken through AR then later the empowered participants can serve as practitioner and engage their peers or colleagues through AR. Once the emancipated participants are able to engage and independently take their peers or colleagues through cyclic and spiral activities of AR then AL is up-to-the-minute. Few definitions of Action Learning are extracted from Mapotse (2017b) and are outlined as follows:

- *Action Learning is a dynamic and deeply reflective way of problem-solving and learning, from and with your peers. Claire Malcolm, Director, New Writing North;*
- *Action Learning is a form of peer learning with a group of colleagues who work on the real live challenges they face as leaders. It is a rigorous and disciplined process that helps people develop the skill of asking insightful questions to unlock thinking.* Ruth Cook, *Director, Action Learning Associates.*

From the above conceptualisation of AL it is clear that there was AR processes first and then followed by AL to their peers or colleagues. An action learning approach to professional development is based on the belief that professionals working together have the ability to ask meaningful questions and searching questions about their own practice and formulate answers to those questions (Revans, 1983).

Technology Education

Western knowledge drives Technology Education in South Africa since it was influenced by curriculum developments in both Australia and New Zealand, which are predominantly Eurocentric (Gumbo, 2017). The South African curriculum was framed from the Constitutions and therefore recognises inclusive principles. Teachers need to understand the Constitution and should have the ability to promote values, such as human dignity, non-racialism, non-sexism, democracy and the rule of law in the classroom (Action Plan 2014).

As a scholar in the field of TE, who has written and read so intensively about this field it should now be brought to the attention of the readership that different countries have their special way of naming this TE subject. What follows is an extract from Mapotse (2017a), about conceptualisation of TE: 'Technology Education is so-called by diverse names in various countries, for example in the USA or New Zealand or Australia or Finland and/or France it is labelled as Technology Education (TE) and in Malawi and/or Bangladesh it is dubbed Science and Technology (S & T), whereas in UK and/or Botswana it is known as Design and Technology (D & T). In Sweden and/or Israel, it is termed Technology, but in South African (SA) context, it has a special name per phase. In the Intermediate Phase, it is entitled Science and Technology; in the senior phase, the subject is branded as Technology and, in Further Education and Training band, it is identified as Technology Field or Education (Mapotse, 2015a, 2017a). It is now over two decades since TE was introduced into the SA curriculum as stated earlier on. There are other countries who gave the subject, Technology, a unique name. For example, in Nigeria, it is called Technical Education, and in Kenya, it is pronounced as Art and Technology, whereas in Tanzania it is named Computer Integrated Education (Mapotse, 2017a)'. The AR practitioner still believes that in some countries in the other part of this world TE will be identified differently as well.

One of the main reason why Technology as a subject was developed, it intent to teach learners the how of technologically solving societal or community problems associated with human needs, wants and desires? If there were no problem to solve, then Technology would not have been developed and/ or needed. The Department of Basic Education in SA provide an explicit definition of Technology as 'the use of knowledge, skills and resources to meet people's needs and wants by developing practical solutions to problems, taking social and environmental factors into consideration' (DBE – Curriculum and Assessment Policy Statement, 2011). Many of the TE teachers still lack Technology pedagogic content knowledge to teach this subject on both senior phase and Further Education and Training band. At Grades 10 to 12 level, where the curriculum allows for specialisation in the form of subjects such as Visual Art and Electrical Technology, a clear problem that perpetuated apartheid era inequalities was the low number of teachers with training in these special subjects in a historically disadvantage areas (Action Plan 2014).

Community Engagement

According to Queensland Department of Emergency Services (2001) community engagement (CE) is a planned process with the specific purpose of working with identified groups of people, whether they are connected by geographic location, special interest, or affiliation or identity to address issues affecting their well-being. The linking of the term 'community' to 'engagement' serves to broaden the scope, shifting the focus from the individual to the collective, with the associated implications for inclusiveness to ensure consideration is made of the diversity that exists within any community.

Cavaye (2001) further declare that CE can take many forms and covers a broad range of activities, and some examples of CE undertaken by practitioners may include but no limited to:

- Informing the community of policy directions the government intend to take;
- Community consultation which serve as a process to build community awareness and understanding or to help the government develop its policy;
- Involve community through a range of mechanisms to ensure they form part of the decision-making process;
- Empowering community to make decisions and to implement and manage change;
- Collaborate with the community by developing partnerships to formulate options and provide recommendations.

The latter form of CE outline by Cayaye speaks well to this chapter as it share the professional development of TE teachers. The AR practitioner collaborated with these TE teachers in CE using AR to formulate new emancipation strategies to be applied in their content delivery.

Since this chapter is focused on teachers, the researcher therefore share the same sentiment of defining CE as outline by Mapotse (2017a) to be in the following manner:

Community engagement is a regular interaction process with incapacitated teachers as a community of co-researchers led by an action research practitioner for empowerment purposes in pursuit of some new teaching skills, methods and approaches within their cultural context to enhance learners' performance and teachers' achievement.

As an Action Research (AR) practitioner, one has realised that it is almost practically impossible to conduct CE activities without the AR components surfacing. A good CE activist will know when to infuse AR strategies within CE processes. This fact is supported by the Principles of Community Engagement Consortium (2011) in defining CE as the process of working collaboratively with and through groups of people affiliated by geographic proximity, special interest, or similar situations to address issues affecting the well-being of those people. It is a powerful vehicle for bringing about environmental and behavioral changes that will improve the socio-economic level of the community and its members. It often involves partnerships and coalitions that help mobilize resources and influence systems, change relationships among partners, and serve as catalysts for changing policies, programs, and practices. CE on the other hand can be taken as a continuum for community involvement, from the principles of CE (2011).

SUMMARY OF THE CONCEPTS

All the four terms used in this chapter: AR; AL; TE and CE are defined as processes, which means they are dynamic by nature and in a state of constant review. The end results of appealing in a process is to improve what is being acceptable at present (Mapotse, 2017b) hence these terms definitions are continually changing. A definition of a process is well articulated by Johnsey (1997) as a way of going about achieving an end and the separate parts of the process can be described as process skills.

- AR (action research) is an interaction process of addressing the empowerment and emancipation issues. In addressing the empowerment and emancipation issues during AR circular and spiral activities, the practitioner and the co-researchers goes through observations, plan, actions and reflections stages.
- AL (action learning) is a question-driven process in addressing the work related challenges of colleagues or peers. In addressing the insight and meaningful questions about own practice, the intention is to unlock thinking as a reflective way to come up with answers to the situation and in that regard advance learning.
- TE (Technology Education) is a design process of addressing community or society needs, wants or desires. In addressing community or society needs, wants or desires the technologist run through the five stages of the technological process which are investigate, design, make, evaluate and communicate.
- CE (community engagement) is a collaborative process of addressing common concerns of a particular group. In addressing common concerns during CE collaborative process, the project leader after identifying and choosing a community then the project leader engage the community by establishing their needs, co-research with them for a solution, and thereafter implement the workable solutions and share the results which are sustainable to that particular community.

In engaging with activities of building and exchanging knowledge, skills, expertise and resources through AR, AL, TE and CE that exercise requires that the AR practitioner have to divide the study into two phases. The ensue section outline the two phases of the study.

PHASES OF THE STUDY

Phase 1: Reconnaissance Study

A preliminary study/reconnaissance phase/initial reflection was undertaken to authenticate the research problem that was acknowledged within the study and hence that has led to the development of themes to be tackled during the AR journey (Mapotse, 2015b). For teachers to maximise the effectiveness of regular classroom teaching it calls for the need to constantly study their own situation in order to understand better the teaching process (Carr & Kemmis, 1986; Kemmis & McTaggart, 1988; Zuber-Skerritt, 1996). During the reconnaissance study, data gathering instruments were administered to the participants so that they could better reflect on their Technology teaching situation. This preliminary study was conducted for a fact-finding mission prior to unfolding the AR process. Ferrance (2000:7) point out that during initial reflection participants examine their own educational practice systematically and carefully apply the techniques of research to bring about change in it. The AR practitioner used only two instruments to collect data, which are reflective questionnaires and non-participatory observation. This is not the focus of this study at this stage, the results of these listed instruments used during reconnaissance study are reported in African Education Review journal, 12 (4), 2015. This study report on one cycle of a preliminary study and it should be noted that this ongoing CE project with three or more schools. Each school was subjected to focus group for data collection.

Phase 2: Main Study

Having read most of AR definitions, as an AR practitioner, one could summarise the four basic themes that AR is overemphasising, those are, 1) empowerment of participants, 2) collaboration through participation, 3) acquisition of knowledge and 4) social change. The four basic themes were swiftly implemented during the one earmarked cycle of AR, which has four stages.

Those stages are: 1) The problem identification or reconnaissance stage: this was done to ensure that the practitioner does not discover later after putting a lot of work, that the research topic being investigated is not a suitable on to embark on (Hofstee, 2006; Kemmis & McTaggart, 1988); 2) The design of the action stage: after the findings of the preliminary study were analysed, the next AR contact session with the co-researchers was to design a way forward so as respond to respond to their concerned. Together (the principal AR practitioner and TE teachers as co-researchers) drafted the programme for emancipation purposes; 3) The implementation of action stage: we (the research team – the practitioner and the participants) did draft an annual schedule to unpack all challenges highlighted by the TE teachers. A consensus was reached that from the results, the hindrances to TE lesson delivery will be linked to the core themes according to the TE teachers yearly workshedule; 4) Evaluation of action_stage: as it was said a bit earlier, this study report only on one cycle of this ongoing CE project where focus group discussion was used. All the AR four basic themes and four basic stages were rolled out to address the research problem which succeed in the next section.

RESEARCH PROBLEM

As echoed in the introduction, Mapotse (2017c:167) also emphasise that, even though the school principals are aware that some of their teachers do not possess any TE qualification the principals went on to ask some of the teachers to volunteer to teach Technology. Many Technology teachers feel inappropriately qualified and under-resourced to be addressing core themes of Technology (Jones, 2007). It is against this background that the study aim was structured as follows:

Aim of the Study

The main aim of this study is to confirm that the use of action research initiative during community engagement can be used to review the TE teachers' professional development. The action research (AR) study with the senior phase Technology Education teachers was conducted from three schools of Mankweng Circuit at Limpopo Province in South Africa. The TE teachers' engagement was motivated by the fact that TE is still a hurdle for most teachers' pedagogy and impediment for learners' comprehension.

Problem Statement and Research Question

I found out when I arrive at Bathabile Secondary School that teachers were in despair to teach this subject. This anxiety was coursed by the fact that these TE teachers are underqualified and/or unqualified to teach TE subject. After reconnaissance study, these TE teachers did welcome the rollout of AR stages with them. As an AR practitioner I ran with these TE teachers for the whole year in unpacking both their TE themes from the district workshedule and guided by Technology policy document.

This focus group session with these TE teachers were solemnly intend to retort to this research question: '*How can Action Research be used as an approach to professionally develop Technology Education teachers' pedagogic content knowledge through Community Engagement?*'

RESEARCH METHODOLOGY

The research methodology comprises primarily of two sections and those are the research design and research methods.

Research Design

This was a qualitative study reporting a case of three schools in their inception into this emancipation community engagement project. The study was deeply immense in action research approach.

Research Methods

Sampling

The three schools in this study are from South Africa, Limpopo Province in the Capricorn region. Two (2) teachers per school, which are teaching Technology, were used as a sample and that makes a total of six (6) TE teachers. These teachers were teaching only Grade 8 and 9, as these are the grades that offer Technology at senior phase level. Four teachers from different schools are teaching Grade 8 and two from different schools also are teaching Grade 9 in these sampled schools. Three schools from diverse backgrounds are used in this study.

Bathabile is a quintile one school (learners don't pay school fee) situated within a needy rural community, and learners' parents are low class income earners. Lamola is a quintile three school (learners' pays minimal school fee). The school is situated in semi-urban area and the parents of these learners are from a mixed class. Kholofelo is a quintile four school (learners are paying their school fee) is situated in an urban environment. The parents of these learners ranges from middle class and up including business owners. This section report on the reconnaissance study conducted with each school every time the TE teachers are in their inception phase, meaning three schools in three years.

Data Collection Instrument

For the tenacity of this study, the researcher will be reporting about the focus group (interviews) conducted to assess general perceptions of the TE teachers in three schools. The reasons of this focus group session was to gather some information about the journey travelled by the TE teachers' in their teaching and learning of Technology' subject in the previous period. The idea is to explore how AR interventions could be used to improve TE teachers' pedagogy and didactics of these three secondary schools.

During the semi-structured interviews with six TE teachers about their classroom practices, few things were also highlighted as part of their AR reflections. The reflection was conducted as a reconnaissance study during the emic phase of the study. This was done to structure the way forward into the participants etic phase space. The main drive was to involve all the parties with interactions in both CE and AR activities. The research dialogue between the participants (TE teachers) and researcher (AR practitioner) were guided by the following series of six (6) predetermined questions. These questions were asked to determine the general perspectives of TE teachers after they were allocated to teach TE without any qualification.

1. How long have you been teaching Technology Education subject?
2. Why are you teaching Technology subject?
3. How do you find the teaching of Technology without any qualification? Share some of your Technology teaching experiences.
4. Department of Basic Education (DBE) in South Africa has introduced TE in its curriculum since democracy. Can you substantiate why DBE embarked on such a move?
5. How many classes are allocated to you this current academic year? In those classes that you teach what is the teacher: learner ratio?
6. The learners in your classes are from diverse cultures and background. How then, as a Technology teacher, do you contextualize your teaching to accommodate those learners?

RESEARCH FINDINGS

AR was reinforced as the desired emancipation approach with these TE teachers. The focus group interviews resulted in thought-provoking dialogues that among other things a team of researchers manages to jointly draft a way forward programme. The TE teachers' were more willing to share their experiences and more open to be engaged at any level of future planning.

The questions asked the participants kept the focus group expedition more focus as teachers share their stories. Most of the teachers' responses are highlighted per question below chronologically all the way through. That is from interview responded (IR) 1 to IR 2 per school:

- How long have you been teaching Technology Education subject?
    - Just about to start

    - One month
    - One year
    - Two years (responded two TE teachers from two different schools)
    - Four years
- Why are you teaching Technology subject?
    - Because the school does not have a Technology teacher.
    - I was ask to teach it based on my background in Science subject.
    - I love it Technology teachers are few.
    - I had no choice as this subject was just allocated to me.
    - It is fun, interesting and compels one to be innovative.
    - I was given to teach this subject on arrival to this school, though History and Afrikaans subjects are my specialization.
- How do you find the teaching of Technology without any qualification? Share some of your Technology teaching experiences.
    - Never taught it before
    - Fun and enjoyable.
    - It is a little bit tough I have even decided to register it with University of South Africa.
    - The Grade 8 book was not user friendly but the Grade 9 book was interesting.
    - Exciting and challenging at the same time.
    - I find it interesting though difficult due to lack of training in the subject.
- Department of Basic Education (DBE) in South Africa has introduced TE in its national curriculum since democracy. Can you substantiate why the DBE embarked on such a move?
    - To develop learners and country citizens in general to reach their international standards.
    - To try to address issues of hand skills since the subject is more practical and skills could be tapped with ease.
    - Technology subject is important as things are rapidly changing, for one to cope with these changes one must be technologically literate and inclined.
    - To improve the understanding of Technology, e.g. saving energy, load shading, saving natural resources, etc.
    - Not sure, but I guess it is meant to tap learners creativity and innovation skills through the technological process.
    - From the ministry side I really do not know, but reading from the definition of Technology I realize the need for it to be implemented.
- How many TE classes are allocated to you during this current academic year? In those classes that you teach what is the teacher: learner ratio?
    - One Grade 8 class, teacher: learner ratio is 1:73.
    - Seven classes of Grade 8 and teacher: learner ratio is 1: 85.
    - Three Grade 9 Classes, teacher/learner ratio is 1: 80.
    - One Grade 8 class, 1: 60 is the teacher: learner ratio.
    - Two classes of Grade 9 and the teacher: learner ratio is 1: 90.
    - Two Grade 8 classes and teacher: learner ratio is 1: 80.
- The learners in your classes are from diverse cultures and background. How then, as a Technology teacher, do you contextualize your teaching to accommodate those learners?
    - My teaching it based on everyday life.

- Yes I do contextualize for instance by using their homes as an example when I teach about structures.
- It seems to be difficult since I am knew on this subject. The examples I give are more real and learners can see them.
- I do talk about water usage to the learners, comparing how ancient people wash in the river and uses the same river water to cook.
- Yes I do, I know most cultures of my learners.
- The example I give during my Technology presentation are cultural based.

DATA ANALYSIS

Few steps that are earmarked by Creswell's (2009) were followed in analysing data. The initial step was to transcribe the focus group discussions onto an MS Word document. The transcripts were thoroughly read by the researcher so that a general sense of the data could be arrived at. Member checking was undertaken with the co-researchers to verify what could have been missed during the discussion process and to ascertain as to whether what was captured during the interviews is a true reflection of our discussion. Of importance was to get the overall depth, and credibility of the information (Creswell, 2009). The researcher approached the co-researchers with predetermined questions since this was assessing the need for emancipation progress and the impact of AR contact sessions could have with the TE teachers. What came out of teachers' responses can be grouped as follows:

1. The sample of the teachers used in this study vary in their teaching period as others are on their incipient stage while others have some years of acumen in their teaching experience of the TE subject.

After TE teachers' profiling review for professional development was done, the next step was to analyse their needs as recommended by Gumbo, Makgato and Müller.

Technology teacher training should be preceded by profiling teachers and analysing needs so that strategic decisions can be made to vary the depth and nature of the training based on the profile and specific needs, otherwise the training may not be beneficial to all (Gumbo, Makgato & Müller, 2012).

2. The review reveals that some teachers are coerced to teach the subject while others just allocated the subject to keep the teaching post within the school. This leaves the researcher with a bit of qualms on how such teachers will be committed to their teaching course.

To make TE relevant, the teacher should always present it at a level that the learners could relate to. It is very important for the teacher to consider the learners' pre-knowledge when involving learners with some scientific knowledge and skills. The study is driven largely by Shulman's (1986) call for increased attention to teachers' subject-matter knowledge, whereas Charalambous and Hill (2012) stress that scholarly interest has focused on exploring the association between teacher knowledge and instructional quality.

3. The novice and the experienced teachers find the teaching of Technology to be enjoyable, fun, interesting, tough and challenging. At least one teacher intent to ameliorate the teaching experience of Technology by enrolling with the Higher Institution of Learning. One teacher commented about the disparity of friendliness between two textbooks of different grades. One still stress lack of prior training.

An Action Plan 2014: towards the Realisation of Schooling 2025, under the goals dealing with the *how of improving schooling* (2012: 108) evince that many and perhaps most, of South African teachers did not receive all the training they need to cope with the responsibilities of teaching and the curriculum changes that have taken place since 1994. The testing programmes conducted on both learners and teachers at primary level reveal that when it comes to subject knowledge, teachers are too often lacking even though they took a minimum of three years training. Whilst improvements amongst teachers are needed with respect to the specifics of the classroom practice and subject knowledge, what is also important is to strengthen the right values as part of the professional development of teachers.

4. The teachers' responses to the question about why Technology is part of the school curriculum or why should it be implemented clearly indicate several things: i) The TE teachers never read the Technology policy document, they do not have the TE policy document in their possession, or it is unbeknownst to all of them. ii) The respond in (b) seem to also apply in this case since in (c) they said all the good things but they might lose passion if they teach just for the sake of teaching.

The changes in the state of the SA national curriculum is hereby supported by Isdale (2016), in confirming that learners in the educational system will benefit if teachers understand the educational transition and contribute towards addressing basic skills shortages. Teachers in SA experience significant difficulties in accessing and receiving support, resources and continuing professional development opportunities close to where they live and work and the large majority of teachers who work in rural areas, the difficulty is even more pronounced (Integrated Strategic Planning Framework for Teacher Education and Development, 2011). Curriculum reform in many education systems appear to have largely failed to also reform teachers' beliefs, values and practices for better student achievement (Al-Daami & Wallace, 2007; Macdonald, 2003).

5. Based on the teacher: learner ratio the situation seem to be anomaly since the standard is supposed to be 1:40 maximum according to UNESCO Education Policy. Some of these schools are from an agrarian set up and least teacher: learner ratio is 1: 60 whereas the most and the worst is 1: 90. This conundrum impede provision of individual attention; create anomic space for assessment and extent the turnaround time for learners' feedback.
6. It seems four out of six co-researchers do not have a glue on how to contextualize their TE teaching. Only the last two interviewees has a background of what contextualizing do involve. The responses are dialectic and calls for AR practitioner to delve and demystify the aspect of TE contextualize teaching.

As Technology is a social phenomenon (Jones, 2003; Pavlova, 2005), in which social views of learning are being emphasised, it is appropriate for research in Technology to be situated in a socio-cultural perspective (Chikasanda, Otrel-Cass, Williams & Jones, 2012). To contextualize Technology in the classroom a teacher should choose examples, problems, projects, assignments and so on that relate directly to or that originate in the environment of a particular learners (Mapotse, 2016). Situations at home, in the school, in the community, in nature, in the market place in business are invaluable sources of examples to contextualise Technology. This is especially true when the situations are linked with the communication of opinions about general interest topics that are being covered by the mass media at a particular time (Flowers, 1994; Ortega & Ortega, 1995).

CONCLUSION

Technology Education (TE) teachers' both novice and experienced ones need to be empowered with the methodology of teaching TE through action research spiral cycles. All the TE teachers from these three schools should be emancipated to teach all the themes of Technology since most are coerced or just allocated to teach the subject. The empowerment cycle programme with these unqualified and/or underqualified teachers should followed the circuit themes per term as per their workshedule for the mere fact that it seems their teaching has been more of trial and error.

It is clear that the TE teachers did not grasp the vision of the ministry of education for implementing TE as a subject in the national curriculum. The teacher: learner ratio exacerbated both the didactic and pedagogy of TE since this subject is not only hands-on by nature but also minds-on. TE as a subject will help the teacher to be a bit flexible for the mere fact that in all its themes they can be taught in rural, urban or semi-rural environment. The incorporation of TE cascading theory with this cohort of TE teachers will boost their morale in presenting TE to their learners, as this will lead to sustaining the of teaching TE in different contexts.

In conclusion, the researcher will recommend that the higher education institutions, especially universities, should embark on community engagement services with Technology teachers from their nearby schools. This partnership will strengthen the teaching of Technology by TE teachers and its learning thereof by TE learners. In this study AR with TE teachers will be structured to close the technological pedagogic content knowledge and didactic gap as raised by the co-researchers. From an Action Plan 2014, Goal 16 state that, 'Improve the professionalism, teaching skills and subject knowledge of teachers throughout their entire careers and that exactly what this study was sort to achieve with Technology Education teachers.

REFERENCES

Action Plan 2014: Towards the Realisation of Schooling 2025. (2012). *The goals dealing with the how of improving schooling.* Problem Statement. Pretoria: Department of Basic Education.

Al-Daami, K. K., & Wallace, G. (2007). Curriculum reform in a global context: A Study of teachers in Jordan. *Journal of Curriculum Studies*, *39*(3), 339–360. doi:10.1080/00220270601057790

Allison, A. (2008). *Critical theory and pre-service art education: one art teacher's journey of equipping art teachers for inclusion*. Austin, TX: University of North Texas.

Buchanan, I. (2010). *Oxford Dictionary of Critical Theory: The Invaluable Multidisciplinary guide to Theory*. New York: Oxford University Press.

Carr, W., & Kemmis, S. (1986). *Becoming Critical: Education, Knowledge, and Action Research*. London: Falmer Press.

Cavaye, J. (2001). *'Community engagement framework project: Scoping and review paper.'* Cavaye Community Development/CEO Committee on Land Resources''.

Charalambous, C. Y., & Hill, H. C. (2012). Teacher knowledge, curriculum materials, and quality of instruction: Unpacking a complex relationship. *Journal of Curriculum Studies*, *44*(4), 443–466. doi:10.1080/00220272.2011.650215

Chikasanda, V. K. M., Otrel-Cass, K., Williams, J., & Jones, A. (2012). Enhancing teachers' technological pedagogical knowledge and practices: A professional development model for Technology teachers in Malawi. *International Journal of Technology and Design Education*. doi:10.1007/s10798-012-9206-8

Cohen, L., Manion, L., & Morrison, K. (2007). *Research methods in education* (6th ed.). London: Routledge.

Creswell, J. W. (2009). *Research Design: Qualitative, Quantitative, and Mixed Methods Approaches* (3rd ed.). Thousand Oaks, CA: Sage.

Creswell, J. W. (2014). *Research Design: Qualitative, Quantitative, & Mixed Methods Approaches* (4th ed.). London: SAGE.

De Villiers, M. R. (2005). Three approaches as pillars for interpretive information systems research: development research, action research and grounded theory. *Proceedings of the 2005 annual research conference of SAICSIT on IT research in developing countries*, 142-151.

Department of Basic Education. (2011). *Education in South Africa*. Available at: http://www.southafrica.info/about/education/education.htm

Department of Education. (2003). *Revised National Curriculum Statement Grades R-9 (Schools): Teacher's guide for developing learning programmes Technology*. Pretoria: Government Printers.

Dick, B. (2010). *Action Research thesis. You want to do an action research thesis? How to conduct and report action research*. Retrieved from http://www.scu.edu.ac/schools/gcm/ar/art/artesis.html

Ferrance, E. (2000). *Action research*. Brown University.

Flowers, J. C. (1994). Applied philosophy Technology Education. *Technology Teacher*, *54*(4), 6–7.

Fullan, M. (1993). *Change forces: Probing the depth of education reform*. Bristol, PA: Falmer Press.

Gibson, R. (1983). *Critical Theory and Explanation*. London: Hodder & Stodder.

Giroux, H. A. (2003). Public pedagogy and politics of resistance: Notes on critical theory of educational struggle. *Educational Philosophy and Theory*, *35*(1), 5–16. doi:10.1111/1469-5812.00002

Guba, E. G., & Lincoln, Y. S. (1994). *Competing paradigms in qualitative research.* Newbury Park, CA: SAGE.

Gumbo, M. (2017). An Indigenous Perspective on Technology Education. In P. Ngulube (Ed.), *Social, Cultural, and Educational Consideration of Indigenous Knowledge in Developing Countries*. IGI Global. doi:10.4018/978-1-5225-0838-0.ch008

Gumbo, M., Makgato, M., & Müller, H. (2012). The impact if In-service Technology Training Programme on Technology teachers. *The Journal of Technology Studies*, *38*(2), 23–33.

Gunther, F., & Chatoney, F. (2016). *Technology Education in France: efficiency of tools from functional analysis in learning process for describing objects. Pupils Attitude towards Technology – PATT 32 Conference.*

Hallström, J., Hultén, M., & Lövheim, D. (2014). The study of technology as a field of knowledge in general education: Historical insights and methodological considerations from a Swedish case study, 18422010. *International Journal of Technology and Design Education*, *24*(2), 121–139. doi:10.1007/s10798-013-9252-x

Hannaway, D. (2016). *Learners' profiles when experiencing Technology-based learning in the Foundation phase.* South African International Conference on Educational Technologies (SAICET), Pretoria, South Africa.

Hofstee, E. (2006). *Constructing a good dissertation: a practical guide to finishing a master's, MBA or PhD on schedule*. Sandton: EPE.

Integrated Strategic Planning Framework for Teacher Education and Development. (2011). *Together, taking responsibility for teacher education and development. Department of Basic education and Department of Higher Education and Training*. Pretoria: Government Press.

Isdale, K. (2016, August). *Smooth, staggered or stopped? Educational transitions in the South African Youth Panel Study*. Labour Market Intelligence Partnership-Human Sciences Research Council (LMIP-HSRC).

Johnsey, R. (1997). Improving Children's Performance in the Procedure of Design and Technology. *Journal for Design and Technology Education*, *2*(3), 201–207.

Jones, A. (2003). The development of a national curriculum in Technology for New Zealand. *International Journal of Technology and Design Education*, *13*(1), 83–99. doi:10.1023/A:1022355410425

Jones, A. (2007). Reflecting on the development of a new school subject: The development of Technology Education in New Zealand. *Waikato Journal of Education*, *13*, 273–294.

Kemmis, S., & McTaggart, R. (1988). *The Action Research Planner*. Victoria, Australia: Deakin University Press.

Love, T. (2015). *An examination of preparation experiences that influence technology and engineering educators' teaching of science concepts*. Presentation at the 102nd Mississippi Valley Technology Teacher Education Conference, Nashville, TN.

Macdonald, D. (2003). Curriculum change and the post-modern world: Is the school curriculum-reform movement an anachronism? *Journal of Curriculum Studies*, *35*(2), 139–149. doi:10.1080/00220270210157605

Mapotse, T. A. (2015a). An Emancipation Framework for Technology Education teachers: An Action Research study. *International Journal of Technology and Design Education*, *25*(2), 213–225. doi:10.1007/s10798-014-9275-y

Mapotse, T. A. (2015b). A reconnaissance study as a *sine qua non* factor for *status quo* change among senior phase Technology teachers. *African Education Review*, *12*(4), 583–599.

Mapotse, T. A. (2016). *Programme in Technology Education. PTECAL-H Study Guide. Centre for Community Training and Development*. Pretoria: Unisa Press.

Mapotse, T. A. (2017a). Development of a Technology Education Cascading Theory through community engagement site-based support. *International Journal of Technology and Design Education*. doi:10.1007/s10798-017-9411-6

Mapotse, T. A. (2017b). *Open and Distance Learning practitioners sustain the teaching of Technology Education through Action Learning: Success stories of ODL institutions. In Open and Distance Learning for Sustainable Growth in Emerging Nations.* IGI Global.

Mapotse, T. A. (2017c). Technology Education teachers' professional development through Action Research. In P. J. Williams & D. Barlex (Eds.), *Contemporary Issues in Technology Education: helping teachers develop research-informed practice*. Singapore: Springer. doi:10.1007/978-981-10-2819-9_11

Ortega, C., & Ortega, R. (1995). Integrated elementary Technology Education. *Technology Teacher*, *54*(5), 11–15.

Pavlova, M. (2005). Social change: How should Technology Education respond? *International Journal of Technology and Design Education*, *15*(3), 199–215. doi:10.1007/s10798-004-5867-2

Pilkington, C., & Gelderblom, H. (2010). Using the Karplus learning cycle to teach learners with ADHD introductory computer programming. *African Journal of Research in MST Education*, *14*(1), 73–84.

Potgieter, C. (2004). The impact of the implementation of technology education on in-service teacher education in South Africa. *International Journal of Technology and Design Education*, *14*(3), 205–218. doi:10.1007/s10798-004-2270-y

Principles of Community Engagement. (2011). *Clinical and Translation Science Awards Consortium; Community Engagement Key Function Committee Task; Force on the Principle of Community Engagement* (2nd ed.). National Institute of Health.

Pudi, T. I. (2005). Educator roles for Technology Education teacher- educator. *Africa Education Review*, *34*(1), 147–167. doi:10.1080/18146620508566297

Pudi, T. I. (Ed.). (2007). *Understanding Technology Education from a South African perspective.* Pretoria: Van Schailk.

Pule, S. (2017). Mixed Media Modelling of Technological Concepts in Electricity. In P. J. Williams & D. Barlex (Eds.), *Contemporary Issues in Technology Education: helping teachers develop research-informed practice*. Singapore: Springer. doi:10.1007/978-981-10-2819-9_6

Queensland Department of Emergency Services. (2001). *Charter for community engagement, Community Engagement Unit, Strategic and Executive Services. The State of Victoria, 1996-2017*. Author.

Shulman, L. S. (1986). Those who understand knowledge growth in teaching. *Educational Researcher, 15*(2), 4–14. doi:10.3102/0013189X015002004

Stevens, A. (2006). Technology teacher education in South Africa. In International Handbook of Technology Education: Reviewing the past twenty years. Rotterdam: Sense.

Stevens, D. (2017). Teaching Bioethics: the intersection of values and the application that advances in Technology-Make-Possible. In P. J. Williams & D. Barlex (Eds.), *Contemporary Issues in Technology Education: helping teachers develop research-informed practice*. Singapore: Springer. doi:10.1007/978-981-10-2819-9_9

Tooley, W. R. (2000). *Political rationality & government mechanisms: Maori education policy in the new millennium* (MA in Education dissertation). Auckland: University of Auckland.

UNESCO (United Nations Educational, Scientific and Cultural Organisation). (2012). Education for All Global Monitoring Report: Education Counts. France: UNESCO.

Williams, P. J., & Gumbo, M. T. (2011). New Zealand Technology pedagogical content knowledge. In Pupils' Attitude Towards Technology (PATT) 25 and Centre for Research in Primary Technology (CRIPT) 8. Goldsmiths, University of London.

Yager, R. E. (2013). Successes and continuing challenges: Implementing the changes in professional development for teachers in the U.S. *International Journal of Education in Mathematics, Science and Technology, 1*(3), 184–189.

Zuber-Skerritt, O. (1996). Emancipatory action research for organization change and management development. In O. Zuber-Skerritt (Ed.), New Directions in Action Research (pp. 83-105). London: Falmer.

Chapter 2
Action Research and Open Innovation:
A Synergy?

Elisabeth Fruijtier
University of Oslo, Norway

ABSTRACT

In today's world, 'global' problems increasingly require global solutions. In order to realize these solutions, innovation processes are 'opening up'. Both of these developments adds a level of complexity to the Action Research process that is traditionally, and perhaps inherently local in nature. However, it also offers opportunities. This chapter explores the case of an Action Research innovation that reached a global scale beyond the initial Action Research process that started it, with the help of Open Innovation strategies. From this case we learn that Open Innovation has a significant potential for sustaining Action Research 'action' beyond its initiation and make them transferable across contexts. At the same time, such 'open' innovations can grow very complex, and therefore so can 'open' Action Research solutions – especially when they concern (free and open source) information systems. The concept of 'Action Research Systems' is introduced as one way in which Action Research can help ground Open Innovation processes in dealing with this complexity.

INTRODUCTION

Especially in the high-tech industry, there is a growing awareness that traditional organizations should be open to 'outside' innovation (Saint-Paul 2003) which furthermore stimulates external exploitation of innovation (Chesbrough & Crowther 2006). In this networking imperative (Enkel, Gassmann & Chesbrough 2009) blurring boundaries between organizations and between them and consumers has led to new ways of innovating, also discussed by Chesbrough (2003) as 'Open Innovation'.

Open Source software development is generally perceived to be an example of Open Innovation in the IS field that has led to highly 'fluid' development processes. Unlike physical or biological matter, software innovation is made up of 'code'; a language that can be accessed from computers from all over

DOI: 10.4018/978-1-5225-2642-1.ch002

the world and is increasingly spoken beyond the traditional information technology domain. Users can access the source code, they can alter it and create their own local alternatives of the product or enter a dialogue about the 'source code' itself and therefore the way the product should function/exist. Building on a rich 'hacker ethics' tradition characterized by values of openness, sharing and collaboration, OSS is argued direct power transferred to corporations as a result of globalisation, back to the people (Bejoy 2010).

The previous example illustrates how consumers as part of more 'open' forms of innovating and co-creating are perceived to be "armed with new connective tools" and "want to interact and co-create value, not just with one firm but with whole communities of professionals, service providers and other consumers" (Prahalad & Ramaswamy 2004 p.14). There is an increasing understanding that design problems faced nowadays require stakeholders from different disciplines and need to include problem-owners (Fischer 2013). To action researchers of information systems (IS), who have specialized in including problem owners in design processes, this paradigm shift is exciting and full of opportunities. However, it is also one in which new challenges present themselves.

Action research (AR) can be generally captured as the collaborative production of scientifically and socially relevant knowledge through a participatory process. In the information systems domain, Action Research cycles often produce socio-technical constructs that combine technology and social intervention in order to address a 'real world problem' and produce scientific knowledge. At the same time, these (scientific process and) practical hands-on interventions also leave behind tangible design. We have imagined Action Research can take place in multiple cycles that spiral in a linear direction, with the advantage that the negotiation of participatory designs remains of a relatively simple nature, and the disadvantage that they are difficult to scale. But have we considered the scenario, partly made possible by open innovating techniques, whereby they fractal out across multiple dimensions?

As pointed out by Waddell et al. (2015), action researchers today like any other change agent are more likely to find their projects entangled in complex global constellations, even when they appear to be – or intend to remain - local. IS increasingly evolve from things into structures; from 'finished' products into customisable platforms. In these environments, platform-based services acquire characteristics of infrastructure[1], while both new and existing infrastructures are built or reorganized on the logic of platforms[2] (Plantin et al, 2016).

This chapter is written with the intention to prepare Action Research for such situations by picturing what this situation and its challenges look like in a particular case and domain which is especially likely to produce 'high maintenance' outputs. Though an Action Research process itself may discontinue to be part of the 'life' of its solution sooner or later, we also imagine how action researchers may still continue to play a valuable role in different chapters of the solutions 'story'. The potential of Open Innovation is explored in an attempt to address the sustainability of such AR actions.

An interest in the growth and sustainability of action research solutions attempts to counter an over-emphasis on method and 'first phase' aspects of Action Research in the literature, in relation to a relatively limited understanding of the change that is created over time. There is a lot of emphasis on the role of Action Research for sustainability, but too little discussion on the sustainability of Action Research outputs. When Action Research generates actions with the potential to serve as a solution for the problems of many, it is important to consider what happens to those actions. Do they need some form of maintenance; do they need to be transferred?

At the same time, researchers of Open Innovation phenomena are still searching for the mechanisms and best practices of this recent era (Enkel et al. 2009) and how 'opening up' innovation should gain shape is still a hot topic of discussion (Chiaroni, Chiesa & Frattini 2011). An important bottleneck for Open Innovation, as well as related participatory design and Action Research approaches, is managing the differences between collaborating partners in terms of organizations and cultures (Du Chatenier, Verstegen, Biemans, Mulder & Omta 2009). Understandings from the journey of Action Research 'actions' that entered this field can reveal the potential value that Action Research can have in addressing the challenges of the Open Innovation process.

The IS that will be explored in this chapter spiralled from an Action Research project that attempted to improve health information management in a particular development country. Over the course of 20 years, the actions that followed needed to be 'opened up' in order for the IS to for it to meet the demands of a diversity of users/ use cases. In other words, new opportunities of the Open Innovation paradigm were used to cope with issues related to scaling a solution and maintaining a scaled/scaling solution. By studying what happens to such a particular Action Research output as it 'lives' on in a time in which innovations are no longer the efforts of single firms but emerge from co-creative efforts of many stakeholders, this chapter aims to answer the question:

(Where) does a fusion between Action Research and Open Innovation lead to a synergy?

In exploring this question, this chapter encourages researchers and practitioners to glance forward and think about two things in particular:

1. The potential of Open Innovation for the scaling, ongoing development and maintenance of Action Research 'action' in ways that ensure the involvement of problem owners in line with Action Research philosophies.
2. The potential of Action Research for dealing with the complex challenges associated with Open Innovation processes – especially when reaching a global scale.

BACKGROUND

In order to meet the previously stated objectives of this chapter in understanding where Action Research and Open Innovation processes can potentially complement each other, this section will first attempt to shed light on the commonalities between these two approaches, their current shortcomings, challenges and potential.

Action Research and Its Challenges

Action research as a scientific method for social inquiry is somewhat caught between its objective to study human practices with the aim of improving them on the one hand, and its own historical attempts to suit the standards of scientific social research on the other (Carr 2006). Action research solutions are born in the midst of these two potentially conflicting forces. In an attempt to add structure to this conflict, Checkland and Holwell (1998) conceptualized the Action Research process as a number of phases in which researchers declare a framework of ideas (F), methodology (M), an area of application (A) and research themes in order to produce two types of research outcomes in the form of action and findings that are informed by reflection on the process. This chapter focuses specifically on this 'action', which is understood to be activity/ efforts in relation to the development of solutions for a 'real problem situation'. Action can be related to Action Research as part of the cycle in Figure 1 (Action Research 'action') but also to the development of IS in general (IS 'action'). In the case of Action Research in the IS domain, these actions may furthermore overlap.

The debate this chapter wishes to enter concerns 'the real problem situation' as the domain, or 'living lab', in which an important part of the Action Research process takes place. After all, it is here that social change needs to be established as a result of action taken. Zooming in on this domain there is one question which the current model for Action Research, designed around a local Action Research project, does not effectively address. In the current innovation climate, how local can we consider Action Research interventions to be, or, to remain?

Globalization, the internet and the deep and widespread integration of communication channels in our society have contributed to an environment in which solutions can be shared easily and extensively. From thinking about creating solutions for a specific setting, the current innovation climate is increasingly urging both practitioners and academics to look at 'global problems' (Gustavsen 2003; Waddell 2007). It has been already briefly mentioned that Action Research projects, even when intending to create a local solution, are bound to encounter global complexities in their surroundings (Waddell et al. 2015). In fact, in some cases, aiming to create a local solution can be problematic in this environment. So called 'pilotitis' in which there are too many small projects that do not scale, is a big problem in the IS field and in particular in low resource settings. Although Action Research can play an important role in transferring solutions across contexts because of its sensitivity for local conditions, this also adds a risk of contributing to this 'pilotitis'.

In addition, like other research approaches, the Action Research process is constricted mostly by time and resources. However, unlike other research approaches, action researchers produce an 'action' in addition to their scientific findings, in the form of a solution for an established problem. It is the

Figure 1. A systematic display of the Action Research process
Source: Adapted from Checkland & Holwell (1998).

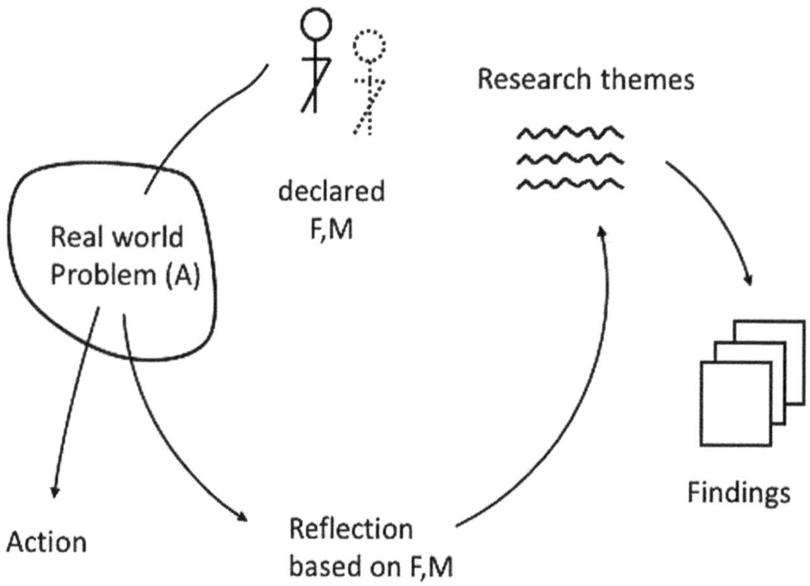

production of this solution that has led to theorizations of the Action Research cycle as endless in the sense that new cycles can start where previous cycles end. This is among others because the real life problem situation that triggered and/or justified the action taken in the first place, may change over time. Aiming for social change, this could mean that Action Research solutions are inherently temporarily in their ability to actually addressing especially complex problems.

Transforming solutions into generic solutions that can deal with diversity across contexts may deal with limited local capacities and resources to maintain solutions locally as well as combat 'pilotitis'. However, making the jump from transferable to transformable solutions adds a whole new level of complexity. Does this complexity concern the action researcher? This chapter argues that is does. Especially in IS, where action researchers are bound to be faced with high interconnectivity in their surroundings, the aim to create social change comes with the responsibility to consider the sustainability of the 'life' of the solutions they co-create - especially when targeting problems of marginalized and vulnerable groups.

In the previously sketched scenario, the problem situation identified as part of the Action Research process has grown in complexity (from a local problem situation to a regional or even global one) that may furthermore change in nature over time. Ironically, the success of an Action Research solution (addressing the problems of many) may contribute to this growth as the solution is transferred. Transforming solutions to address this complexity requires a re-thinking of the current model through which action is produced.

Open Innovation and Its Challenges

Open Innovation offers such a model. Open and networked forms of innovation are emerging from changes in the landscape of software development that challenge notions that any given company can create value unilaterally and that value resides exclusively in the company's or industries products and services (Prahalad and Ramaswamy 2004). Open source software development are often perceived as example of an Open Innovation approach in the IS field.

In some ways, Open Innovation is nothing new. As discussed by Huizingh (2011) the roots of Open Innovation go back far in history (Chesbrough 2003). A literature review conducted by Dahlander and Gann (2010) furthermore unravels how within the field of Open Innovation many ideas reflect concepts such as 'absorptive capacity' (Cohen & Levinthal 1990) in the innovation literature and the notion of 'lead users' (Von Hippel 1986) to name a few. And yet in other ways, the Open Innovation paradigm is unlike anything we have seen before. As stated by Badawy (2011 p.66):

The open innovation paradigm embraces the non-traditional activities relating to creativity, invention, innovativeness, and product/ service differentiation from the currently existing and accepted mold. As such, 'newness' is the name of the game.

What Action Research in IS and Open Innovation efforts have in common is a recognition of the value of user participation and 'co-creation'. This understanding stems from the ontological assumption that knowledge is practice-based and accordingly equivocal, dynamic and context-dependent by nature (Newell et al 2009). In Action Research in the IS domain, action researchers often act as designers, or on behalf of designers and or users. In Open Innovation, an organization will take the lead in organizing this mediation. In some forms of co-creation there may not be one organization, but several organizations that partake in the co-creation of solutions (Prahalad and Ramaswamy 2004).

In all the previous approaches, the innovation emerges as a compromise of these participatory processes. However, as the notion 'compromise' indicates, designer-user relationships can be challenging (Suchman 1988; Bødker, Ehn, Kammersgaard, Kyng & Sundblad 1987). In firms, negotiation processes are often 'top-down' in the sense that ultimately, the company decides how it groups, extracts and values input from its users (Pollock et al. 2007). Co-creation and Open Innovation processes challenge the previous top-down approach with 'bottom-up' strategies. However, this does not mean that these processes are necessarily smooth.

Though challenging the previous top-down approach with more 'bottom-up' strategies, this does not mean that the co-creation process surrounding Open Innovations is necessarily smooth. This becomes clear when we take Free and Open Source Software (FOSS) development as an example. The FOSS approach, whereby the so called 'source code' of a software product is made freely available, is highly suitable for development of software where participation by various users in the design is important. However, as it focuses on general use, this approach is also bound to addressing a selective range of user requirements, given its aim is to cover universal aspects of local requirements (Camara & Fonseca 2007).

In the terminology used in system development, tension occurs between the 'trunk' and its 'branches'. In short, creating a 'trunk' involves that networked development focuses on creating a generic technical solution for widespread use. A tension between these generic features and local requirements may lead to the creation of branches which offer local alternatives. However, for various reasons, the creation of branches (also referred to as forking of the core code) can be undesired. Problems occur when, in adapting the software for it to address a local problem, the 'forked' software solution may not be brought back into the generic core. The disadvantage of developing such software 'forks' is that the user eventually can no longer benefit from later versions of the system and the advantage of consulting the wider community. These disadvantages compete with the situation of the user that may require urgent adaptation. Consequently the open source software development process too requires a constant negotiation of conflicting interests in order to develop a generic product. Especially when an open source community involves many different stakeholders, interests may differ greatly. This is likely to be the case as solutions scale.

In addition, the required speed of developing and maintaining IS solutions adds further pressures on designer and user-relationships. The information system domain is characterized by the way technology and software in particular 'lives' or, perhaps even more specifically; lives *fast*. As pointed out by Kankanhalli et al. (2003), the production of software artefacts happens in an environment where the rate of innovation and the speed of new product development is crucial. At the same time, the production of support and the building of use capacity need to be tailored to local requirements and often in a timely fashion (ibid.) In comparison, we can imagine hardware to 'live' too; they are subjected to fashions, to the environment and technological improvements. However, these processes move slower and it takes more time adapt objects to changes in the physical environment that simply does not change as fast as virtual environments do. Software is highly dependent on this virtual environment that is furthermore highly competitive too: there are many that speak the language and information travels quickly. The tricky part hides in designing for such a fast life, especially if it is to be a long one, when design processes are complex collaboratives.

THE DHIS2 'NETWORK OF ACTION'

In previous sections we have identified a number of challenges that are new to the IS domain in which both action researchers and practitioners are grappling with enabling co-productive designer-user relationships as solutions increase in scale and complexity. This section will discuss an example of such an Action Research solution, the District Health Information System version 2 (DHIS2).

DHIS2 is a free and open source based software platform. It is widely used as a tool for collection, validation, analysis, and presentation of aggregate and transactional data, tailored (but not limited) to integrated health information management activities in low resource settings. Being a platform, it is highly configurable and customizable, and has extensive interfaces allowing for integration with other systems (Nielsen & Sæbø, 2016). It is also a platform for innovation, enabling the development of apps by a global network of software developers. While apps are developed to meet particular local needs, they can be made available for all users by being published in the DHIS2 app store. Apps may also be implemented in or result in changes in the generic software core (Gizaw et al, forthcoming).

Originating from Action Research efforts, the development of DHIS2 nowadays is the result of collaborations of a global network, the Health Information Systems Programme (HISP), which is established and coordinated by the Department of Informatics at the University of Oslo (UIO). The HISP network comprises of various entities, including Universities, Ministries of Health, non-government organizations (NGO's), international agencies like WHO and NORAD and many in-country implementing agencies commonly referred to as HISP nodes.

DHIS2 found its origins in South Africa as DHIS1, as collaboration between a PhD researcher from UIO who had received some funds and local members of a committee formed by the Ministry of Health with an interest in tackling weaknesses of the country's paper-based health information system. Following participatory Action Research approaches, whereby de-signers closely collaborated with ministry staff and health practitioners in the field, they aimed for the implementation of a digital district health information system. The success of their solution led to the export of the software to other countries that faced similar challenges, as illustrated in Figure 2.

Over the course of 20 years, the system scaled to use in over 60 countries (in 2016) with a majority of implementations in Africa and Asia. In this time, local developers have organized themselves as part of a network of 'nodes' that gradually emerged around the customization and implementation of DHIS.

Figure 2. DHIS expansion
Author: Ola Titlestad (2014) power point slide. Labels added.

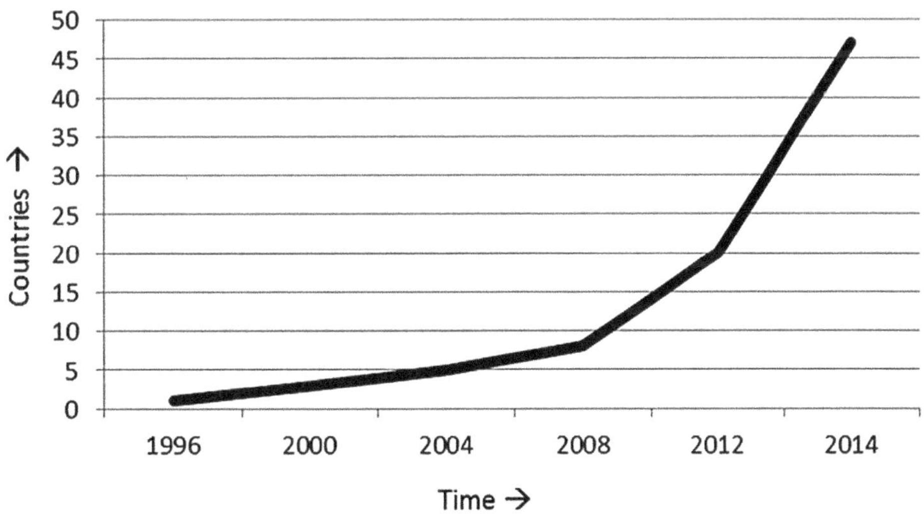

Currently, decentralizing design processes toward regional or local HISP nodes and other skilled users (and thereby diffuse the development of the software's core) is considered to be of significant importance in order to cope with the growing user base.

These developments will be discussed in more detail in the remaining of this section, which derives from the insights of the longitudinal involvement in the activities of the HISP UIO node within the network over a period of 1,5 year. Insights provided are based on interviews, observations, field conversations and document analysis was conducted during this period. In addition, frequent meetings (averagely held on a 2-week basis) with the community coordinator of the DHIS2 community (at the time also in charge of coordinating the academies) were used to reflect upon various elements of the research process.

Analysis aimed at developing a broad understanding of the distribution of development processes, mapping of key actors involved and their roles and regional activity over time. Part of the data was collated and used to create a timely reconstruction of key events (the development of the object, in this case DHIS2 and its support structures). This timeline, a summary of which is provided in Figure 3, enabled an overview of the various participatory activities in the network; provides an historic overview of the case and creates an image of the complexity and scale of the network as it develops over time. These developments will be discussed in more detail in the remainder of this section.

After developing a solution from scratch in South Africa in 1996, other Action Research projects followed in which participation took the form of the introduction of a pre-developed and flexible system that was then customized in collaboration with the developers. Following a highly intensive process, referred to as participatory customization (Kimaro & Titlestad 2008), developers and local health management worked together in setting up the overall structure of the database to collect health data at lower levels. Developers then needed to work together with the health workers to customize the system so it could be used in local contexts. Because this approach did not adapt well to the differences between countries and use cases, an international version of DHIS was released in 2004 under the name DHIS 1.4. HISP UIO would take the lead in the development of the system from here onwards.

Figure 3. Timeline
Source: Author.

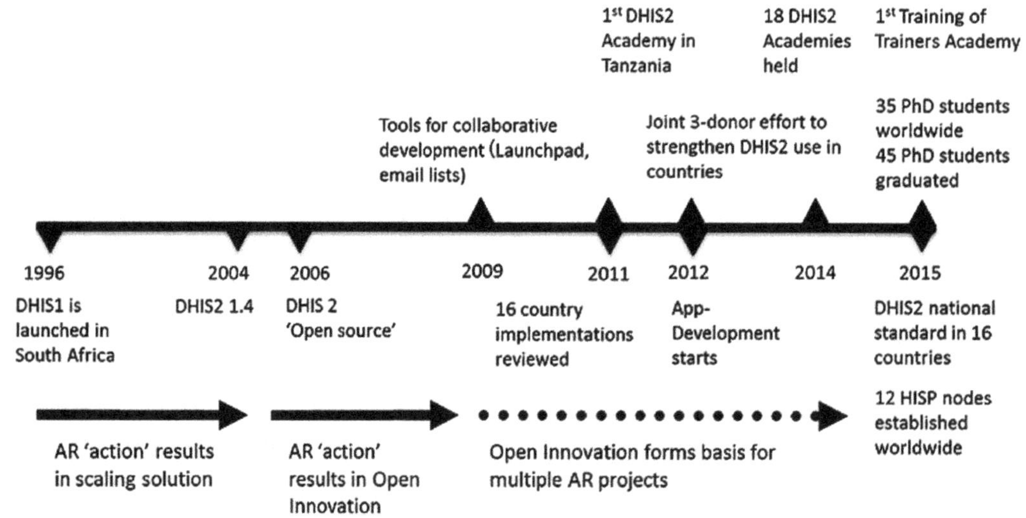

The following participant recalls how the system, inspired by the South African case, was piloted in his country. He became aware of DHIS during a course in another country where it was promoted as a solution for related issues, which he describes as follows:

Record people at the Health Unit would record at a monthly basis, then send that by taxi or whatever to the District level. Then the District HMIS focal person would get into a taxi or with bus or whatever and take the paper based reports to the Ministry of Health. It was very difficult to meet the time limit (...) so we thought the DHIS would improve on the reporting. – Actor local donor funded NGO

In this example, DHIS as a solution answers to a pre-existing local concern about the existing situation of health systems that transfers across contexts. In this case, the DHIS version 1.4 was piloted in one district by a local NGO as an alternative to the previously discussed existing paper-based system:

We piloted it at [district x] (…) and at that time it was only [district x] that had timely reporting. It was an eye-opener for the entire country, and that is how DHIS came to be [in country x]. Because [donor x], which supported the Ministry of Health to roll out DHIS, took an example of the DHIS 1.4 from one of the districts where we had installed it. So they came and looked at it, because at that time the Ministry representatives were bragging about how [name] district reports are always timely and they are always send to them by email, and all that kind of thing. So the [donor x] folks came over to that district, looked at how DHIS 1.4 worked, and then a few months later we were called for a training in DHIS. – Actor local donor funded NGO

In 2006, researchers at UIO changed the DHIS 1.4 architecture to a Java-based platform, which made the software free and open source. This version, now called DHIS2, introduced two important changes; 1) for the developers at UIO it required identifying generic needs across settings and stakeholders and 2) it allowed other developers to participate in contributing to this design by submitting 'blueprints' (code) that could then be integrated in to the software 'core' code, while for users it meant they have to update regularly to new versions that are released quarterly if they want to benefit from any of these improvements. In 2012, the platform approach furthermore enables those users who need customized requirements that cannot be integrated to the core to extend the software to develop their own applications (in 2012) with the opportunity to share them with the wider user community (Braa and Nielsen 2015).

These changes furthermore required putting in place various technological tools and social interventions to enable collaboration and stimulate capacity building (Braa & Nielsen 2015). These capacity building efforts, which started in 2009, have gradually made the development process more interactive through development platforms (Launchpad and Github) and user- and developer email lists. In addition, regional training academies for basic and advanced users became structural events (see Figure 4).

The following excerpt from an email to the DHIS2 community illustrates the way DHIS2 academies are able to increase the capacity of users:

[Ministry of Health country x] has left the pilot phase since three months (...) with the integration of different programs [x y z] in a common DHIS2 database. This work has been done due to the skills we have acquired at the DHIS2 academy Level 2&3 at [country] (Thanks to [academy organizers]) and a tool that we developed for importing existing Excel and Access files in technical directions and health programs in DHIS2. – Email Ministry of Health representative

Figure 4. DHIS2 academies
Author: Matthieu Pinard.

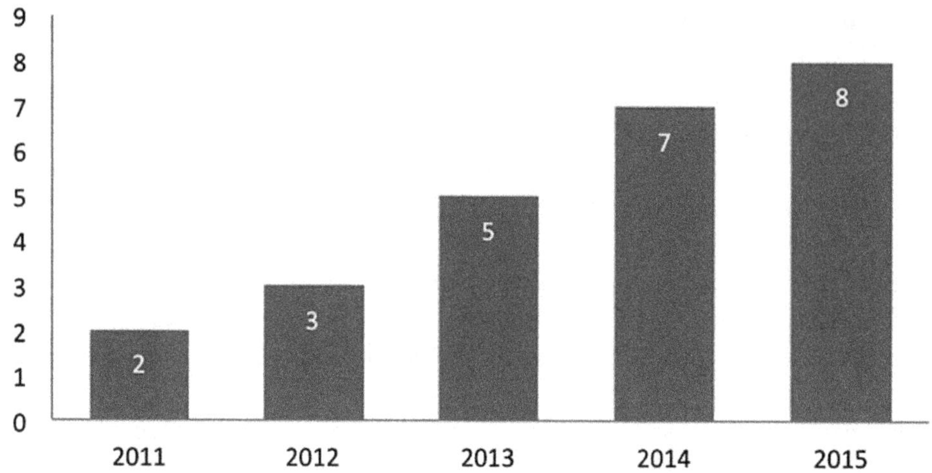

A network of HISP nodes, local organizations of varying nature, has emerged over time to provide in-country or regional support. In 2015, core researchers at the UIO state in a publication [(Braa & Nielsen 2015 p.10) that this network of nodes is increasingly being formed strategically for the purpose of decentralizing aspects of the development process; to make use of the expanded size and skills available in a 'maturing' community of developers as well as external knowledge beyond the health domain and limited resources of the core team. The importance of such nodes is also emphasized in the evaluation report (2011 p. 24; 27 and 28) stating:

The two strongest successes for HISP are the HISP [country x] and HISP [country y], the two oldest and most influential nodes within the network in terms of institutionalization of DHIS (referring to government linkages) and capacity to provide support.

While the new possibilities created by architectural changes instigated a rapid increase in the number of users, all DHIS2 solutions had (and still are required) to 'speak' to the core version of the software for other actors (users of the software) to benefit from the network, During a regional academy, one of the researchers from HISP UIO explains this to the participants as follows:

Whenever we learn something from you guys [audience of participants from mainly local NGOs and Ministries] from a new use case, it will be generalised and taken into the core, and become available in the next release for all the other countries. – Spokesperson HISP UIO at regional DHIS2 academy

Following this strategy, new innovations that appear when user-representatives solve a problem on the ground are integrated into the DHIS2 platform, making it available for other users through quarterly releases.

This generalization process (also referred to as generification (Pollock et al. 2007) however is full of challenges, and only a selection of use cases make it into the core. These challenges result from a diversity of perspectives between users as well as between (South based) users and (North based) developers, and the way certain actors (among which those considered the software's 'founders') have much more influence on prioritizing proposed features then others (Gizaw et al, forthcoming p 28). For instance, funding sources tend to play an important part in the prioritization of new features (ibid.). At the same time, Gizaw (ibid.) reminds us that the open source development approach provides a permanent opportunity for users to 'fork out' and create parallel development processes, stressing that 'this creates a necessary tension convincing the community to remain open, collaborate and stay together.'

From a developers point of view, there is also a tension between allocating time and resources toward integrating use requirements to meet certain expectations versus keeping up with the latest technological requirements and developments to stay ahead of others. This is reflected in the following conclusion that was drawn in an evaluation report in 2011 (p. 39) as follows:

The need for developers to be flexible and independent for meeting their partner country needs and the reduction of development redundancy are at odds.

Maintaining this balance was an important motive to develop the system into a platform, enabling users to address local needs through developing apps and extensions, which are furthermore attempted to be made generic and available for others in an DHIS2 'app store'.

The initial approach which relied on Action Research efforts for the software's implementation and development remains to be an important drive within the network. Through Action Research, researchers at UIO secure their involvement and support in local implementations to either further develop or make improvements to the system. An evaluation report from 2011 (p. 19, 20, 45) underlines the way 'researchers have had direct impact on the technology development and the new deployments of DHIS and other HIS' and 'the important role of UIO graduates for example in HISP node [country x] as well as nearly all country deployments'. Nowadays, the latter is no longer the case as developers at UIO express there are now more DHIS2 deployments than they are even aware of.

Today, the overall aim of the program is formulated in a strategic document (2014- 2016 p.1) as follows:

The overall goal of the Health Information Systems Programme (HISP) at the University of Oslo is to enable and support countries to strengthen their health systems and their capacity to govern their Health Information Systems in a sustainable way to improve the management and delivery of health services.

Strategic aims of the program in achieving this overall aim involve the development and improvement of DHIS2 which followed the DHIS1; enabling the use of DHIS2 as a data warehouse to counter fragmentation, and providing training in the use of DHIS2 (strategy document 2014- 2016 p.1). In a research publication (Braa & Nielsen 2015) researchers at the UIO assign the following roles to various actors in the 'network of action' which are defined around the programs aims (see Table 1).

In addition, in the strategic document donor roles are described as follows (Strategy document 2014-2016 p 2):

- ...'Provide core funding for core development of the DHIS2 platform and staff for coordination and support of implementation in countries'
- ...'Supporting the implementation of DHIS 2 in countries'
- ...'Fund further development of DHIS 2 to support specific reporting requirements within the scope of country-owned health information systems.'

Table 1. Actors and their roles

Actors	Roles
University of Oslo	Academic research, strategizing, software development, prioritising requirements for generification, coordinating software development, demonstrations, piloting, engaging countries and donors, generating funding, facilitating Academies and related knowledge sharing, develop and participate in Master programs in the South, offer PhD program, provide API's, integrate with other software etc.
Developer nodes in the South	Software development, explore new domains
Software providers (e.g. openLMS)	Offer complementary functionality, integration with DHIS
Ministries of Health in the South and local NGO's	Training, implementation, knowledge sharing, data use in management
Regional health organisations	Facilitate the establishment of regional standards and regional implementers of DHIS
Universities in the South	Offer Master programs, participate in implementation projects
Independent consultants	Implementation and software development support, training
Academy trainers	Facilitating learning and knowledge sharing
Hosting partners	Virtual server provision or DHIS as software as a service
Health workers	Participate in training and use DHIS

Braa and Nielsen 2015 p 10.

DISCUSSION

As pointed out by Walsham and Sahay (2006) scaling a system in scope and size requires growing diverse networks that surround a technical solution. In previous sections we have seen how experiences of developing the initial DHIS version '1.4' inspired a 'networks of actions' approach in previous work by Braa, Monteiro and Sahay (2004). In the DHIS2 case, the Researchers formed the beginning of this network with the aim to build knowledge on implementing health information systems while building systems on the ground and addressing local problems collaboratively.

According to this approach, researchers mediate between users and their institutions – in this case Ministries of Health and later NGO's– in several places. Put differently, the researchers apply Action Research in one place and then take their lessons learned to the next. They become a thread that spins between different organizations. This is also possible because they are the one building the solution in collaboration with different actors in various locations and at different moments in time. Where these 'threads' would be of a temporarily nature (in the sense that they are broken when the researcher leaves) these cases could be considered individual Action Research projects.

However, in an alternative scenario that follows the networks of action approach, these threads between the researchers and their separate Action Research projects remain to exist over time (Braa & Nielsen 2015) due to the required efforts to sustain them. We could say that these 'threads' in a sense 'institutionalized' with the introduction of DHIS2, as all users now depend on a central product that ties them to each other and to the researchers. The response of the researchers to this new situation leads to a more 'Open Innovation' approach. The cornerstones of this approach are:

- A team of 'core' developers, established to coordinate and secure continuous and speedy development of the solution;

- A network of user-representatives, inspiring improvements, new use and innovation
- Open sourcing the software, enabling collaborative development of the solution together with user representatives in various countries and organizations that have employed the system. Although this also enables creating local 'branches' (that exists separately from the generic solution) the core aim of this strategy is to enable addressing of common needs shared by members within the network;
- Developing the DHIS2 into a generic platform, enabling flexible customization and extension of the solution to address individual needs within the network.
- Capacity building strategies in the form of as DHIS2 Academies become a prominent and formalized part of the networks activities and offer the network physical opportunities to meet.

As the development of the solution becomes the effort of a team of developers and a network of practitioners, the role of the researcher changes. The initiating researchers are now supervising and introducing new action researchers and Action Research projects to support implementers and problem owners to improve the solution.

Action Research Systems

Looking back, we could say the traditional Action Research process ended when DHIS moved from South Africa onward to other countries, where the solution became part of new Action Research projects that started and ended. Once DHIS2 became a standard in the world of HMIS in low resource settings, more actors became involved in its development and funding. However, Action Research still had an important role to play as networked Open Innovation efforts continued to take shape. No longer emerging from one, the solution began to draw in new Action Research projects. One could say that from driving the solution, the solution now began driving research. The involvement of action researchers would contribute to informed action by participating in a wide array of design and implementation efforts ('actions') undertaken in different parts of the network.

As part of this new model, action researchers and practitioners are no longer working together from the initial phase of the AR cycle, but instead the action researcher enters an on-going 'action' process that is the combined effort of a variety of practitioners. This effort can be seen as an Open Innovation effort by an organization, in this case HISP UIO, that takes an Action Research solution forward, inviting action researchers to join a vast network of practitioners in their effort to tinker and improve aspects of it as problem owners are attempting to implement the solution to address real world challenges.

In Figure 5, this process is conceptualized as an Action Research System, in which (aspects of) Open Innovation processes are reflected upon by action research projects which, though temporarily in nature, are part of a permanent structure that informs the Open Innovation process as it attempts to negotiate a compromise between the needs and efforts of a variety of stakeholders. The following sections explore how Action Research Systems may contribute to current debates in relation to both Open Innovation and Action Research related domains.

Currently, these systems take the shape of individual Action Research projects that inform action in a specific part of the network involved in the Open Innovation process. For instance: AR researchers collaborate with Ministry X and HISP node Y to enable a new use of DHIS2 on mobile phones in country X. After a successful Action Research pilot, the solution can be made generic and available for everyone and action researchers share their lessons learned with others in the network. Such contribu-

Figure 5. Action research system as part of Open Innovation models
Adapted from Checkland & Holwell 1998.

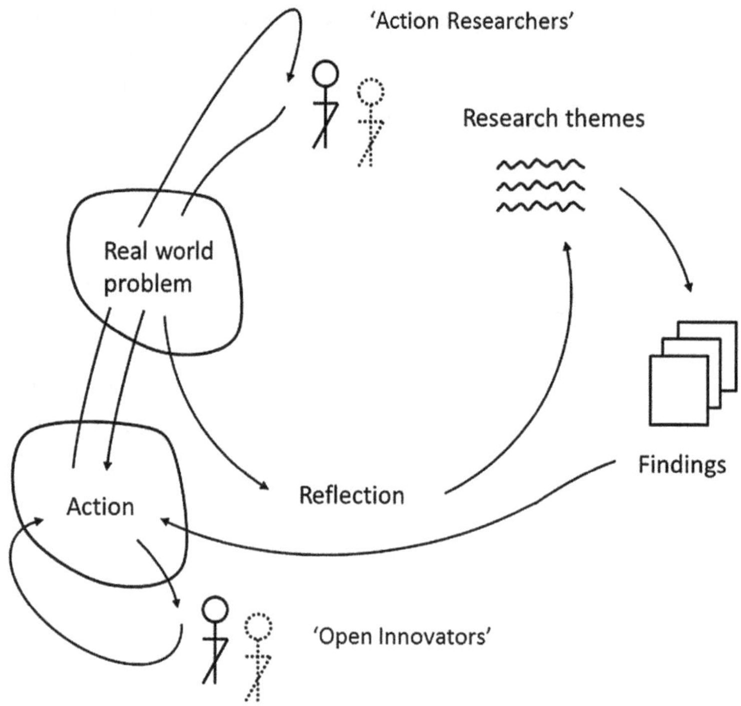

tions certainly also emerge outside of these Action Research projects. However, with involvement of the Action Researcher, the innovation process is extensively studied and reflect upon inspired by theory and comparative cases. Being part of a 'system', the role of the action researcher makes it easier to consult AR research colleagues active in other projects within the network and research perceptions of the systems core designers as well as the perceptions of the end users who will be using it. In their role as a researcher, they furthermore document and share this knowledge. This way, AR research contributes to new innovations, collaborations and knowledge production in the network.

Action Research Systems have similarities to the 'soft systems methodology' (SSM). Checkland and Poulter (2010) developed this action-oriented method after studying the initial response to messy and complex situations that (among others) IS may find themselves subjected to / be designed to respond to. Action Research Systems are also developed for these situations, in which solutions required to respond to such complexities are dealt with continuously.

As part of the soft systems methodology, users find out about a complex problem situation through discussion or debate about the situation, identify how it might be changed in the form of models and take action to improve it (Checkland & Poulter 2010). Their conceptualization of SSM, which is the output of an Action Research process, focuses on the way participants of action oriented change processes have different world views of the problem situation. They base this approach on the assumption that 'the complexity of problematical situations in real life stems from the fact that not only are they never static, they also contain multiple interacting perceptions of 'reality'' (ibid. p 192). Because the problem situation may change, the SSM process needs to be repeated with the aim of securing adaptive and holistic systems of *purposeful action.*

The concept of Action Research Systems stems from the same assumptions, however focuses on the way the solutions that follow from these non-static problem situations are required to 'shape-shift' along. The assumption is therefore not so much that problem situations change (and also world views themselves) but that the solution also changes as a result. Whereas Checkland and Poulter (2010) focuses on the process that went before this solution, Action Research Systems in some sense could be seen to emerge 'after it' to ground (changing) models used and emerging from these or other processes in research. It pictures an active role for researchers as part of holistic change systems such as Open Innovation initiatives / networks. As such, the aim is to contribute to *informed action* as part of these systems. Grounding innovation processes in research is considered important given the fast developing nature of both IS and knowledge in this area. The involvement of action researchers can enable dialogue and actions to be informed by knowledge of best practices both in and outside the network.

The Relevance of Action Research Systems for Action Research and Action Oriented Approaches

The previous conceptualization of Action Research Systems not only offers an alternative interpretation of the notion that scaling efforts of action researchers are unique in that they have the potential to transcend 'roll-outs' (of solutions) in terms of replication and generalizability (Gearty et al. 2015). It also has implications for the role of action researchers.

In early studies of Action Research in information systems (such as Braa, Bratteteig & Øgrim 1996 and Bødker et al. 1987), action researchers would take on the role of mediators between firms (their managers) and workers or users of a technological solution (for instance for the workplace) in the negotiation of purposeful design. Their Action Research solutions, though developed as much as possible together with users, are adopted by the firm. As mediators, action researchers tend to act as coordinators in bringing together various stakeholders and fine-tuning aspects of the design cycle.

In some aspects this traditional role of the action researcher is left unchanged in Action Research Systems. They still mediate for the needs of individual 'parts' of the network in order to enabling improved design for the network as a 'whole'. However in other aspects, the role of the action researcher changes significantly. Whereas practitioners in the network become 'mediators' in the sense that they act as coordinators in bringing together various stakeholders and fine-tuning aspects of the design cycle, the action researcher becomes a mediator between the practitioner and the scientific community. The Action Research targets specific parts of the network, as a result of which both the knowledge about the network as well as the knowledge within the network grows.

This conceptualization of Action Research systems furthermore requires a renegotiation of the role of the action researcher, who is no longer seen to engage participants and stakeholders as 'co-researchers' (Gearty et al. 2015 p. 46) but is engaged by stakeholders in 'action' instead. This new role has a potential of addressing what are considered weaknesses of the action research process. What Greenwood (2002 p. 130) points out as one of the most 'unsettling' features of AR literature is the way the term action research is used without attaching 'any serious meaning to the concept of research', adding that "throughout these [research] processes the collaborative process of reflection is the guiding thread that integrates the work." This critique is perhaps especially valid in the IS domain.

Following Greenwoods perception that the collaborative process (unique to Action Research) is *guiding*, Action Research Systems have a potential to restore this 'unsettling' feature into one of the strongest assets of action researchers. Being engaged by practitioners in a collaborative intervention rather than driving it frees up the hands of the action researcher to reflect on knowledge gained from being deeply embedded in the change process. In a way, we could say that as part of Action Research Systems, the action research process 'opens up' as the actual 'action' becomes the main responsibility of a collaborative in which the action researcher takes on the role of a mediator and a counsellor based on his knowledge of both worlds – theory and practice.

In fact, perhaps the greatest benefit of Action Research Systems hides in a challenge pointed out by Greenwood (2002 p 133) which is very central in the case examined, namely: who owns action research products? This is where differentiations between Action Research within the IS domain and Action Research elsewhere becomes even more visible. Whereas Greenwood (2002) writes mostly about reports and documentation being created, in the IS field, these products concern technical solutions that require future development. Yet, as Greenwood (ibid.) suggests, 'a collaborative project belongs to the collaborators'.

In the case of DHIS2, researchers and developers at UIO claim responsibility for the software's development. Nevertheless, others are and remain free to fork out, maintain their own instance and remain part of the collaborative development process when they desire to. At the same time however, the negative consequences of not being able to benefit from later versions outweigh are likely to outweigh the positive for most actors involved. This is why the ability to stay part of collaborative development of the solution is so important, and where Open Innovation strategies offer the most potential. In the case of Open Source products, the answer to Greenwoods question about who owns the AR product is: no-one, and everybody does. And that answer is potentially the closest solution to solving the issue raised.

The Relevance of Action Research Systems for Open Innovation and Social Innovation Approaches

In an attempt to deal with scale and retain (local) participatory processes, action researchers at UIO have searched for solutions in Open Innovation strategies. These solutions in turn gave life to a new role for action researchers in studying them while contributing to them in line with the overall Open Innovation approach. As seen in the case discussed in this chapter, Open Innovation strategies have a great potential in dealing with complexity; however these approaches come with their own challenges.

Action Research Systems may complement the Open Innovation process that characterize as spontaneous, free-wheeling, and organic (Badawy 2011) by making sure its processes continue to serve the needs of problem owners and subjecting solutions and practices to a vast body of research. Depending on the context, Action Research may further continue to infuse the Open Innovation process with creativity and novelty as well as share such cases with a wider network of scholars. As we have already established, Open Innovation and Action Research have many common grounds, and Action Research could take forward values of Open Innovation in relation to the sharing of knowledge in bridging activities of research and practice. In addition, Open Innovation is not merely enabled by developments in technology. For instance, as seen in the case discussed in this chapter, physical meetings and capacity building efforts may be an important part of its processes. Action researchers as part of Action Research Systems can stimulate and/or support networks or organizations in setting up such community-structures.

In doing so, Action Research Systems could furthermore help address two key areas in the Open Innovation literature that struggles to 1) understand the relevance of Open Innovation beyond high-tech industries and 2) study how firms implement Open Innovation in practice (Gassmann 2006).

Although this chapter has focused on Open Innovations as a wide spread and increasingly known phenomena, it must be said that Action Research Systems would perhaps in fact be most beneficial to the (sub-)domain of 'Digital Social Innovation'. This domain has risen along with developments in the area of Open Innovation and is referred to in a report about this approach for the European Committee (2015 p.9) as:

A type of social and collaborative innovation in which innovators, users and communities collaborate using digital technologies to co-create knowledge and solutions for a wide range of social needs and at a scale and speed that was unimagined before the rise of the internet.

As hinted at in this definition, the field of Digital Social Innovation is a relatively new field with 'little existing knowledge on who the digital social innovators are, what types of actions they are involved in and how they are using digital tools to achieve a social impact' (European Union, 2015 p. 23). Accordingly, Digital Social Innovation presents an almost entirely untouched area in both IS, Open Innovation as well as Action Research literature.

According to the current Action Research Systems approach, the ability of the action researcher to contribute to social change is limited to (multiple) local AR projects. By triggering and studying action in the networks' 'parts', these projects together contribute to a greater understanding of the bigger 'whole'. However, the concept of Digital Social Innovation in which co-creative Open Innovation processes that steer toward social change inspires a possible expansion of the relevance of Action Research Systems in this domain. Action Research Systems could potentially carve out a role for action researchers in enabling structured dialogues as part of (often unstructured) Open Innovation efforts about its social purpose. As explored by the SSM approach, this could lead to the development of shared models with the intention of enabling *purposeful action* – something which aligns with the intention of Action Research to contribute to meaningful change while learning from the process. Future research is encouraged to explore further how these two approaches can complement each other in realizing Digital Social Innovations and understanding how this is practised.

CONCLUSION

In the current innovation climate, Action Research interventions are no longer 'local' by default. However, our Action Research methods are currently limited in designing for solutions that address more complex problem situations. Open Innovations appreciate and increase opportunities for personalizing and customizing them to our own individual needs and tastes, and as such offer important opportunities for Action Research solutions to scale. Nevertheless, action research 'action', as one DHIS2 expert referred to it, is 'a bit like building a mud house; when you are busy trying to hold one part together somewhere else another part is slipping away'. This metaphor illustrates well how Action Research solutions are not well-defined objects with a beginning and an ending and clear boundaries; they gather an array of often unpredictable processes and actors.

Given that 'co-creation' is key in both Open Innovation and Action Research approaches, this chapter has looked at their challenges as well as the way they can complement each other in overcoming their individual weaknesses; Action Research solutions are difficult to scale as part of the research process, and the chaos of the Open Innovation process may compromise the ability of innovators to reflect on its processes. Nonetheless, when Open Innovation initiatives gain the ability to span across over a dozen countries and trickle into ministries and multinational organizations of resource-poor countries there is no room for short sightedness. Scenarios such as the case discussed in this chapter enable us to think beyond local Action Research cycles toward Action Research Systems that can create and or/become part of the on-going development of real world solutions as they increasingly open up to the involvement of problem owners.

REFERENCES

Badawy, M. (2011). Is open innovation a field of study or a communication barrier to theory development?: A perspective. *Technovation*, *31*(1), 65–67. doi:10.1016/j.technovation.2010.09.006

Bejoy, T. (2010). Participation in the knowledge society: The free and open source software (FOSS) movement compared with participatory development. *Development in Practice*, *20*(2), 270–276. doi:10.1080/09614520903566509

Bødker, S., Ehn, P., Kammersgaard, J., Kyng, M., & Sundblad, Y. (1987). A utopian experience: On design of powerful computer-based tools for skilled graphic workers. In G. Bjerknes, P. Ehn & M. Kyng (Eds.), *Proceedings of the Aarhus Conference on Computers and Democracy—A Scandinavian challenge* (pp. 251-278). Academic Press.

Braa, J., Monteiro, E., & Sahay, S. (2004). Networks of Action: Sustainable Health Information Systems across Developing Countries. *Management Information Systems Quarterly*, *28*(3), 337–362.

Braa, K., Bratteteig, T. & Øgrim, L. (1996). Organizing the redesign process in system development. *Journal of Systems and Software, 33*(2), 133–140.

Braa, K., & Nielsen, P. (2015). Sustainable Action Research: the networks of actions approach. In *Proceedings of the 13th International Conference on Social Implications of Computers in Developing Countries* (pp. 331 – 343). IFIP.

Bråten, S. (1973). Model Monopoly and Communications: Systems Theoretical Notes on Democratization. *Acta Sociologica*, *2*(16), 98–107. doi:10.1177/000169937301600202

Camara, G., & Fonseca, F. (2007). Information Policies and Open Source Software in Developing Countries. *Journal of the American Society for Information Science and Technology*, *58*(1), 121–132. doi:10.1002/asi.20444

Checkland, P., & Holwell, S. (1998). Action Research: Its Nature and Validity. *Systemic Practice and Action Research*, *11*(1), 9–21. doi:10.1023/A:1022908820784

Checkland, P., & Poulter, J. (2010). Soft Systems Methodology. In M. Reynolds & S. Holwell (Eds.), *Systems Approaches to Managing Change: A Practical Guide* (pp. 191–242). Springer-Verlag London Limited. doi:10.1007/978-1-84882-809-4_5

Chesbrough, H. (2003). *Open Innovation: The New Imperative for Creating and Profiting from Technology*. Harvard Business School Press.

Chesbrough, H., & Crowther, A. K. (2006). Beyond high tech: Early adopters of open innovation in other industries. *R & D Management*, *36*(3), 229–236. doi:10.1111/j.1467-9310.2006.00428.x

Chiaroni, D., Chiesa, V., & Frattini, F. (2011). The Open Innovation Journey: How firms dynamically implement the emerging innovation management paradigm. *Technovation*, *31*(1), 34–43. doi:10.1016/j.technovation.2009.08.007

Cohen, W. M., & Levinthal, D. (1990). Absorptive Capacity: A New Perspective on Learning and Innovation. *Administrative Science Quarterly*, *35*(1), 128–152. doi:10.2307/2393553

Du Chatenier, E., Verstegen, J., Biemans, H., Mulder, M., & Omta, O. (2009). The Challenges of Collaborative Knowledge Creation in Open Innovation Teams. *Human Resource Development Review*, *8*(3), 350–381. doi:10.1177/1534484309338265

Enkel, E., Gassmann, O., & Chesbrough, H. (2009). Open R&D and open innovation: Exploring the phenomenon. *R & D Management*, *39*(4), 311–316. doi:10.1111/j.1467-9310.2009.00570.x

European Union. (2015). *Growing a Digital Social Innovation Ecosystem for Europe*. Retrieved from https://www.nesta.org.uk/sites/default/files/dsireport.pdf

Fischer, G. (2013). Learning, social creativity, and cultures of participation. In A. Sannino & V. Ellis (Eds.), *Learning and Collective Creativity: Activity-Theoretical and Sociocultural Studies* (pp. 198–213). Routledge.

Gassmann, O. (2006). Editorial: Opening up the innovation process: Towards an agenda. *R & D Management*, *36*(3), 223–228. doi:10.1111/j.1467-9310.2006.00437.x

Gearty, M. R., Bradbury-Huang, H., & Reason, P. (2015). Learning history in an open system: Creating histories for sustainable futures. *Management Learning*, *46*(1), 44–66. doi:10.1177/1350507613501735

Gizaw, A., Bygstad, B., & Nielsen, P. (in press). Open Generification. *Information Systems Journal*.

Greenwood, D. J. (2002). Action research: Unfulfilled promises and unmet challenges. Concepts and Transformation. *Concepts and Transformation*, *7*(2), 117–139. doi:10.1075/cat.7.2.02gre

Gustavsen, B. (2003). New Forms of Knowledge Production and the Role of Action Research. *Action Research*, *1*(2), 153–164. doi:10.1177/14767503030012003

HISP UIO Strategy 2014-2016. (n.d.). Retrieved from http://www.mn.uio.no/ifi/english/research/networks/hisp

Holwell, S. (2004). Themes, iteration and recoverability in action research. In B. Kaplan (Eds.), *Action Research* (pp. 353–362). Kluwer Academic Publishers. doi:10.1007/1-4020-8095-6_20

Huizingh, E. K. R. E. (2011). Open innovation: State of the art and future perspectives. *Technovation*, *31*(1), 2–9. doi:10.1016/j.technovation.2010.10.002

Kankanhalli, A., Tanudidjaja, F., Sutanto, J., & Tan, B. (2003). The role of IT in successful knowledge management initiatives. *Communications of the ACM*, *46*(9), 69–73. doi:10.1145/903893.903896

Kimaro, H. C., & Titlestad, O. H. (2008). Challenges of user participation in the design of a computer based system: The possibility of participatory customisation in low income countries. *Journal of Health Informatics in Developing Countries*, *2*(1), 1–9.

Newell, S., Robertson, M., Scarborough, H., & Swan, J. (2009). *Managing knowledge work and innovation*. Hampshire, UK: Palgrave Macmillan. doi:10.1007/978-0-230-36641-1

Nielsen, P., & Sæbø, J. (2016). Three Strategies for Functional Architecting: Cases from the Health Systems of Developing Countries. *Information Technology for Development*, *22*(1), 134–151. doi:10.1080/02681102.2015.1026304

Plantin, J., Lagoze, C., Edwards, P., & Sandvig, C. (2016). Infrastructure studies meet platform studies in the age of Google and Facebook. *New Media & Society*. Available at: http://nms.sagepub.com/content/early/2016/08/02/1461444816661553.abstract

Pollock, N., Williams, R., & DAdderio, L. (2007). Global Software and its Provenance: Generification Work in the Production of Organizational Software Packages. *Social Studies of Science*, *37*(2), 254–280. doi:10.1177/0306312706066022

Prahalad, C. K., & Ramaswamy, V. (2004). Co-creation experiences: The next practice in value creation. *Journal of Interactive Marketing*, *18*(3), 5–14. doi:10.1002/dir.20015

Saint-Paul, G. (2003). *Information sharing and cumulative innovation in business networks. CEPR Discussion Paper Series (No. 4116)*. London: Centre for Economic Policy Research. Available at https://ssrn.com/abstract=480643

Suchman, L. (1988). Designing with the user: book review of Computers and democracy: A Scandinavian challenge. ACM Transactions on Information Systems, 6(2), 173–183.

Von Hippel, E. (1986). Lead users: An important Source of Novel Product Concepts. *Management Science*, *32*(7), 791–805. doi:10.1287/mnsc.32.7.791

Waddell, S. (2007). Realising Global Change: Developing the Tools; Building the Infrastructure. *Journal of Corporate Citizenship*, *2007*(26), 69–84. doi:10.9774/GLEAF.4700.2007.su.00011

Waddell, S., McLachlan, M., Meszoely, G., & Waddock, S. (2015). Large scale change action research. In H. Bradbury (Ed.), *Action research handbook* (pp. 538–548). Sage Publications Ltd. doi:10.4135/9781473921290.n55

Walsham, G., & Sahay, S. (2006). Research on information systems in developing countries: Current landscape and future prospects. *Information Technology for Development*, *12*(1), 7–24. doi:10.1002/itdj.20020

ENDNOTES

1. Infrastructures are essential, widely shared socio-technical systems ranging from electric power grids to communication networks (Plantin et al., 2016).
2. Platforms are computing devices and software environments that are extended with application software built upon them, ranging from game design to content-sharing websites and social media applications (Plantin et al., 2016).

Chapter 3
Action Learning Coaching:
A Practical Model for Novice and Emergent Leader Development

Angie Danielle Carter
University of Georgia, USA

ABSTRACT

Action learning coaching is advocated as an aspirational means to develop novices and emergent leaders that builds personal capacity as well as ability to meet new challenges. In this chapter, action learning coaching will be considered in three different ways. The first marks a broad overview of action learning and coaching, along with its integrated practice. The second way presents a practical model of action learning coaching as an innovative method to fortify novices and emergent leaders in the workplace. Finally, insights are drawn from an action learning action research (ALAR) case study and an object lesson to show as to how five novices in human resource development (HRD) in roles such as instructional design and training and development gained heightened capacity and efficacy in performing their roles through action learning coaching.

INTRODUCTION

An ever-changing competitive landscape and heightened business volatility compel organizations to think differently about their approach to human resource development in the workplace. The approach to and utilization of human resource development (HRD) practices, including training and development initiatives, plays an important role in developing employees in organizations (Noe, Clarke & Klein, 2014). Integration of consistent approaches or well developed methodologies for managing and developing employees can prove to be illusive (Axelrod, Handfield-Jones, & Michaels, 2002; Boudreau, 2004). As a result, many firms continue to rely on formal training opportunities which focus narrowly focus on surface-level skills that prepare employees with current skills and competencies but not tomorrow's (Cullen et al., 2002; Matthews, 1999; Winch & Ingram, 2002).

DOI: 10.4018/978-1-5225-2642-1.ch003

Management theorists have long concluded that sustained success in the workplace is only possible when new forms of learning, knowledge creation, and organizational change emerge (Drucker, 1993; Senge et al., 2005). Contemporary trends in workplace learning now focus on how individuals ideally transcend their existing limits (Illeris, 2003). In an effort to source innovative methods for developing employees and specifically leaders, human resource development practitioners continuously seek ways to address competency development needs, and build current as well as future organizational capacity (Volz-Peacock, Carson & Marquardt, 2016). One method that has been demonstrated to be an effective approach and used universally to develop leaders is the action learning coach-facilitated methodology (Volz-Peacock, Carson & Marquardt, 2016).

Action learning coaching produces learning for innovation in complex, ambiguous situations (O'Neil & Marsick, 2014). Its subjects, though responsible involvement in some real, complex and stressful problem, achieve the intended changes to improve their observable behavior through participation in small groups called "sets" (Revans, 1982, p. 626-627). Similarly, coaching within professional or business settings is an informal dialogue that facilitates new skills, possibilities and insights in the interest of individual learning and organizational advancement (Bacon & Spear, 2003). The International Coaching Federation website defines professional coaching as "partnering with clients in a thought-provoking and creative process that inspires them to maximize their personal and professional potential" (icf.com, n.d.). Of late, the incorporation of coaching with action learning has begun to gain traction in scholarly circles, albeit with little practical application (Carter, 2015).

The use of action learning and professional coaching as an integrated practice actually has its roots in business literature as far back as the 1950's (Ellinger, 2014). Action learning coaching can be the first step for participants in a journey toward greater self-insight and a greater capacity to learn from experience (Marsick & O'Neil, 1999), ultimately allowing participants to work on the problems occurring in their own work settings and reflect with peers who offer valuable insights into each other's workplace problems (Cho, 2013). Not only is action learning an effective tool for learning, problem solving and leadership development, the addition of coaching elements and the personalized attention that coaching brings (Ellinger & Kim, 2014) can serve as a means to develop employees' leadership abilities and personal capacity in the workplace.

Participants largely missing from the action learning coaching discussion include novice and emergent leaders, also known as high-potential employees. Novices are individuals who are new entrants in the workplace and emerging leaders, also known as high-potentials, and are recognized as an organization's likely future leaders (Cope, 1998). Novices and high-potentials are critical to ensuring current and future businesses success and are playing an increasingly important role in an organization's competitiveness (Lamoureux, 2012). Few studies have specifically investigated the learning of employees who are not yet leaders, or who are new to the organization or industry. Privileged instead is the research that heavily favors an action learning approach used by leadership development programs and teams within organizations to resolve larger business or team-based issues. (Carter, 2015).

In this chapter, we discuss the facilitated action learning coaching as an effective way to support novices, emergent leaders and new professional members in the workplace, and further, assist them in building capacity for the future. The beginning of the chapter engages in a broad discussion of action learning and coaching, seeks connections to empirical data of learning by novices and emergent leaders as under-researched groups. The middle part of this chapter presents a practical model depicting action learning and business coaching as a way to build capacity in emergent leaders and novices in the workplace. Following the model is an examination of an action learning action research (ALAR) case study. Derived from a dissertation study, an action learning intervention was proffered for five novices in human resource development (HRD) practitioners, who after participating in action learning coaching subsequently gained greater capacity and efficacy in performing their roles. The chapter concludes with a review of the implications of the intervention and offers lessons for others who are interested in instituting action learning coaching as an innovative coaching method to fortify those who are yet to become leaders in the workplace.

INTEGRATION OF ACTION LEARNING AND COACHING

The melding of action learning and coaching can bolster participants' abilities, leaving them better fortified for dealing with future problems (Carter, 2015), which is the ultimate goal of learning—to have access to information for a wide set of purposes so that the learning will transfer to other circumstances (Bransford, 2000). We will discuss the blending of these two for the benefit of novices.

What Is Action Learning?

Action learning originated in the 1940's, out of Reg Revans' interest in the convergence of systems thinking, social justice, education and societal reform (Revans, 1972). Revans (1983) is credited as saying, "there can be no learning without action and no (sober and deliberate) action without learning". In action learning, participants use work on an actual project or problem as the way to learn, with participants working in small groups to initiate steps to solve their problems and know how to learn from that action (O'Neil & Marsick, 2007).

Action learning is widely used for workplace problem solving, its utility well established for organization change strategy, organization development, and leadership (Boshyk, 2002; Marquardt, Leonard, Feedman, & Hill, 2009; O'Neil & Marsick, 2007; Raelin, 2009; Pedler, 1991; Dilworth & Boshyk, 2010). However, significant value can be derived from application in individual learning and development contexts (De Loo, 2002, 2006; Pedler, Burgoyne & Brooks, 2005), professional education and management development training (Zuber-Skerritt, 2002; Kramer, 2008), and adult learning (Waddil & Marquardt, 2003). Previous reviews of action learning literature focused on myriad categories: gathering and exploring accounts of practice and examples of application (Mumford 1994; Smith & O'Neil 2003), theoretical schools (Marsick & O'Neil, 1999), steps (Pedler, 2003; Marquardt, 2004), distinguishing between action-oriented, learning-oriented and balanced action learning practices (Cho & Egan, 2009) and action learning coaching (O'Neil & Marsick, 2014).

As an aim, action learning endeavors to improve or change work practices in order to create knowledge or understanding (Marquardt, 1999; Pedler, 1997; Revans, 1982). According to Zuber-Skerritt (2002), action learning is a process by which participants "learn from concrete experience and critical reflection on that experience – through group discussion, trial and error, discovery, and learning from and with each other" (p. 114-115).

Empirical studies have tended to show a propensity for deep learning within action learning sets, which is what is needed for transfer of skills derived from participation into the larger work context. One study sought to determine the learning that takes place within action learning sets and found that the learning group learns more during than after the action-learning sessions, mainly by exploring issues in depth and receiving personal feedback (deHann, 2005). Other, more recent studies have centered on how action learning programs significantly enhance both individual and organizational outcomes as compared to traditional executive education formats. Learning outcomes from action learning cannot be taught in books (Tushman, et. al., 2007), but develop substantially through personal and professional participation (Harris, 2012). Leaders develop through action learning and increase empathetic listening, formulate more informed actions and initiative and exhibit higher readiness to take responsibility (Raudenbush & Marquardt, 2008), and also exhibit self-confidence, problem-solving, expanded professional network, coaching skills, and listening (Cowan, 2013).

So called soft skills are formed through action learning, including increased empathetic listening, an enhanced ability to formulate more informed actions, and higher readiness to take on responsibility and initiative (Marquardt, 2008) as evidenced in research. Participants increase their ability to listen, accept criticism and openness with one another (Davey, Powell, Powell and Cooper, 2002). Compelling research from Acker-Hocevar, Pisapia and Coukos-Semmel, (2002) indicates that strong leadership skills may be developed through action learning that develops basic leadership skills in a safe environment; reinforces managerial, transformational, political and professional aspects of leadership; and allows participants to understand themselves as developing leaders.

Professional Coaching

The practice of professional, structured, and systematic coaching in the workplace came about during the late 20th century as part of targeted leadership and management development (Maltbia, Marsick & Ghosh, 2014). As "a human development process that involves structured, focused interaction and the use of appropriate strategies, tools, and techniques to promote desirable as well as sustainable changes for the benefit of these coaches and potentially for other stakeholders" (Bachkirova, Cox and Clutterbuck, 2014, p. 1), coaching just might be the new face of leadership for the 21st century (Marsick & Maltbia, 2009). In the workplace, coaching involves helping on various levels of a team's work, including the task, project, process, or the team dynamics level (Reddy, 1994).

The benefits of coaching are widely recognized (Coutu & Kauffman, 2009) in that it addresses "people issues" (Ladegård, 2011; Moen, 2011; Slåtten, Svensson, & Sværi, 2011; Wasylyshyn, 2003); goal setting (Clutterbuck, David & Megginson, 2013); individual fulfillment (Ellinger & Kim, 2014); and can develop high-potential employees (Ready, Conger, & Hill, 2010). In addition, coaching promotes a number of practices that are generally associated with learning and with good communication, including listening, clarifying, and questioning, as well as encouraging reflection, criticality and collaborative learning (Cox, 2013). However, there seems to be some confusion from practitioners and scholars alike over exactly what coaching is, what it helps with, and how it is practiced along with the theories that inform it. Some even call it atheoretical (Ellinger, Hamlin, & Beattie, 2008).

Coaching creates value by providing leaders with a "developmental process for focusing on growing the emotional and relational aspects of their leadership capacity, by addressing specific relationship challenges and how the leader might handle them differently" (Hawkins, 2008, p. 30). Towell's (2013) study suggests that as individual empowerment grows through the utilization of a coaching style, it can even help facilitate a change in organizational culture. Employees surveyed confirmed that this approach was more motivating and created more trusting relationships. The key to coaching seems to be about the latent ability to unlock potential, maximizes the individuals' performance, and helps participants learn rather than to teach them (Whitmore, 2003).

Action Learning Coaching

O'Neil and Marsick (2014) assert that melding action learning with coaching takes it to another level—that of learning-how-to-learn" (p. 206) and helps participants learn rather than teach them (Whitmore, 2003). According to O'Neil and Marsick (2014), bringing coaching into an action learning process can aid self-development and problem identification, utilizing such tools as 360° feedback, personality inventories and interviews. Action learning coaching equips participants with important tools to become better employees and individuals, provides a safe place to experiment and provides opportunity for incremental learning and practice (Carter, 2015).

Action learning coaches seek to open minds to a deeper level of questioning that may follow, and are freer from the power and organizational dynamics that contextualize the coaching relationship because their role is less tied to explicit results or predefined objectives (O'Neil & Marsick, 2014). The coaching role enables a team to build a learning climate; create openness, trust, and critical thinking; encourage teamwork; enhance listening, questioning, and reflection; and stress the importance of actions to support the team as well as the three levels of learning; making certified coach-facilitated action learning more effective for the development of leadership skills (Volz-Peacock, Carson & Marquardt, 2016).

Empirical study on action learning coaching is scarce. However, the few studies found tend to confirm that action learning coaching practice is worthwhile and imparts benefits for both the coach and participants. For instance, when Vaartjes (2005) applied action learning practice to strengthen the form and structure of executive coaching engagement, she established that when delivered within a supportive, client-centered coaching framework, such integration can support rigor, theoretical depth and the achievement of intangible and tangible outcomes. Abbott's (2006) study on expatriate acculturation to El Salvador using action learning coaching prepared expatriate managers to face the challenges, provided a medium to transfer and apply knowledge, and allowed them to understand themselves and others through reflective practice.

To be sure, positive working relationships with the coach aided transfer and motivation for the participant to continue the process, and the coaches reported similar experiences of self-efficacy with regard to facilitating learning outcomes. Finally, a Sankaran and Brown (2012) micro case study used the group coaching methodology rooted in action learning to help managers improve their capacity to think more strategically and expand the managers' creativity and innovation for mutual collaboration.

Advocacy for Novices and Emergent Leaders

Given the important role that leadership plays within an organization, understanding how to facilitate the development of effective leaders and leadership is critical (Bialek, 2014). Emergent leaders consistently display superior levels of performance indicating a strong capacity to grow and succeed in their roles more quickly and effectively than others in their peer groups (Ready, Conger & Hill, 2010). Despite evidence of such topics as induction, onboarding, professional socialization and development, or work-role transitions, few studies specifically highlight problems novices confront in their new roles, and little is known about the learning process novices undertake as they gain competency.

Since novices have less experience with real situations, they must rely on the rules they have learned in their preparatory education in order to function. In contrast, the expert professional intuitively knows how to assess and grasp situations and can accurately zero in on the problems without wasteful consideration of unfruitful possible problem situations (Benner, 1982). However, adding an experiential component loosens the rigidity and allows them to develop based on the experiences they encounter in their work setting so what they learn in one setting can transfer to their work environment (Daley, 1999).

Novices face both technical and adaptive challenges, often do not have the skills necessary for competency, and may not feel efficacious in their new roles. New employees learning how to deal with rapidly changing environments need help in building internal capacity to face new challenges and developing skills, including those that are needed to think outside of the box, cross boundaries, integrate perspectives, experiment with feedback (O'Neil & Marsick, 2014), and learn how to balance their work with the learning from that work (O'Neil & Marsick 2007).

Fortunately, the addition of coaching to action learning takes the process to new levels, particularly for novice and emergent leaders. Action learning coaching can, for novice and new employees, help assuage feelings of confusion and doubt, deal with rapidly changing environments, think outside of the box, cross boundaries, integrate perspectives, experiment with feedback, and ultimately build internal capacity to face new challenges. Through iterative planning, reflection and action cycles, action learning coaching could be the mechanism novices seek to deal with gaps in knowledge as they successfully transition to competency and efficacy. Novices can be supported through the mixture of action learning that enables individuals to deal with rapidly changing environments and use critical thinking skills (O'Neil & Marsick, 2014) and coaching increases consciousness and helps them "know themselves better, live more consciously, and contribute more richly" (Sherman & Freas, 2004, p.85). Seemingly, novices are ideal for this manner of improvement.

Capacity building is a process that improves the ability of a person, group, organization, or system to meet its objectives or to perform better (Brown, LaFond & Macintyre, 2001). In fortifying individuals for the future, capacity building involves making individuals the best forms of who they are so that they might develop personal mastery, contribute their gifts, and find balance in their lives (Reinelt, 2006). Marsick (2003) advocates action learning as a means to stretch individuals farther than they could reach themselves through deep questioning, assumption identification, and problem reframing that results in increased individual abilities for enhanced informal learning and metacognitive skill development. According to Sheckley and Keeton (1999), individuals develop proficiency by working in challenging and supportive environments, self-monitoring, engaging in deliberate practice, and solving ill-defined problems.

In this section, the ideals of action learning, professional coaching, and their integrated use have been linked to related literature and empirical studies. The literature review supports the view that action learning coaching could be beneficial for novices, emergent leaders, and new professional members to an organization. There is a paucity of literature on the topic of integration of action learning and coaching to build future capacity within novices, and missing from the conversation is a practical model for action learning coaching that integrates the two practices. The next section endeavors to show the practical application and operationalization of action learning coaching through such a model.

ACTION LEARNING COACHING MODEL

Similar to action learning, set members use action learning coaching to work though real problems experienced in their workplaces. Members divide the time among themselves, and each person is helped by the others to think differently about the situation, the proposed solutions and the results from any accompanying action (Marsick & O'Neil, 1999). Participants share professional experiences, discuss work-based problems, interrogate case holder assumptions and reflect on the resulting action. (O'Neil & Marsick, 2014). The addition of coaching amplifies and enhances the session with a strong focus on supportive development of individual skills and through growth opportunities. Coaches help draw out the participants' reflections and learning (Carter, 2015), and intuitively work to fulfill the needs of case holders by slowing down the coachee's thoughts and providing time for reflection (Allen, Eby & Lentz, 2006). In addition, the coach maximizes learning by using activities to spur action within sets, such as role-plays and timely reading material as well as the ability to relate and reflect in the moment.

A model is included here as a way to show action learning coaching in operation. Within action learning coaching sets, members share personal and professional problems that occur at work. The steps in action learning coaching typically begin with an initial welcoming meeting which focuses on tools that identify each set member's personal or professional problem, after which, the set members bring a significant work problem to the group for consideration. As they present their cases, set members use questioning insights to help case holders discover their own "solutions". Ultimately, members draw their own conclusions about the problem through reflection and group perspectives, and are offered support towards effective actions. Any learning derived from the participants' sharing of professional experiences, interrogation of case holder assumptions and reflection upon the resulting action is not limited to just the case holder. In fact, all participants—including the coach—learn through each other's learning and reflection during and after the set.

To operationalize action learning coaching, a model is shown in Figure 1 followed by an explanation of its steps:

Figure 1. Action learning coaching model

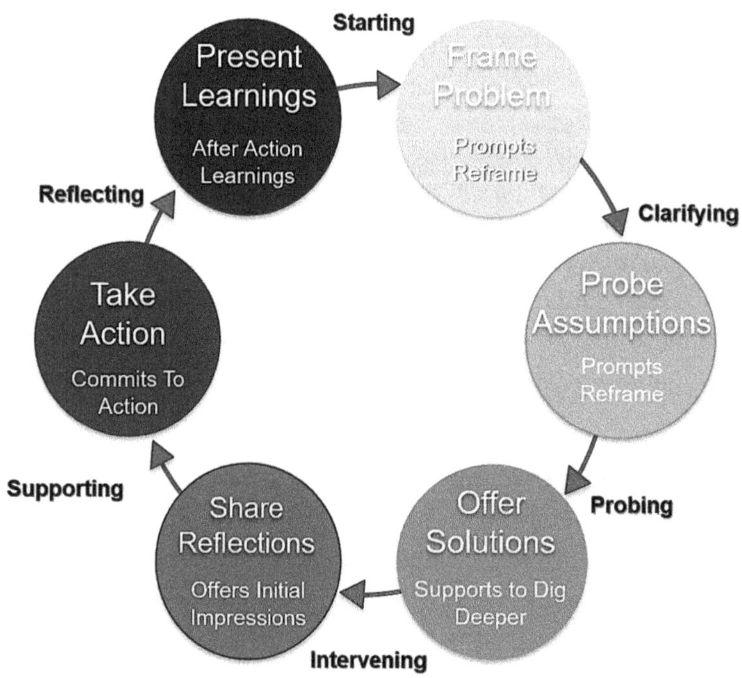

1. **Framing Problem:** A first meeting is held to identify each set member's personal or professional problem that will become their focus during the set meetings Volz-Peacock, Carson and Marquardt (2016). Prompted by the coach, participants define an area of personal and/or professional development (can be derived through 360° assessments, personality testing, and/ or reflection). As "homework", and as a precursor to presenting the problem to the group, each individual also reflects upon the problem and further develops through the use of an Action Learning Toolkit (Watkins & Marsick, 2011). The coach lays out the format for upcoming set meetings and the members decide on the rules for the action learning set meetings.
2. **Clarifying Questions:** During an action learning coaching meeting, one case holder will present and provide context to their previously discovered personal or professional problem, as described in their Action Learning Toolkit (Watkins & Marsick, 2011). The coach opens a dialogue and models clarifying questions posed to the case holder; the set works together to ask the case holder challenging questions. After a round of questioning, the coach can pause to ask the case holder to consider the sets' input and identify if the problem has aptly been recognized. If not, another round of questions and reframing ensues.
3. **Probing Assumptions:** After the problem is adequately identified for the case holder and the set, the coach resumes the set by asking questions in an effort to uncover hidden assumptions the case holder may be making about the situation that brought them to present the case. The coach may intervene whenever she sees an opportunity to help members develop insight, improve upon their problem-solving process, or provide just-in-time practice of a particular skill, such as role play (Volz-Peacock, Carson & Marquardt, 2016).
4. **Offering Solutions:** By this point, the case holder, through questioning, likely exhibits greater understanding and comes to his or her own conclusions about the problem through deep reflection. The coach may again pause the prescribed process to ask the set to offer an array of solutions for the case holder. The case holder gets to choose the path of action that feels right for them, and commits towards taking actions on the problem in their work setting.
5. **Sharing Reflections:** At the end of the session, the case holder followed by volunteers from the set offer initial reflections on their learning through the process of action learning coaching. The coach encourages further reflection and action, and concludes the set.
6. **Presenting Learnings:** After presenting their case, the case holder reflects and takes action on their workplace problem and reports back to the set any insight or further understanding culled from implementing that action, and any following action reflections. The coach seeks offers for a new case holder and initiates another action learning set meeting.

As we have seen, action learning coaching shows promises to allow individuals to gain competency while working on problems occurring in their own work settings, and reflect with peers offering insights into each other's workplace problems (Cho, 2013). There have been a few models that specifically show this level of integration between action learning and coaching. A model has been shown here that not only imagines the steps that lead to action learning coaching facilitation and participation, but also includes potential verbiage to envisage the process.

A Case Study of Novice HRD Practitioners

The difference between cleverness and wisdom is a small but important distinction. Reg Revans, the man credited with the creation of action learning said, "The clever man will tell you what he knows; he may even try to explain it to you. The wise man encourages you to discover it for yourself even though he knows it inside out" (McNiff, 1988, p. 52). In the workplace, human resource development (HRD) professionals are being trained to be clever, instead of being groomed to be wise. Certainly, a clever person is sharp, skillful, witty and inventive (Carter, 2015). However, as a person becomes wiser, they understand that in order to most effectively change a group, company, environment or the world, the individual must first change from within (Garrison, 2012). This case study examined just that—how novice HRD professionals could best make internal changes, gain competency and build capacity using action learning action research (ALAR) to enhance their professional practice and personal development.

Methods

ALAR is an integration of action learning, used in employee development and action research, used in social work, education and higher education (Zuber-Skerritt, 2015). In ALAR, members share their experiences with the intention of learning about themselves and from each other (Kuany, 2012) and have been utilized particularly when referring to research that involves improvement of individual practice (Zuber-Skerritt, 2009).

Both action learning and action research are based on a similar learning cycles, focus on learning in action and emphasize on collaborative relationships. Where they are different is in their goals. Action learning is an educative process with its focus on learning, while action research is focused on research (Coghlan & Coughlan, 2007). As both action research and action learning are related processes (Dick, 1997), there are a variety of approaches to action research (Aniekwe, et. al., 2012). Zuber-Skerritt's (2009) ideation of ALAR involves iterative cycles of identifying the problem, making plans for innovative improvements, taking action, interpreting results and making plans for the next cycle, as shown in Figure 2.

The following case study gleans insights from a group of novice human resource development (HRD) practitioners convened as an action learning coaching set as part of a dissertation study. Their aim was to investigate, for their personal benefit and for the benefit of the systems in which they interacted, a deeper understanding of how novices might best acquire practice competency and learn through the learning of others (Carter, 2015). The steps taken below show the deployment of the action learning action research (ALAR) study, the framing of the case study research, defined the participants, and indicated actions taken as well as outcomes derived from the research study. This chapter chronicled the (ALAR) case study. Running concurrent to the research story of 10 novice participants were the action learning action research (ALAR) cycles of identifying the problem, making plans for innovative improvements, taking action, interpreting results, and making plans for the next actions.

ALAR Step 1: Identifying the Problem

The Association for Talent Development (ATD), a professional organization dedicated to the promotion of talent and training development, was a sponsor of this study. ATD's membership performs human resource development (HRD) duties and hold diverse titles such as instructional designer, corporate trainer, business coach, and organization development consultant. ATD, formerly ASTD, American

Figure 2. ALAR model
Zuber Skerritt (2009).

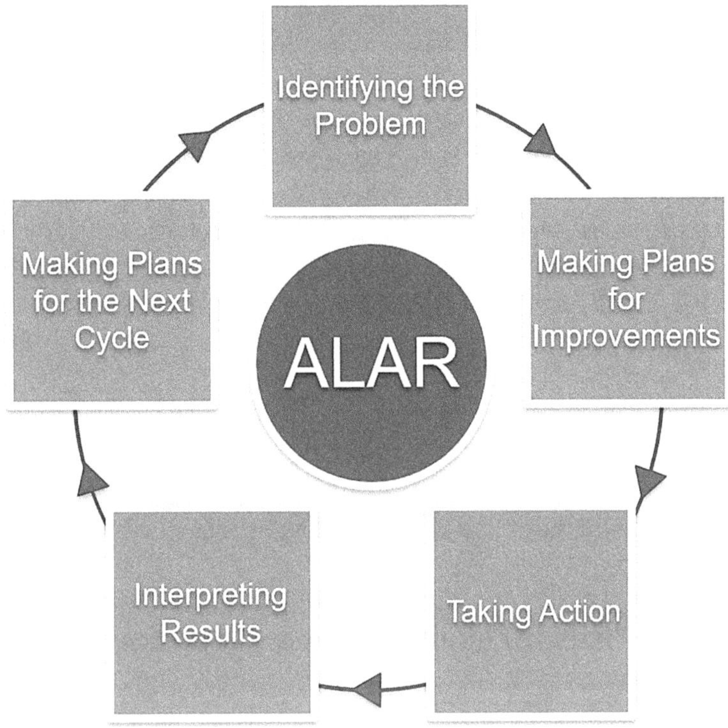

Society for Training and Development, was established in 1944, and is a non-profit association formed to study the dynamics of the workplace to propagate the advancement of the workplace HRD professional. ATD seeks as its mission to empower professionals to develop knowledge and skills successfully, and to recognize the impact learning has within a global and democratic context. (Association for Talent Development, 2014).

As a member of ATD, the researcher had occasion to observe and discuss the lack of socialization and induction opportunities for novices. Ideally, ATD fosters socialization into the profession, transmits knowledge and cultural understanding, provides legitimization of credentials, and aids personal individual development (Jarvis, 1987). The association intends for learning to be derived from participation in formal activities such as conferences, workshops, and seminars, and informally through mentorships, networks, and committee participation (Rusaw, 1995). Formal training was used rather inadequately as the sole means of helping novice trainers find their way in their new professions. Scholars argue that formal training provides impractical learning because it does not incorporate live or real-world experience into the learning process, and does not induce sufficient reflection (Coghlan & Coughlan, 2007). It is widely known that adults learn best from live projects, from the support and constructive criticism of colleagues, from rigorous self-reflection leading to serious reinterpretation of their previous experiences, and from a willingness to test their hypotheses in action (Garratt (1991, p. 24-25).

The framework that influenced and guided the study consisted of novice capacity building using a blended method of action leaning action research, and business coaching (Carter, 2015). The combination of action learning and action research under the acronym ALAR was chosen as an ideal platform for engaging novices in improving processes, and applying practical means of learning (Chapman, 2002). ATD became the site for which the research would be enacted, using professionals who belonged to the association and who were new to the field of workplace learning. As a first step in the process, the purpose and problem of practice were identified.

The purpose of the study was to explore the use of action learning coaching as a means of personal and professional development for novice HRD practitioners. The participants, gleaned from ATD chapters across the southeastern part of the United States, experienced problems of practice that novices within ATD and the HRD field faced. To ascertain what they learned from participation, and determine how, if at all, novice HRD professionals apply that learning to their workplaces, their professions, and their lives, three key questions guided the study and provided for the exploration of relationships between action learning coaching and the professional development of novices in the field of human resource development:

1. How do novice human resource development practitioners construct the challenges they face in their new roles?
2. How do these novices make meaning of their participation in action learning coaching?
3. What systemic changes will they make in their professional practice as a result of participation in action learning coaching?

ALAR Step 2: Making Plans for Innovative Improvements

After deciding on the research problem and questions, the next step of the ALAR process was to start formulating plans to improve upon the problem. To do this, the researcher convened five participants who were novice HRD professionals from different companies, after they answered recruitment information posted on the Association for Talent Development (ATD) website. Three women and two men (Daria, Lisa, Marie, Rasheen, and Travis) were new to the fields of instructional design, training, and/or talent development and were all within 1.5 years of starting new roles, as is reflected in the demographic Table 1.

ALAR Step 3: Taking Action

The third step in the ALAR process involved taking action. Early on in the first meeting, Rasheen, Marie, Travis, and Lisa, introduced themselves and chose a pseudonym in keeping with the privacy and anonymity promised in the study. The coach explained the research process, primed the group to offer feedback and reflections and discussed action learning coaching methodology. Each set member was

Table 1. Study demographics

Daria	Female	36	Healthcare	Education Specialist	4 Months
Lisa	Female	23	Pharmaceutical	HR Assistant	2 Months
Marie	Female	24	Manufacturing	Master's Student and HR Intern	4 months
Rasheen	Male	30	Non-Profit	Training Specialist	1 Year, 2 months
Travis	Male	26	Manufacturing	Trainer	1 year

asked to identify an initial problem to present to the group that was of reasonable difficulty, or worrisome to the individual in their work.

After the set worked through the steps of action learning coaching, the coach asked for reflections, as they would begin the actual process the next week. Rasheen thought the sessions were both practical and exciting. He noted, "Theories are fantastic, but I need something practical. And so even if the practical stuff doesn't work or it fails a hundred fifty million times, at least the constant generating of ideas thrills me". Travis, too, spoke about theory and practical implementation as a novice: "There are certain theories that you might learn in school, but until you get into that specific role you can't really apply [them]. And also, it's not [that] it's not relevant but it's just a whole other ball game".

Before the sessions began in earnest, the participants filled out the Action Learning Toolkit (Watkins & Marsick, 2011), and prepared to interrogate the problems they had being novices in their new roles. The coach began by framing the study's questions and asking for feedback, then asked what the real problem was in the context of them being novices in their new roles. Travis spoke about application, noting, "Sometimes there are certain theories and stuff that you might learn in school, but until you get into that specific role you can't really apply that stuff". Rasheen did not previously have an understanding of exactly the tasks he would be expected to perform, and Lisa felt she did not know the terminology well. Daria said that the particulars of the role were eluding her: "I've done this kind of a role before, but . . . it's like learning a totally new language".

Travis' Case

Travis had previously volunteered to be first to present his case, and he began by asking how he could make others within the organization understand his role. After questioning, Travis realized his problem was actually intrinsic. He had a "problem sometimes overcoming those expectations that people have of me". Further questioning from the group led to Travis understanding some of the assumptions he was making, such as:

1. Travis assumes people want him to be the expert.
2. Travis assumes he cannot fill the role.
3. Travis assumes he has no clout.

As the final exercise, Travis came up with an action plan of proactively meeting with managers and to report the next week on the actions he actually did take.

The third week of sessions started in the same manner as previous sessions. The coach asked Travis to reflect on the last week:

I have noticed that I've been nicer to some of the employees when I do my training. I'm not so stone-faced or so monotone. My training seems better actually, believe it or not, because I have my shoulders down. I'm not waiting for somebody to say something out of the way to me. I'm able to connect with them better on another level because I don't think that they're all against me.

Travis was not the only one who had reflections on his case, as Rasheen also confessed that he was confronted with an issue similar to Travis' and though he was not impervious to the necessity, he felt challenged to act in the moment. The next time when he was confronted with a similar situation, he pledged to act. Rasheen applied learnings from Travis' case to his own work setting: "at the time of Travis's [case] I remember that in that conversation, there were some things that I pulled out and tried to apply within my job."

Marie's Case

Marie's presentation of her case began with a discussion of her perceived problem with having little technical knowledge in her new role. Marie felt her coworkers were giving her subtle hints that led her to believe she was not up to her role. Reframing her initial question, Marie understood her problem to be, how can Marie quiet her feelings of internal pressure to be perfect and need to stay in control?

With the group's help through questioning, she reframed her initial inquiry. She discovered that the problem was not in others' expectations, but that she was putting too much pressure on herself. Her assumptions played into her harsh view about herself. Assumptions she may have been making were, first that Marie assumes she is incompetent because she does not know the technical information, second, she assumes people think the way she does about herself, and third, that she assumes she represents the entire field of HR/OD. The coach prompted a discussion on anxiety, and asked if anyone had methods they used to counteract it. Daria discussed her use of mindfulness. At the end of the session, Marie pledged to take action by seeking out a mentor, giving herself a break and trying out the mindfulness technique. Marie reflected:

I know over time I'll learn, but it's got to be something I have to be willing to ask and get more details and just being embarrassed that I may not know the information. Because the end goal is to help and I've got to get over my pride a little bit and maybe seek out that help.

As a closing reflection, Lisa felt a kindred spirit with Marie in wanting to try mindfulness. In the interim, the coach reflected on the session and sensed that something else was at play in the novices' hesitance to take up their roles, deciding to send an article about Imposter Syndrome to the group so as to read prior to the next meeting.

A discussion on mindfulness started during week four. Lisa researched the technique and was confused. Daria explained that it helps her to focus on something else so that she can go back and approach the problem with a calmer demeanor.

Daria's Case

Initially, Daria wanted the group to help her decipher how she could survive in a job in which she was overqualified. Her initial problem was framed as, how can I demonstrate that I am ready to assume more/greater responsibility in my new role? After the action learning process, Daria reframed her problem and realized that she was afraid of getting back out into the job market so soon after her long term unemployment. After questioning from the group, her problem became: Should Daria leave the job she's overqualified for or approach her boss about a role change?

Daria worked through several assumptions; including the idea that she thought job-hopping was inherently bad, that Daria assumes she will lose her job if she tells them she is overqualified and bored, and that Daria assumes her job search will be long, or that she has to leave this job before she finds another. Daria admitted she was ultimately afraid of being unemployed again and was thus trying to maintain a low profile. With a promise to make some industry contacts, and look for opportunities to expand role as her action items, Daria looked to make actionable her action learning case. To start the week, Daria reported to the group that she had indeed taken action over the previous week in reaching out to a few industry contacts to try to set up some informational interviews.

Rasheen's Case

In Rasheen's case, he brought the issue of his micromanaging boss to the group for assistance. He discussed her increasingly dominant management style and was comforted that the group could "feel what I feel." Rasheen's boss was due to be off on an extended medical break, and he felt (and the group agreed) that waiting for them to leave was an action in itself. He also thought he could, in the interim, be very selective and deliberate in his communications with his boss. In this case, the coach did not prompt for assumptions, opting instead for a technique called question-storming to help him generate alternatives for communication with his boss, including:

1. Continue to seek opportunities to learn from your boss.
2. Be intuitive
3. Self-care
4. Wait them out- it will get better after extended medical leave

The coach reviewed learning from the previous week and asked for reflections or actions in the interim. Daria reflected on the challenge of being new in the workplace: "I think it's just about knowing yourself and what you are willing to accept".

Lisa's Case

Lisa began to frame her problem about her boss was going out of town and left Lisa, a novice, in charge of a hiring team. She had never done anything like this and was inevitably very nervous. The coach suggested the group role play in order to help her practice. Although Lisa was nervous, she and the rest of the group thought it would be an excellent idea. With that, the group started to act as if they were Lisa's interview team, each naturally taking on a role and Lisa guided the team to make a "decision" on a simulated candidate. At the end, the set was very complimentary of Lisa's handling of the learning opportunity, and she was relieved, opining that the activity was worthwhile. Her action items were the following:

1. Practice
2. You can admit that you are new, helps to clear the air
3. Look at it as a great learning experience
4. Ask for feedback, as the more comfortable you get at it, the more comfortable it will be

Lisa came back to the group in this concluding meeting and reported that she had, in fact applied her newfound skills learned during role play to a new situation at work. She said that her colleague at first thought the exercise strange, but that her confidence in using the interview techniques grew exponentially.

I did go up to the hiring manager who's responsible for making decisions about the position that we were hiring, and I told him I wanted to learn more about this position. I kind of put myself as the candidate to kind of see what it is that he was looking for in the candidate. So I role played with him to see what he was expecting. And so based on that, I wrote out the questions that I felt like would be necessary to ask during the close out meeting.

When asked how she felt after her try at role-playing, Lisa reflected, "he said that he had never seen this before, which makes me feel good, because he's learning something from me. It's not just me learning from him." Her capacity increased, evidenced by her own admissions: "I definitely think that I will behave differently—I have been behaving differently. I try to incorporate it".

Daria, Lisa and Travis were also excited about what they learned during the previous weeks.

Daria: *I really just enjoyed hearing everybody's own individual issues because it made me feel that I'm not alone. I feel the same way that everybody else feels and when I try to do something new or take on a new task or whatever, we all have the same set of similar issues in that we don't want people to think that we're not qualified to do what we're doing.*

Lisa: *I think that I really like this whole activity, just because I felt like I was not alone and also it's a reminder that it's okay to be a novice and it is okay to make mistakes and just learn from the mistakes. It is also okay to not be afraid to talk with your boss because I did have a hard time bringing up this issue or problem to my boss. But really, communication is the key to a lot of things, so I feel like this is a great reminder for that.*

Travis: *I think that's great. When a different situation arises, you've gone through that situation before and you have something in your background to help you through that. I think that's the winning piece of it all.*

ALAR Steps 4 and 5: Interpreting Results and Making Plans for the Next Cycle

The final stages of the ALAR process involve interpreting the results of the experience and making decisions on how to conclude the research. In a nutshell, the research yielded a great knowledge and learning experience for the participants. Specifically, the research produced the following results:

1. The first research question asked, "How do novices construct the challenges they face in their new roles?" The problems the novices expressed were collected into three dominant themes relating to stress, self-concept, and their jobs. Problems were evidenced in feeling pressure, being overwhelmed, and with setting expectations, both on their own and from others.
2. The second research question asked, "How do novices make meaning of their participation in action learning coaching?" Novices learned from within and through others as a key finding. The ability to learn, reflect, and transfer learning to novel situations was a key outcome of the study and a hallmark of the action learning process.
3. The last question asked, "What systemic changes will novices make in their professional practice as a result of participation in action learning?" Participants learned skills they could use in their work and in their lives. Participants came away empowered through action learning, and applied their what they learned to their roles and used their knowledge in new and novel situations. Participants reported learning skills they could use in their work and in their lives (Carter, 2015).

Finally, one of the tenets of ALAR is that there is a public display or presentation of study findings. Zuber-Skerritt (2002) offers oral and written presentations as vehicles for individual and team learning, reflection and conceptualization (pg. 146). Action learning sessions that forgo public presentation produce learning that is transitory, unnoticed, unappreciated and unrewarded (Zuber-Skerritt, 2002). When the presentation and recommendations were sent out to the set members for feedback and accompaniment to the meeting, one participant responded: "I believe if I were taught and informed that it is ok not to know everything, my transition into the work environment would have been easier. We all want to be experts, but have to realize it takes time".

CONCLUSION AND FUTURE RESEARCH DIRECTIONS

As a practice in its own right, action learning coaching borrows first from action research programs, where creativity remains one of the core values. Practitioners engaging in the practice should be willing to 'take risks, be flexible and innovative, and to encourage others to be likewise' (Zuber-Skerritt, 2002, p.149). The risk is well worth the reward, as creativity allows for the design and development of programming that meets participant's emergent needs.

Original permutations of action learning omit coaching or at least discourage its use. Reg Revans has famously denounced reliance on an action learning coach, "decry[ing] models that stray[e]d too far from his conceptualization, but healthy experimentation and critique help it grow" (Marsick and O'Neil, 1999, pg. 160). In fact, Revans was adamant that the initial facilitator give way to the group process and groups should not be hampered or dependent on a coach (Johnson, 2012; Pedler, & Abbott, 2008). Further, Revans felt that reflection was a fundamental component in participants' learning. However recently several theorists have come to embrace and even elevate the learning coach to facilitate the process. Volz-Peacock, Carson & Marquardt (2016), O'Neil and Marsick (2014), and Dilworth (2010), are all enthusiastic ambassadors of the coaching aspect of action learning. At the same time, coaching is both rising in stature and use in the workplace. It stands to reason that action learning coaching should be a tool in the professional coach's arsenal to aid and develop employees, particularly those who are not already leaders.

Action learning coaching takes cues from established tradition of action learning that has as an outcome the ability to deal with the inevitable change that occurs in life (Stark, 2006). In organizations, much of the initial training for novices is front-loaded and includes formal education and/or training that need to be completed by entrants to an occupation before they can be regarded as qualified workers. Usually taking place in classrooms remote from the workplace and not conducive for proper transfer, this 'front-loaded' training may not be sufficient, and can actually mask the importance of learning from practice (Beckett & Hager, 2002). Within action learning is the premise that proficiency is developed by working in challenging and supportive environments, self-monitoring and engaging in deliberate practice (Sheckley & Keeton, 1999). Action learning coaching lends support to learning that takes place through problems that are ill-defined and 'wicked' (Sankaran & Brown, 2012) and solving such problems allows for meaningful learning opportunities.

Setting the stage for learning is an important part of action learning coaching. It is important to retain the idea of learning through action (Vaartjes, 2005) that is inherent in action learning programs. As stated previously, for this study, action learning was coupled with business coaching, which involves unlocking a person's potential to maximize their own performance, and helps them to learn rather than teaching them (Whitmore, 2003). Coaching was integral as an in-the-moment intervention that served to deepen the learning of the participants.

The researcher-as-coach dynamic deepened the learning and outcomes of the participant's cases and served to guide the participants' on their quest for self-improvement. Action learning coaching can be a first step for participants in a journey toward greater self-insight, greater capacity to learn from experience and better awareness of the political and cultural dimensions of organizational change (Marsick & O'Neil, 1999). Certainly, the time is ripe to explore action learning coaching specifically as a means to develop future individual capacity.

REFERENCES

Abbott, G. N. (2006). *Exploring evidence-based executive coaching as an intervention to facilitate expatriate acculturation: Fifteen case studies*. Australian National University.

Acker-Hocevar, M., Pisapia, J., & Coukos-Semmel, E. (2002, April). *Bridging the abyss: adding value and validity to leadership development through action learning—cases-in-point*. Paper presented at the Meeting of the American Educational Research Association, New Orleans, LA.

Allen, T. D., Eby, L. T., & Lentz, E. (2006). Mentorship behaviors and mentorship quality associated with formal mentoring programs: Closing the gap between research and practice. *The Journal of Applied Psychology*, *91*(3), 567–578. doi:10.1037/0021-9010.91.3.567 PMID:16737355

Aniekwe, C. C., Hayman, R., Mdee, A., Akuni, J., Lall, P., & Stevens, D. (2012). *Academic-NGO collaboration in international development research: A reflection on the issues*. A Working Paper.

Axelrod, B., Handfield-Jones, H., & Welsh, T. (2001). War for talent, part two. *The McKinsey Quarterly*, *2*, 9–11.

Bacon, T. R., & Spear, K. I. (2003). *Adaptive Coaching: The art and practice of a client-centered approach to performance improvement*. Palo Alto, CA: Davis-Black.

Baron, L., & Morin, L. (2009). The coach-coachee relationship in executive coaching: A field study. *Human Resource Development Quarterly*, *20*(1), 85–106. doi:10.1002/hrdq.20009

Beckett, D., & Hager, P. (2002). *Life, Work And Learning: Practice in Postmodernity*. London: Routledge.

Bialek, T. K. (2014). *An examination of how participation in a cohort-based leadership development program for high-potential employees contributes to the development of leaders at a major professional services firm in the United States* (Unpublished doctoral dissertation). University of Minnesota.

Boshuizen, H. P., Bromme, R., & Gruber, H. (2004). *Professional learning: Gaps and transitions on the way from novice to expert*. Dordrecht: Kluwer Academic. doi:10.1007/1-4020-2094-5

Boudreau, J. (2004). Talentship: A decision science for HR. *Strategic HR Review*, *3*, 28–31.

Bransford, J. (2000). *How people learn: Brain, mind, experience, and school*. Washington, DC: National Academy Press.

Bransford, J., Pellegrino, J. W., & Donovan, S. (1999). *How people learn: Bridging research and practice*. Washington, DC: National Academy Press.

Carter, A. D. (2015). *Building novice capacity through action learning coaching* (Doctoral dissertation). University of Georgia, Athens, GA.

Chapman, J. (2002). *Monitoring and Evaluating Advocacy, PLA Notes No. 43*. London: IIED.

Cho, Y. (2013). What is Action Learning? Components, Types, Process, Issues, and Research Agendas. *Learning and Performance Quarterly*, *1*(4), 1–11.

Cho, Y., & Egan, T. M. (2009). Action learning research: A systematic review and conceptual framework. *Human Resource Development Review*, *8*(4), 431–462. doi:10.1177/1534484309345656

Clutterbuck, M. D., Megginson, M. D., & David, M. S. (2013). *Beyond goals: Effective strategies for coaching and mentoring*. New York: Routledge.

Coghlan, D. & Coughlan, P. (2007). *Action learning and action research (ALAR): A methodological integration in an inter-organizational setting*. Springer Science+Business Media, LLC.

Cope, F. (1998). Current practices: Current issues in selecting high potentials. *Human Resource Planning*, *21*(3), 15–16.

Cowan, C. A. (2013). *An exploration of significant leadership development factors in action learning: A comparison of three action learning programs* (Doctoral dissertation). The George Washington University.

Cox, E. (2013). *Coaching Understood: A pragmatic enquiry into the coaching process*. London: SAGE.

Cullen, J., Hadjivassiliou, K., Hamilton, E., Kelleher, J., Sommerlad, E., & Stern, E. (2002). *Review of current pedagogic research and practice in the fields of post-compulsory education and lifelong learning*. London: The Tavistock Institute.

Daley, B. J. (1999). Novice to expert: An exploration of how professionals learn. *Adult Education Quarterly*, *49*(4), 133–147. doi:10.1177/074171369904900401

Davey, C., Powell, J. A., Powell, J. E., & Cooper, I. (2002). Action learning in a medium-sized construction company. *Building Research and Information*, *30*(I), 5–15. doi:10.1080/09613210110090403

David, S., Clutterbuck, D., & Megginson, D. (2014, August). Goal orientation in coaching differs according to region, experience, and education. *International Journal of Evidence Based Coaching and Mentoring*, *12*(2).

de Haan, E., Bertie, C., Day, A., & Sills, C. (2010). Clients Critical Moments of Coaching: Toward a Client Model of Executive Coaching. *Academy of Management Learning & Education*, *9*(4), 607–621. doi:10.5465/AMLE.2010.56659879

de Haan, E., & de Ridder, L. (2006). Action learning in practice: How do participants learn?. *Consulting Psychology Journal: Practice and Theory, 58*(4), 16- 31.

Dick, B. (1997). *Action learning and action research*. Retrieved from http://www.scu.edu.au/schools/gcm/ar/arp/actlearn.html

Dilworth, R. L., & Boshyk, Y. (Eds.). (2010). *Action learning and its applications*. Basingstoke, UK: Palgrave Macmillan. doi:10.1057/9780230250741

Drucker, P. F. (1993). *Post-Capitalist Society*. New York, NY: HarperCollins.

Ellinger, A. D., Hamlin, R. G., & Beattie, R. S. (2008). Behavioural indicators of ineffective managerial coaching: A cross-national study. *Journal of European Industrial Training*, *32*(4), 240–257. doi:10.1108/03090590810871360

Ellinger, A. D., & Kim, S. (2014). Coaching and human resource development examining relevant theories, coaching genres, and scales to advance research and practice. *Advances in Developing Human Resources*, *16*(2), 127–138. doi:10.1177/1523422313520472

Garratt, B. (1991). The power of action learning. In Action learning in practice (pp. 15-30). Gower Publishing, Ltd.

Garrison, G. (2012). Clever vs. wise. *Journal of Healthcare Contracting*. Retrieved from http://www.jhconline.com/wisdom-vs-cleverness.html

Hawkins, P. (2008). The coaching profession: Some key challenges. *Coaching: An International Journal of Theory, Research and Practice*, *1*(1), 28–38.

Illeris, K. (2003). Workplace learning and learning theory. *Journal of Workplace Learning*, *15*(4), 167–178. doi:10.1108/13665620310474615

Janosik, S. M., Carpenter, S., & Creamer, D. G. (2006). Beyond Professional Preparation Programs: The Role of Professional Associations in Ensuring a High Quality Workforce. *College Student Affairs Journal*, *25*(2), 228–237.

Johnson, M. (2012). *Business Consulting: What you Need to Know For IT Operations Management*. Emero Publishing.

Ladegård, G. (2011). Stress management through workplace coaching: The impact of learning experiences. *International Journal of Evidence Based Coaching and Mentoring*, *9*, 29–43.

Lamoureux, S. T. (2013). *Five Trends in Leveraging Leadership Development to Drive a Competitive Advantage*. Bersin Research Bulletin.

Maltbia, T. E., Marsick, V. J., & Ghosh, R. (2014). Executive and organizational coaching: A review of insights drawn from literature to inform hrd practice. *Advances in Developing Human Resources*, *16*(2), 161–183. doi:10.1177/1523422313520474

Marquardt, M. (2004). *Optimising the power of action learning: Solving problems and building leaders in real time*. London: Davies-Black.

Marquardt, M., Leonard, S., Freedman, A., & Hill, C. (2009). *Action Learning for Developing Leaders and Organizations through Action Learning*. Washington, DC: American Psychological Association Press.

Marquardt, M. J. (1999). *Action learning in action: Transforming problems and people for world class organizational learning*. Palo Alto, CA: Davies-Black Press.

Marsick, V. J. (2003). Invited Reaction: Informal Learning and the Transfer of Learning: How Managers Develop Proficiency. *Human Resource Development Quarterly*, *14*(4), 389–395. doi:10.1002/hrdq.1075

Marsick, V. J., & ONeil, J. (1999). The many faces of action learning. *Management Learning*, *30*(2), 159–177. doi:10.1177/1350507699302004

Matthews, P. (1999). Workplace learning: Developing a holistic model. *The Learning Organization*, *6*(1), 18–29. doi:10.1108/09696479910255684

McLagan, P. (2015). Preface: Making sense out of complexity. *Advances in Developing Human Resources*, *17*(2), 139–144. doi:10.1177/1523422315572617

McNiff, J. (1988). *Action research: Principles and practice*. London: Routledge. doi:10.4324/9780203305676

Moen, F. (2011). Executive coaching and the effect on causal attribution. *International Journal of Coaching in Organizations*, *8*(2), 90–105.

Mumford, E. (1994). New treatments or old remedies: Is business process reengineering really socio-technical design? *The Journal of Strategic Information Systems*, *3*(4), 313–326. doi:10.1016/0963-8687(94)90036-1

Noe, R. A., Clarke, A. D. M., & Klein, H. J. (2014). Learning in the twenty-first century workplace. *Annual Review of Organizational Psychology Organizational Behavior*, *1*(1), 245–275. doi:10.1146/annurev-orgpsych-031413-091321

O'Neil, J., & Marsick, V. J. (2007). *Understanding action learning*. AMACOM Div American Mgmt Assn.

ONeil, J., & Marsick, V. J. (2014). Action Learning Coaching. *Advances in Developing Human Resources*, *16*(2), 202–221. doi:10.1177/1523422313520202

Pedler, M. (1997). Interpreting action learning. *Management learning: Integrating perspectives in theory and practice*, 248-264.

Pedler, M. (2003). *Action learning in practice* (3rd ed.). Aldershot, UK: Gower.

Pedler, M., & Abbott, C. (2008). Am I doing it right? Facilitating action learning for service improvement. *Leadership in Health Services*, *21*(Iss: 3), 185–199. doi:10.1108/17511870810893010

Pedler, M., Burgoyne, J., & Brook, C. (2005). What has action learning learned to become? *Action Learning*, *2*(1), 49–68. doi:10.1080/14767330500041251

Raudenbush, L., & Marquardt, M. (2008). Growing leaders at the US Department of Agriculture: A case study of leadership development using action learning. *Public Administration Quarterly*, 147–173.

Ready, D., Conger, J., Hill, L., & Stecker, E. (2010). The anatomy of a high potential. *Business Strategy Review*, *21*(3), 52–55. doi:10.1111/j.1467-8616.2010.00685.x

Reinelt, C. (2006). A Framework for Leadership Investment and Evaluation. In *Investing in Leadership* (vol. 1). Grantmaker's Framework for Understanding Nonprofit Leadership Development.

Revans, R. W. (1972). Action learning: A management development program. *Personnel Review*, *1*(4), 36–44. doi:10.1108/eb055215

Revans, R. W. (1982). *Origins and Growth of Action Learning*. Bromley: Chartwell Bratt.

Sankaran, S., & Brown, S. (2012). Coaching collaborative creativity and innovation: An action-based method for sustainable innovation learning and development in business organization. In O. Zuber-Skerritt (Ed.), Action research for sustainable development in a turbulent world (pp. 127–150). Bingley: Emerald Publishing.

Senge, P. M., Schamer, C. O., Jaworski, J., & Flowers, B. S. (2005). *Presence: Exploring profound change in people, organizations, and society*. New York, NY: Doubleday.

Sherman, S., & Freas, A. (2004). The wild west of executive coaching. *Harvard Business Review*, *82*(11), 82–93. PMID:15559448

Slåtten, T., Svensson, G., & Sværi, S. (2011). Service quality and turnover intentions as perceived by employees: Antecedents and consequences. *Personnel Review*, *40*(2), 205–221. doi:10.1108/00483481111106084

Smith, P., & ONeil, J. (2003). A Review of Action Learning Literature 19942000 Part 2. *Journal of Workplace Learning*, *15*(4), 154–166. doi:10.1108/13665620310474606

Stark, S. (2006). Using action learning for professional development. *Educational Action Research*, *14*(01), 23–43. doi:10.1080/09650790600585244

Towell, M. (2013). *How can coaching improve effectiveness of performance management conversations?* (Unpublished doctoral dissertation). Oxford Brookes University, Oxford, UK.

Tushman, M. L., OReilly, C. A., Fenollosa, A., Kleinbaum, A. M., & McGrath, D. (2007). Relevance and rigor: Executive education as a lever in shaping practice and research. *Academy of Management Learning & Education*, *6*(3), 345–362. doi:10.5465/AMLE.2007.26361625

Vaartjes, V. (2005). Integrating action learning practices into executive coaching to enhance business results. *International Journal of Evidence Based Coaching and Mentoring*, *3*(1), 1–17.

Volz-Peacock, M., Carson, B., & Marquardt, M. (2016). Action Learning and Leadership Development. *Advances in Developing Human Resources*, *18*(3), 318–333. doi:10.1177/1523422316645884

Waddill, D., & Marquardt, M. (2003). Action Learning and the Five Schools of Adult Learning. *Human Resource Development Review*, *2*(4), 406–429.

Wasylyshyn, K. M. (2003). Executive coaching: An outcome study. *Consulting Psychology Journal: Practice and Research*, *55*(2), 94–106. doi:10.1037/1061-4087.55.2.94

Watkins, K.E., Marsick, V.J., & O'Neil, J. (2011). *Action Learning Toolkit*. Academic Press.

Whitmore, J. (2003). *Coaching for Performance*. London: Nicholas Brealey Publishing.

Winch, A., & Ingram, H. (2002). Re-defining the focus of workplace learning. *International Journal of Contemporary Hospitality Management*, *14*(7), 361–367. doi:10.1108/09596110210440666

Yorks, L., Lamm, S., & ONeil, J. (1999). Transfer of Learning from Action Learning Programs to the Organizational Setting. *Advances in Developing Human Resources*, *1*(2), 56–74. doi:10.1177/152342239900100205

Zuber-Skeritt, O. (2002). The concept of action learning. *The Learning Organization*, *9*(3), 114–124. doi:10.1108/09696470210428831

Zuber-Skerritt, O. (2015). *Action Research*. Palgrave Macmillan.

Zuber-Skerritt, O. D. (2009). *Action learning and action research: Songlines through interviews*. Sense Publishers.

KEY TERMS AND DEFINITIONS

Action Learning Coaching: An aspirational process by which an individual works on real personal or professional problems with the help of a small group called a set and a learning coach. Set members learn from each other through questioning insight, take action, and reflect on their actions. Coaches help individuals in the moment to work on and develop capabilities and skills for future growth.

Development: The acquisition of individual knowledge or skills for personal growth.

Emergent Leaders: Also called high potentials (HiPos), these are individuals who have the potential to develop into leaders in their workplace.

Formal Training: The learning and development undertaken at the individual level on the job, usually provided face to face by an instructor.

Human Resource Development Practitioners: Those whose role is to train and develop others in the workplace.

Novices: Employees new to the workplace, or to their roles.

Practice: To engage in a profession in a corporate or workplace environment.

Professional Coaching: The belief that family is central to well-being and those family members and family issues take precedence over other aspects of life.

Workplace Learning: The process used by individuals who are engaged in training programs designed to help them acquire skills and competencies to perform current and future roles in the workplace.

Chapter 4
Synergising Action Research and Technology Education:
A Pedagogical Perspective

Mishack Thiza Gumbo
University of South Africa, South Africa

ABSTRACT

Design process is a dominant method (backbone) for teaching the Technology Education subject in schools which demands critical and creative thinking skills. Moreover, the design process is research orientated, which could jell well with action research. This chapter explores the synergy between Technology Education and action research, and ultimately contributes a theoretical pedagogical framework which could help in teaching Technology by consciously pulling in action research principles into the design process, thus helping also to approach design process as a cyclical rather than a process.

INTRODUCTION

Can the design process, which is the method for teaching Technology, be driven by action research? The answer is yes. Both design process and action research involve problem solving approaches, which entail identifying and investigating the problem, designing a solution to the problem, (making the solution in Technology Education), evaluating the solution and making adjustable improvements if need be (Ferrance, 2000; Eggert cited in Williams, 2011). Though Technology Education is a school subject and action research a research design under research methodologies, it is important to note that the method of teaching Technology, that is to say design process, is inherently research orientated. Most importantly, the design process engages the above steps in a cyclical manner, which needs learners' reflective and critical thinking skills. On the other hand, action research also involves steps which should be executed cyclically. These similarities lead to synergy between Technology Education and action research (Mapotse, 2012; Merrill, 2017) and ultimately contribute towards enhancing the teaching of Technology. In this chapter I explore this synergy and contribute a framework which can concientise Technology Education teachers to enrich their pedagogical strategies by consciously involving action research in their teach-

DOI: 10.4018/978-1-5225-2642-1.ch004

ing activities. This jells well with the theme of this book. The framework in question is formed on an understanding of Technology, Technology Education, design process, action research and constructivist learning, which are discussed subsequently. An indigenous perspective is entertained in the discussion due to the diverse nature of learners in many educational contexts.

CONCEPTUAL FRAMEWORK

Technology

The word "technology" seems to be overused but with little insight (Mapotse, 2012:15), hence it is uneasy to define it (Heidegger, 1977; Dugger 2008; Li-Hua, 2009; Misa, 2009; Mapotse, 2012). Its overuse is caused by the fact that life on earth seems unliveable without technology – technology surrounds and 'lives' with us: it is evident in our eating, bathing, sleeping, traveling, working, etc. It is also evident in the processes and methods we follow and the tools, knowledge and materials we use. Why then is it uneasy to define technology? The high speed at which technology changes, different cultures and different versions of technology applied in different contexts are all behind this uneasiness. However, attempts have been made to define technology. A few are considered subsequently.

Technology generally refers to what has been described as high technology in an industrial context, for example computers, superconductivity, chips, genetic engineering, robotics, magnetic railways, etc. (Li-Hua, 2009, p. 18). Technology can also be defined as the technical means that people use to improve their surroundings (Kalanda, 2005). According to Ankiewicz, De Swart and Stark (2000, p. 40), technology concerns technological knowledge, skills and technological processes and it involves the understanding of the impact of technology on the individual and society. The definition by Ankiewicz, De Swart and Stark lies very close to that of the (South African) Department of Basic Education (DBE): "the use of knowledge, skills, values and resources to meet people's needs and wants by developing practical solutions to problems, taking social and environmental factors into consideration" (2011, p. 8).

Then there is an indigenous perspective to the definition of technology. Obikeze (2011) defines technology as any human-made or culture-generated devices, formulations or organisations that may be used for purposes of producing or creating needed goods and services. This production happens by employing tangible (material) and intangible (non-material) forms of technology, which root technology in a cultural context (Ogungbure, 2011). Tangible and intangible devices, formulations and techniques fulfil some need or provide some service for humankind in a given environment (Moalosi, Popovic, Hudson & Kumar 2005; Obikeze, 2011).

Three categories of devices include the following:

1. Material (physical) technology, for example bows and arrows, ploughs, looms, laboratories, machines, electronic devices, knives, fishing nets and explosives.
2. Social technology, for example methodologies, techniques, organisational and management skills, bookkeeping and accounting procedures, negotiating and counselling techniques and social institutions like patriarchy and women's league, songs, jokes, ideas and skills.
3. Communication technology, for example language, signs and symbols, drumming and the internet.

Organised into goods and services, these cultural products can in turn be applied in the current context as follows:

1. Material goods, such as soap, food items, houses, ornaments, aeroplanes and television sets.
2. Social goods or services, such as values, norms, customs, motherhood, priesthood, friendship, concerts and plays, football games, health and healing systems and belief systems.
3. Intellectual goods, such as ideas, abstract concepts, names, terminologies, cognitive knowledge and idioms.

The indigenous perspective on technology creates awareness that not all technology is tangible, a mistake which many make – each time the term "technology" is mentioned people look for a device or tool they can see, touch or use. The social services category of technology brings an invisible dimension to the term, which can be perceived better by looking at the processes or services rendered: knowledge, processes and techniques that are designed to address problems or satisfy needs or wants.

Since the indigenous definition of technology situates technology in a culture, it becomes important, then, to briefly define the term "culture". Culture is the totality of the way of life evolved by a people in their attempt to meet the challenges of living in their environment. Their natural instinct to question the unknown in order to find a rational explanation for the way things are in the universe, to seek understanding of the environment, to harness the resources therein and to make use of the earth's possessions has been the principal force in the development of humankind and society (Ogungbure, 2011, p. 88). It can be noticed from this definition that technology is an integral part of the cultural system, as is evidenced by the fact that cultures (people or communities) are in a constant mode of harvesting the natural environment to meet their needs or wants and even to solve problems. Chasing game for a kill can, for instance, be a problem; if one's hunger is not satisfied as quickly as it is desired, one may not succeed due to feeling lethargic. A design of a snare or trap thus resulted from the need to catch game without having to run for kilometres after it.

From this indigenous perspective, technology is intertwined with indigenous knowledge. The fact that indigenous knowledge is mostly evident in practical activities such as agriculture, food preparation and conservation, health care and education, it is befitting to refer to it as technology (Senabayake, 2006). This claim ties in well with the intangible form of technology. Indigenous knowledge is uniquely and closely related to a particular culture or society and can thus be referred to as local/traditional knowledge, folk knowledge, people's knowledge, traditional wisdom or traditional science (Senabayake, 2006). The holistic approach to life and nature, which is most common in indigenous communities, supports this soft side of technology (Watson & Chambers, 1989; Gumbo, 2016a).

If the different perspectives explored this far are taken into account, technology can now be defined as the amalgam of knowledge, processes, methods and resources engaged by a given culture to manipulate the environment in an effort to realise and sustain its livelihoods; such technology is evidenced in tangible and intangible forms.

Of much interest in this chapter is the scientifically held view of technology, that is to say technology-as-applied-science (TAAS), which is widely debated by scholars such as De Vries (1996), Davies (1997), Dugger and Naik (2001), Gil-pe'rez, Vilches, Fernana'ndez, Cachapuz, Praia, Valde and Salinas (2005) and Gumbo (2016b). This view helps in drawing the similarities between Technology Education and action research. The logic is that action research is one of many research designs or approaches in the field of science. It applies scientific principles even though it was met with critical opposition – a brief discussion on this is entertained under the section on action research below. Technology is taught to some extent through research as investigation principles are applied. However, this is seldom mentioned in research on Technology Education.

According to Davies (1997) the TAAS view claims that technological capability grows out of scientific knowledge. However, he entertains the opposing view as well, which claims that designers are open-minded and draw on knowledge from whatever other sources which can help them design solutions to technological problems. Technology is indeed not solely dependent on science; measuring applied in a bridge construction project is a mathematical activity, the technology of cooking needs some knowledge from consumer studies, the technology of marketing the designed and/or realised product needs some business studies and economics knowledge. That is why technologists working on a project normally see a need to team up with the geoscientists, social scientists, land surveyors, artists, etc. Davies (1997, p. 107) argues that science has a tendency to depend heavily on technology, both for ideas and apparatus; he bags this argument with the fact that Galileo would most probably not have succeeded in developing a model of the solar system without the telescope with which he observed the moons of Jupiter. But instead of being caught up in a chicken-and-egg debate of technology and science, energy can rather be spent on acknowledging how these two complement each other. The synergy between technology and science is illustrated in Figure 1.

Gil-pe'rez et al (2005, p. 312) contend that science draws on technological developments and achievements to understand the universe; there is, for instance, a need to conceive and construct experimental designs in order to test a hypothesis in research. For Gil-pe'rez et al (2005, p. 312), this is in itself technological work, that is to say to solve technological problems that humans encounter. Conversely, technology develops a solution or meets a need or want by developing technical objects, systems, products or processes, and for that it uses scientific principles, laws and theories (Québec Education Program,

Figure 1. The model of technology as it relates to scientific inquiry
Adapted from the Atlantic Province Education Foundation, 2001, p.6.

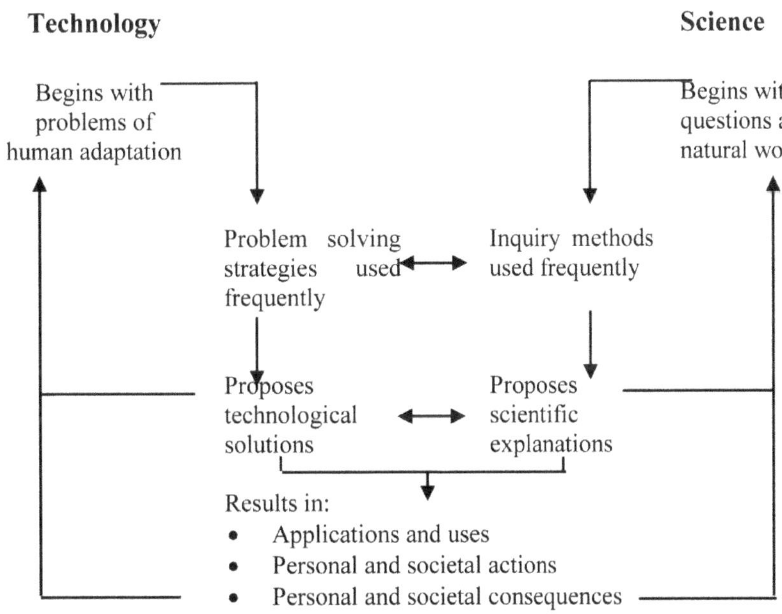

2016). Gil-pe'rez et al (2005, p. 314) remind us: "We should bear in mind that research is an answer to problems that are often linked to human needs and so to the search for adequate solutions to previous technological problems". The complementarity of technology and science can be seen in their respective practical and design approaches to the physical world (Québec Education Program, 2016, p. 2). In both cases direct engagement with the problem is central to defining and solving it (Massachusetts Science and Technology/Engineering Curriculum Framework, 2006, p. 8).

This discussion mirrors the synergy illustrated in Figure 1. Technological inquiry starts with identifying and explaining the technological problems existing in the social environment out there. This environment happens to be part of the natural environment which a scientist attempts to understand by posing questions worth investigating. A researcher could use certain research methods to address the problem, for example reflective notes, interviews or observation. Problem solving strategies could also include some of these methods to gather information which will enable the designer to come up with an informed solution. Both the technologist and researcher must come up with solutions to the respective problems they have identified. These solutions are intended for use by the society and should help to improve its situation or to bring about certain consequences. If the latter are not favourable, critical explanations and suggested improvements will have to be provided.

The foregoing definition and discussion on technology provide the bridge to the concept of Technology Education, which is looked at in the next section.

Technology Education

Let me begin this section by clarifying that I am not talking about educational technology. It is important to bring this clarity because "there is much confusion today when attempting to understand the differences between technology education and educational technology" (Dugger & Naik, 2001, p. 31). Technology Education is a subject taught in schools just like subjects such as Mathematics, Setswana, Science and so forth. The aim of the subject is to help learners understand the concept of technology so they will ultimately qualify as engineers, artisans and suchlike. Educational Technology, sometimes referred to as Instructional Technology, is a specialised field which has to do with the technological resources which can aid the teaching of subjects. Just as a teacher may need to use certain technological resources to teach Science, so may Technology teachers need technological resources to teach Technology.

In Technology Education learners are taught about the theory and practice of a range of material processes for metal, wood, plastic materials, textile, leather and food materials, with the greater and more important component being that of gaining practical experience (Kumar, 2002, p. 125). The subject encompasses, but is not limited to, learners learning with respect to design, making, problem solving, technological systems, resources and materials, criteria and constraints, processes, controls, optimisation and trade-offs, invention and other topics that concern human innovation (Dugger & Naik, 2001; Kwaira & Gumbo, 2014). Learners should know technology (conceptual knowledge) and be able to do technology (procedural knowledge and/or tacit knowledge) (Ankiewicz, 2003; Gumbo, 2015). In addition, learners will acquire a range of skills such as how to investigate, design, make, evaluate, communicate, create, criticise, collaborate, work in a group or team and many more.

In direct line with the definition of technology above, Technology Education is about studying technology which provides an opportunity for learners to learn about technological processes and the knowledge needed to solve problems and to extend human capabilities (ITEA, 2000, p. 242). According to the DBE (2011, p. 8), Technology Education aims to contribute towards learners' technological literacy by giving them opportunities to:

- Develop and apply specific design skills to solve technological problems,
- Understand the concepts and knowledge used in Technology Education and use them responsibly and purposefully, and
- Appreciate the interaction between people's values and attitudes, technology, society and the environment.

From these definitions it is clear that Technology Education is a subject that provides learners with opportunities to acquire the knowledge and skills they need to understand and apply processes using different materials and methods to effectively manipulate the environment for intended purposes. By its very nature the subject is highly practical, thus it should involve learners actively in learning tasks while ensuring the relevance of its content and pedagogy to diverse cultural contexts represented by learners. Technology Education is a very important subject in the education of every learner because every country's development rests on it (Pudi, 2006). There is a dire need to prepare learners to fit in with the industrial demands (Gumbo, 2010) – and not only those in high tech (urban) environments, but the indigenous ones as well. I am saying this because design as a concept and central methodological aspect of teaching Technology, is a culturally spiced concept. The next section is dedicated to defining and discussing the design concept.

Design Process

"Design is fundamental to the study of technology" (Warner & Morford, 2004, p. 33). This fundamentality has been captured succinctly by McCracken (2000):

As a human soul is to the body, design is to technology. It is important to understand the interdependence and complementarity nature of technology and design. Like the inseparable relationship between body and soul, technology is incomplete without design. Design cannot be fully appreciated without an understanding of technology. If technology is to be fully understood, then the concepts of design need to be understood. (p. 87)

In line with this relational view, design means the process of devising something – a creative, iterative and often open-ended process of conceiving and developing components, systems and processes (Warner & Morford, 2004; Asunda & Hill, 2007; Williams, 2011). Thinking about a problem/needs/wants and situation, imagining a solution to the problem, proposing the solution, and putting an action to this process defines teaching and learning in Technology Education. The problem solving nature of the design process is therefore in tandem with the action research or learning principles as can be noticed in the design process stages at the end of this section.

The concept of design can be understood like a pendulum which swings between Technology Education and Design Engineering. Asunda and Hill (2007, p. 3) claim that the "National Center for Engineering and Technology Education (NCETE) has proposed that the field of technology education adopt the interpretation of design based on the engineering definition". The procedural terminology used in Engineering and that in Technology Education is pretty much the same, for example "formulating a problem, generating alternatives, analysing and evaluating" (Eggert cited in Williams, 2011, p. 398). This, however, does not suggest that Technology Education and Engineering Education are the same. Engineering depends so intensely on the body of knowledge from science and mathematics (Williams

2011) that if the Engineering definition had to be adopted in Technology Education it would mean that learners should be taught a lot of Science and Mathematics. As it is, the knowledge areas from Science and Mathematics are not applied as intense in Technology Education as in the case of Engineering, which affords every learner the opportunity to learn Technology. Engineering design makes use of mathematical analysis whereas Technology Education makes more use of conceptual design (Williams, 2011, p. 399), with design as the backbone of teaching Technology. This difference is illustrated through the construction of a model bridge:

In Design and Technology, after developing an understanding of the design factors, students will construct a model bridge and then test it to destruction. Then they will analyse the model and the testing process to further develop their understanding, and then possibly construct another model as a result of the new information they have discovered. In Engineering, students will develop an understanding of the design brief requirements, and then construct the model. If the testing of the bridge indicates that it does not meet specifications, then the design has failed. (Williams, 2011, pp. 399-400)

In an attempt to explain the synergy between Technology Education and action research in this chapter, it is important to draw this distinction between Technology Education and Engineering Education, to create an understanding that the synergy may not work as desirably between action research and Engineering Education as it would between action research and Technology Education. So, Technology Education cannot be viewed as being the same as Engineering Education because Engineering Education's mathematical approach prevents creativity and flexibility, which can be restricting to involving action research in its teaching from a design point of view. I concur with Williams (2011), who is not comfortable with the fact that in Engineering Education the opportunity for learners to be divergent and creative is closed up by more deterministic design criteria. A range of criteria which necessitates research and the development of ideas are crucial for realising design success in Technology Education (Williams, 2007).

Design is a concept and can thus flavour learners' creative ideas in their collaboration on design projects from their cultural milieus. In this regard the logic of Moalosi et al. (2005) is almost the same as that of McCracken:

The concept of culture and design in the cultural perspective complement each other and one is inconceivable without the other. Therefore, design does not take place in isolation but it is embedded in users' culture. Designers need to be mindful of the interdependence between culture and design because it might enable them to improve their concepts in responding appropriately to users' needs, wants and desires. (p. 1)

From a cultural context, innovation commences with people, not with enabling technologies (Custer 1995; Moalosi et al., 2005; Moalosi, 2007). Artefacts are the expression of culture (Custer, 1995). According to Moalosi et al (2005, p. 2), culture "is revealed in all human actions and artefacts they design". Moalosi et al. write further that "observing the design of artefacts produced and consumed in a society often reveals the cultural situation and the people's lives, education, needs, wishes, hopes and fears" (2005, p. 2). The culture informed design expressed in technology is evident in the nature of the technologies which certain communities practice on daily basis. The technology of lekgotla (method of discussion), for example, is a prevalent concept in many African communities. From a sociological point of view, it is a collective problem solving technology in which each one is offered the chance to contribute his or her ideas while the principle of eldership is respected within the confines of botho/ubuntu – the most senior (usually the king) speaks first and passes the matter to the next most senior; on and on it goes until the least senior is covered by completing the circle – the circle activity is characterised by problem identification, certain propositions and reflections which may trigger action depending on the nature of the problem; thus drawing from the design and action research principles. The contribution of the least senior cannot be undermined because of their lack of status. This is a process which is endowed with knowledge, intense wisdom and problem solving technicalities. The conventional design process illustrated in Figure 2 should include the lekgotla technology. The stages in the design process (Figure 2) could also be viewed as participants, seated in a circular shape, tackling the problem at hand. This provides an innovative way of understanding the design process. With that in mind, learners from diverse cultural contexts are taught to adopt the lekgotla principle when they tackle technological problems.

The stages of the design process in Figure 2 are briefly explained below.

Stage 1 - Identify the Problem: A brief description of the problem is given in clear, short sentences.

Stage 2 - Design the Brief and Write Specifications: A plan containing clear, very specific terms is generated to solve the problem. Specifications should spell out exact details, for example "three centimetres long", not "long enough".

Stage 3 - Research the Problem: At this stage data should be gathered to collect information for the design of the solution. Existing solutions should be investigated.

Stage 4 - Generate Ideas: Ideas inspired by gathered data are fleshed out. These should be the learners' own creative ideas; they may not be copied from other sources.

Stage 5 - Develop Generated Ideas: The generated ideas need to be developed further, particularly the optimum idea which was selected. The ideas should be developed fully, just like a completed

Figure 2. Design process

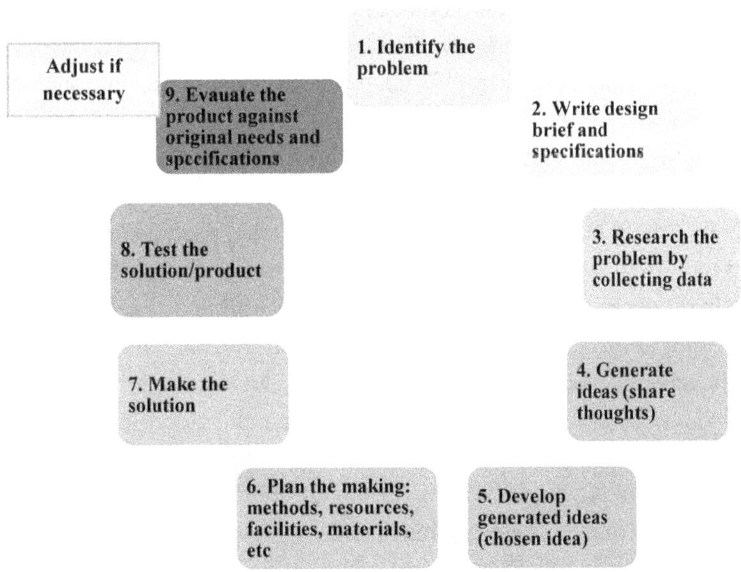

house plan waiting for the builder to execute it. It should be presented through basic and exploded drawings, showing the details in measurement, shape, materials, etc.

Stage 6 - Plan the Making of the Solution: Making methods, required facilities, resources and materials are identified and prepared for the making. Normally facilities, resources and materials are obtainable from a Technology workshop.

Stage 7 - Make the Solution: The solution (product) is made practically in the form of a prototype as per the developed ideas. This requires working in a resourced environment such as the Technology workshop.

Stage 8 - Test the Solution/Product: The finished product is tested to determine if it can perform the intended function. If it is, for example, a lift to transport shoppers from one level of a shopping mall to another, the weight that the prototype (not the real lift) can carry should be determined.

Stage 9 - Evaluate the Final Solution/Product: During this stage the final product is evaluated according to the original needs and specifications. The designer has been engaging his or her critical thinking on decisions and choices made throughout the process, including this stage. The following questions should be asked: What worked well and/or what did not work well? If something did not work well, what can be done to rectify or improve on it? The designer, who wears an evaluator's cap in this stage (or who could be someone else, for example a teacher and/or peers in the learning situation), should ensure that he/she evaluates all important design aspects such as function, value, ergonomics, aesthetics, manufacturing and materials.

Like in a lekgotla, the design process should be as democratic as possible. In a learning situation this means that learners working on certain activities of the design project should make an effort to accommodate each other's ideas and resources where possible. Each must show interest in the creative ideas of the others. Action research lends itself handy to facilitate the planning and execution of this kind of activity where learners co-participate and co-research the problem and suggest solutions.

Action Research

Action research is a process in which participants examine their own educational practice systematically and carefully. They use research techniques to devise ways to make things work differently if it seems that they are not working as desired (Ferrance, 2000, p. 1). The systematic approach does not mean shunning flexibility and openness to others' ideas; this is substantiated by its principles which promote active participation of the team. Systems thinking is heralded as a crucial aspect in action research wherein a synergy between a phenomenon and the whole, i.e. associated environment/context is valued. According to Flood (2010, p. 269), "the whole is greater that the sum of its parts" and "synergy is the sexy label for it". This thinking by Flood is useful to understand action research as it applicable to the teaching of Technology, and Technology Education in terms of its action related teaching and learning activities. Furthermore, the teaching and learning activities in Technology Education, which are practice-based, form a perfect synergy between understanding Technology Education against the social reality in which problems are identified and solved, and applying action research principles is the way to go. Adopting an action researcher's role, the Technology teacher can position him- or herself to engage learners in what Davison, Martinsons and Kock (2004) refer to as action learning. Action learning in Technology Education is demonstrated through a scenario that also shows a synergy between Technology Education and action research in Table 1.

Action research actively involves those regarded as participants in the research, that is to say they become co-researchers; it is emancipatory and overt political (Charles & Ward, 2007). A much more helpful way of defining action research is by concentrating on its process and goals, and for that matter it:

- Is participatory and is undertaken by or with insiders;
- Is collaborative, thus, it should involve all those who have an interest in the outcome of the research;
- Democratises research by changing the role and relationship of the researcher to that of the participants; all are responsible and co-own the research process;
- Enables both the participants and researcher to co-generate the knowledge which informs practice;
- Is a reflective and systematic process which involves ongoing intervention in the research setting, and as such it is an emergent and flexible process; and
- Is aimed at effecting change or action as agreed upon by the participants, targeting issues of social justice or improving the quality of life of the participants. (Greenwood & Levin, 1998; Reason & Bradbury, 2001; Kemmis & McTaggart, 2003; Herr & Anderson, 2005; Charles & Ward, 2007)

The above definition of action research creates an understanding that action research is a democratic process. Even though it may be one person's (researcher's) initiative, the researcher and participants are co-owners of its process. The democratic nature of action research is seen as a working partnership process by Somekh (2006, p. 1). She declares that it "is a means whereby research can become a systematic intervention, going beyond describing, analysing and theorizing social practices to working in partnership with participants to reconstruct and transform those practices". As a relationship-based enterprise it thus "implies joint research actions, within a relationship where equality between partners is promoted. It allows for flexibility and openness – something often required when out in a research field, working with human participants" (Ferreira, 2016, p. 2). In a teaching context, I want to conceive of action research as a teacher's reflection on the teaching or learning activities and processes in which he or she works collaboratively with learners to identify problems, needs or wants, and involves learners in suggesting the solutions, planning and implementation of such solutions.

"Implicit in the term action research is the idea that teachers will begin a cycle of posing questions, gathering data, reflection, and deciding on a course of action" (Ferrance, 2000, p. 1). Figure 3 shows the stages and cyclical nature of action research. Many action research projects start with a problem which needs to be solved; this may spiral into a cycle of new problems which are each sparked by the solution provided for the previous problem. This spices educational practices with new perspectives and improvements.

The steps of action research are subsequently briefly discussed according to Ferrance (2000, pp. 10-13).

1. **Identify the Problem:** First a doable problem worth investigating must be identified from the number of questions that one may have. It should be a higher order question stated in common language; it should be concise and meaningful and it should not already have an answer.
2. **Gather Data:** Gathering data helps to decide what action should be taken. Data can be gathered from multiple sources such as journals, individual files, logs of meetings, videotapes, case studies, surveys, records (tests, report cards, attendance registers), self-assessments, samples of student work, projects, performances, interviews, portfolios, diaries, field notes, audio tapes, photos, memos, questionnaires, focus groups, anecdotal records and checklists.

Figure 3. Action research cycle
Adopted from Ferrance, 2000, p.9.

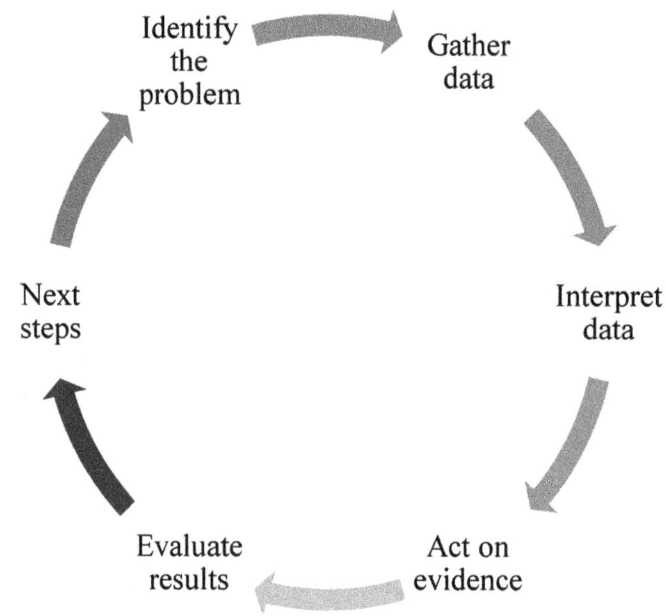

3. **Interpret Data:** At this stage data are analysed, for example classroom data, individual data and group data. Some of the data can be easily quantified without the help of a statistician and opinions, attitudes, checklists, etc may be summarised in a table form. Others can be viewed holistically and themes be noted.
4. **Act on Evidence:** Backed by the gathered data and literature review, an action plan can be designed which will enable you to make changes and to study these changes. It is important to document the action process and collect data on the performance.
5. **Evaluate Results:** It is important to assess the impact of the action or intervention to verify if any improvement has occurred and to ascertain that the data provide supporting evidence. If no change has occurred other actions should be tried to bring better results.
6. **Next Steps:** Were additional questions raised by the data? If so, plan for additional improvements, revisions and yet next steps.

The term "action research" was coined by a United States-based scholar, Kurt Lewin, in the 1940s. He did not separate investigation of the problem from acting on the evidence in pursuance of a solution (McFarland & Stansell in Ferrance, 2000; Charles & Ward, 2007; Manfra, 2009), but he involved a cyclical process which is based on a non-linear planning pattern: action, observation and reflection on changes in the environment of practice (Ferrance, 2000; Charles & Ward, 2007). Stephen Corey of the Teachers College at Columbia University is hailed as being among the first who took action research into the educational field. He believed that action research would bring about change as a result of teachers being involved in research and application of their knowledge at the same time. To him the value of action research was the changes that resulted from the input of research and action. He advised researchers and teachers to work together.

Action research was confronted in the 1950s for being allegedly unscientific, close to common sense activities and the work of amateurs (Ferrance, 2000, p. 8). But action research saw its own resurrection from the claws of opponents; it has endured since the 1970s and is used more for professional teacher development (Ferrance, 2000, p. 8). Now action research is used interchangeably with teacher research or practitioner research (Johnston, 2005; Manfra, 2009). While action research is practiced by researchers and teachers, I submit that teachers can be co-researchers with their learners in a teaching-learning situation. Technology Education presents opportunities for them to engage in action research where the teacher facilitates learners' problem solving activities. In support of Stephen Corey, a university-based experienced researcher can act as a support system for the Technology Education teacher concerned.

The design process and action research should be used within constructivist teaching and learning environments.

Constructivist Learning

A Technology Education teacher, who approaches his or her teaching by employing the principles of the design process and action research, is a teacher who will nurture the co-construction of knowledge in learners. This claim is supported by De Jager (2011):

As learners come into the classroom from different backgrounds and indeed different learning styles, the primary responsibility of the teacher is to create and maintain a collaborative problem solving environment where learners are allowed to create their own knowledge in their own way and their own time. (pp. 1-2)

Constructivism is an epistemic theory which is fundamental in learning. People generate knowledge and meaning from an interaction between their experiences and their ideas (Hill 2002). In the educational setup learners construct knowledge by interpreting their experiences of the external world to form a perception of it (Brunning, Schraw & Ronning, 1999). This is important in their learning of Technology as they attempt to make meaning of the technological environment and to solve problems in it. It is also important in action research for the research team to identify and solve their problems as a collective.

Researchers, learners or others engage in active and social cognitive learning. Social cognitive constructivism is premised on Piaget's cognitive developmental approach and the socio-cultural approach of Vygotsky (in Subban, 2006). The cognitive developmental approach focuses on the learner's own construction of knowledge, while the socio-cultural approach mainly focuses on the social process of the learner's interaction with others and participation in constructing shared knowledge.

Vygotsky (in Subban, 2006) has influenced the teaching approaches which supposedly support learner-focused learning environments. He (Vygotsky) supports learning that is enriched by collaborative and learner engaged approaches. His theory raises a need on the part of the teacher to cater for the learners' diverse learning styles in a sociocultural learning context (Subban, 2006, p. 936). Teaching Technology through the design process, which is action research-inspired, can ensure accommodation of learner diversity through collaborative design activities.

According to Shackelford and Maxwell (2012, p. 230), constructivism includes three elements which ensure learner interaction, namely cognitive, social and teaching presences. Cognitive presence is about a learner's ability to construct meaning through sustained communication in the learning community context (Shackelford & Maxwell 2012). Cognitive presence is ensured through a learner's collaboration as he or she explores, constructs, resolves and confirms his or her understanding of the content (Garrison in Shackelford & Maxwell 2012). Social presence is the ability of a learner to project him- or herself socially and emotionally through communication (Shackelford & Maxwell 2012). For cognitive and social presences to materialise teaching presence is needed, which is about the teacher's ability to project him- or herself (Swan 2003) through subject design and organisation, discourse facilitation and direct teaching (Shackelford & Maxwell 2012).

Social cognitive constructivism owes its origin to the gurus, Jean Piaget, Jerome Bruner, Lev Vygotsky and John Dewey, who claimed that the learner actively constructs new ideas or concepts based upon current and past knowledge or experience and direct interaction with the environment (Subban, 2006). The teacher's role is to facilitate learning, which encourages learners to discover principles for themselves and to construct knowledge by working to solve realistic problems (technological problems in the context of this chapter). As per author, Dewey asserted that meanings and truth are achieved through the application of ideas into actions and experiences (Fox, 2001). Dewey emphasised the idea that schools should bring real-world problems, which occur in a social context, into the school curriculum (Taylor, 2011, p. 11). The teacher's role is then to facilitate learning, which encourages learners to discover principles for themselves and to construct knowledge by working to solve realistic problems (technological problems in the context of this chapter). Thus, teaching should be facilitated such that it encourages learners to analyse, interpret and predict information by using open-ended questions and promoting extensive dialogue and reflection.

ACTION RESEARCH-DRIVEN DESIGN

In this section a framework for engaging action research in a school term-based design project is presented in the form of an example (Table 1). The teacher engages Grade 7 Technology learners in a term design project in which they identify and solve a food problem in a squatter camp, a community of which the majority of these learners are members of. This activity happens at the school where a few squatter campers are invited by the teacher to help expound on the situation. The school is situated only a kilometre away from the squatter camp. Food is not properly preserved at the squatter camp and as a result it goes bad quickly, causing diseases and deaths. The learners' lives are threatened by this condition, so much that they passionately contributed ideas about the identified problem. The framework in Table 1 shows the stages of the design process (not linear as the table may suggest, but cyclical as depicted in Figure 1), targeted content, concepts and skills, the teacher's activities, learners' activities, resources or learning support materials (LSM) and where and how action research features in the process. The framework should not be implemented rigidly. It should also have a degree of flexibility to accommodate learning styles and knowledge construction by learners from diverse cultural backgrounds and adapted to other grades. It needs the teacher to be creative with critical reflection. The term dates column has, for example, not been included; the teacher should consider including that as well as the assessment column.

Table 1. A synergy pedagogical framework for technology education and action research

<table>
<tr><td colspan="6">Knowledge construction: Diverse learner backgrounds and learning styles, collaborative problem solving, social cognitive learning, sociocultural learning, learner-centred, learning community, active learning, discovery learning, meaning making, dialogue and reflection, iterative, innovative, creative and critical thinking, group work, team work, participatory, flexible, democratic, reflective, transformative, emancipatory, open discussions.
Technology Education's targeted knowledge: conceptual knowledge, procedural knowledge and tacit knowledge.
Action research's targeted knowledge: knowledge of action research (concepts and principles) and knowledge of methods which should be applied.</td></tr>
<tr><th>Technology Education (design process)</th><th>Content, concepts and skills</th><th>Teacher's activities</th><th>Learners' activities</th><th>Resources / equipment technological facilities, consumables</th><th>Evidence of action research</th></tr>
<tr><td>Identify the problem: Depending on the grade, the teacher or learners can identify a problem to be solved.</td><td>Identifying and explaining food problem in a squatter camp</td><td>Facilitates problem identification in a squatter camp

Organises the literature and invites five squatter camp representatives</td><td>Identify food problem faced in a squatter camp

Read on and discuss the problem with squatter camp representatives</td><td>Literature</td><td>Identify the problem: Teacher identifies the problem to be solved (could be a technological problem) together with or by involving learners (and refugees) as co-researchers – participatory, open discussion, sociocultural learning.</td></tr>
<tr><td>Design brief: Teacher sets parameters for solving the problem, i.e. specific design brief (what needs to be done exactly), specifies materials, function, cost, safety measures, etc.</td><td>Designing the brief with clear specifications; stating clearly what the solution must to do.</td><td>Invites five squatter camp representatives to talk about their food problem
Teaches food processing, preservation and packaging

Refers learners to the design briefs they wrote in the previous terms and asks them to write this design brief in teams.</td><td>Discuss with squatter campers – 1 squatter camper per learner group

Design clear brief</td><td>Exercise books

Food preservation and packaging samples

Data projector

Laptop

Internet / Wi-Fi connection</td><td>(Discuss and negotiate plan to solve problem): Teacher discusses and negotiates plan with learners (and squatter campers). This can be design brief about which the teacher asks questions such as: What should we do exactly? What must the product be able to do? What do we need? – participatory, team work, collaboration, problem solving, creativity, critical reflection, learning community, innovation, learner-centred.</td></tr>
<tr><td>Investigate the problem: Investigate the existing solutions, noting challenges experienced by end-users</td><td>Investigating the problem of unpreserved food in a squatter camp:

What situations commonly cause food not to be preserved in a squatter camp?

What initial problems do squatter campers face about food that is not preserved in terms of:

-health hazard?

-hunger?</td><td>Designs knowledge (squatter camping, food preservation, etc), concepts (squatter camp, investigation, etc), skills (investigate, discussion, etc) questions.

Identifies and organises learning resources (literature, video, invites squatter campers, etc).

Group learners into teams

Discussions with learners</td><td>Answer questions set by the teacher

Watch video and take notes

Speak to squatter campers</td><td>Literature

Data projector

Computer

Techno-lab or suitable room</td><td>Gather data: Engage research methods and strategies to investigate the problem, e.g. interviews or surveys for end-users, observations, literature survey, device analysis, etc.
Interpret data: Make sense of data by analysing and interpreting them – learner-centred, problem solving, co-researching, collaboration, negotiation of ways and strategies, teacher engages in discussions with learners in confirming and refining the problem, teacher designs interview questions for squatter campers and observation sheet with learners, knowledge co-construction.</td></tr>
</table>

CONCLUSION

The concept of design process as a method of teaching Technology can be driven by the principles of action research due to the similarities between these two. This chapter has explored this synergy and contributed a framework which demonstrates the featuring of action research in the teaching of Technology. Going forward, Technology teachers should develop an understanding of action research as it applies in this context. Thus, action research practitioners should make it their responsibility to develop teachers in this regard and extend research to the practical engagement of action research in the teaching of Technology.

REFERENCES

Ankiewicz, A. (2003). *Technology education at school: Illusion or reality? Inaugural address.* Johannesburg, South Africa: University of Johannesburg.

Asunda, P. A., & Hill, R. (2007). Critical features of Engineering Design in Technology Education. *Journal of Industrial Teacher Education, 44*(1), 25–48.

Atlantic Provinces Education Foundation. (2001). *Foundation for the Atlantic Canada Arts Education Curriculum.* Halifax, Canada: Atlantic Provinces Education Foundation.

Brunning, R. H., Schraw, G. J., & Ronning, R. R. (1999). Cognitive psychology and instruction (3rd ed.). Prentice-Hall.

Charles, L., & Ward, N. (2007). *Generating change through research: Action research and its implications.* Retrieved September 15, 2016 from: www.ncl.ac.uk/cre/publish/discussionpapers/pdfs/dp10.pdf

Custer, R. L. (1995). Examining the dimensions of technology. *International Journal of Technology and Design Education, 5*(3), 219–244. doi:10.1007/BF00769905

Davies, D. (1997). The relationship between science and technology in the primary curriculum: Alternative perspectives. *The Journal of Design and Technology Education, 2*(2), 101–111.

Davison, R., Martinsons, M. G., & Kock, N. (2004). Principles of canonical action research. *Information Systems Journal, 14*(1), 65–86. doi:10.1111/j.1365-2575.2004.00162.x

De Jager, M. L. E. (2011). *Teachers' understanding and implementation of the design process as it relates to teaching Technology Education in the intermediate and senior phases (Grades 4-9) in the General Education and Training Band in KwaZulu-Natal* (Unpublished PhD thesis). Durban University of Technology, Durban, South Africa.

De Vries, M. J. (1996). Technology Education: Beyond the Technology is applied science paradigm. *Journal for Technology Education, 8*(1), 7–15. doi:10.21061/jte.v8i1.a.1

Department of Basic Education. (2011). Curriculum and assessment policy statement grades 7-9: technology. Pretoria, South Africa: Government Printers.

Dugger, W. (2008). The perspective of technology education. *Proceedings of the Japanese International Symposium of Educational Cooperation for Industrial Technology Education.*

Dugger, W. E. Jr, & Naik, N. (2001, Sept.). Clarifying misconceptions between Technology Education and Educational Technology. *Technology Teacher*, 31–35.

Ferrance, E. (2000). *Action research.* Retrieved September 14, 2016 from: www.google.co.za/?gws_rd=ss1#q=themes+in+education+action+research+by+eileen+ferrance

Ferreira, R. (2016). *Action research as research design.* Pretoria, South Africa: University of Pretoria.

Flood, R. L. (2010). The relationship of systems thinking to action research. *Systemic Practice and Action Research, 23*(4), 269–284. doi:10.1007/s11213-010-9169-1

Fox, R. (2001). Constructivism examined. *Oxford Review of Education*, *27*(1), 23–35. doi:10.1080/03054980125310

Gil-pe'rez, D., Vilches, A., Fernana'ndez, I., Cachapuz, A., Praia, J., Valde, P., & Salinas, J. (2005). Technology as 'applied science': A serious misconception that reinforces distorted and impoverished views of science. *Science and Education*, *14*, 309–320.

Greenwood, D. J., & Levin, M. (1998). *Introduction to action research: Social research for social change*. Thousand Oaks, CA: Sage.

Gumbo, M. T. (2010). *A critical reflection on curriculum review of Technology Education in South Africa*. Paper presented at the 4th Annual Teacher Education at a Distance Conference, Lanseria, South Africa.

Gumbo, M. T. (2015). Indigenous technology in Technology Education curricula and teaching. In P. J. Williams, A. Jones, & C. Buntting (Eds.), *Contemporary issues in Technology Education: The future of Technology Education* (pp. 57–75). Singapore: Springer. doi:10.1007/978-981-287-170-1_4

Gumbo, M. T. (2016a). Pedagogical principles in Technology Education: An indigenous perspective. In G. Emeagwali & E. Shizha (Eds.), *African Indigenous Knowledge and the Sciences: Journeys into the past and present* (pp. 13–32). Rotterdam, The Netherlands: Sense. doi:10.1007/978-94-6300-515-9_2

Gumbo, M. T. (2016b). The question of Technology Education: A pertinent issue. Inaugural Lecture of Professor Mishack T Gumbo, Pretoria, South Africa.

Heidegger, M. (1977). *The question concerning technology and other essays*. New York: Harper & Row.

Herr, K., & Anderson, G. L. (2005). *The action research dissertation*. Thousand Oaks, CA: Sage.

Hill, H. C. (2002). *Learning: A survey of psychological interpretations* (7th ed.). Boston: Allyn & Bacon.

ITEA. (2000). *Standards for technological literacy: Content for the study of technology*. Retrieved February 9, 2016 from: http://scimathmn.org/wp-content/uploads/2013/08/tech_standards.pdf

Johnston, M. (2005). The lamp and the mirror: Action research and self studies in the social studies. In K. Barton (Ed.), *Research methods in social studies education: Contemporary issues and perspectives* (pp. 57–83). Greenwich, CT: Information Age Publishing.

Kalanda, K. (2005). *Factors influencing college students' attitudes toward Technology in Lesotho* (Unpublished MEd dissertation). UNISA, Pretoria, South Africa.

Kemmis, S., & McTaggart, R. (2003). Participatory action research. In N. K. Denzin & Y. S. Lincoln (Eds.), *Strategies of qualitative inquiry* (pp. 151–183). Thousand Oaks, CA: Sage.

Kumar, K. L. (2002). *Impact of new technology on teaching and learning in technology education*. Paper presented at the Design and Technology Association International Research Conference.

Kwaira, P., & Gumbo, M. T. (2014). Applying Design and Technology Education in addressing farmers' problems in the Makonde Rural District, Zimbabwe. *Journal of Human Ecology (Delhi, India)*, *48*(1), 103–113.

Li-Hua, R. (2009). Definitions of technology. In J. K. B. Olsen, S. A. Pedersen, & V. E. Hendricks (Eds.), *A comparison to the philosophy of technology* (pp. 18–22). Malden, MA: Wiley-Blackwell. doi:10.1002/9781444310795.ch2

Manfra, M. M. (2009). Action research: Exploring the theoretical divide between practical and critical approaches. *Journal of Curriculum and Instruction*, *3*(1), 32–46.

Mapotse, T. A. (2012). *The teaching practice of Senior Phase Technology Education teachers in selected schools of Limpopo Province: An action research study* (Unpublished DEd thesis). UNISA, Pretoria, South Africa.

Massachusetts Science and Technology/Engineering Curriculum Framework. (2006). *Purpose and nature of science and technology/engineering*. Malden,MA: Massachusetts Department of Education.

Merrill, C. (2017). *Action research and Technology Education*. Retrieved May 18, 2017 from: https://www.iteea.org/File.aspx?id=85497&v=6eec1be

Misa, T. J. (2009). History of technology. In J. K. B. Olsen, S. A. Pedersen, & V. E. Hendricks (Eds.), *A companion of the philosophy of technology* (pp. 7–17). Malden,MA: Wiley-Blackwell. doi:10.1002/9781444310795.ch1

Moalosi, R. (2007). *The impact of socio-cultural factors upon human-centred design in Botswana* (Unpublished PhD thesis). Queensland University of Technology, Queensland, Australia.

Moalosi, R., Popovic, V., Hudson, A., & Kumar, K. L. (2005). *Integration of culture within Botswana product design*. Paper presented at International Design Congress, Yunlin, Taiwan.

Obikeze, D. S. (2011). *Indigenous knowledge systems and the transformation of the academy in Africa: The CULPIP model*. Unpublished paper.

Ogunbure, A. A. (2011). The possibilities of technological development in Africa: An evaluation of the role of culture. *The Journal of Pan African Studies*, *4*(3), 86–100.

Pudi, T. I. (2006). *Technology education (Study Guide)*. Pretoria, South Africa: UNISA Press.

Reason, P., & Bradbury, H. (2001). Introduction: Inquiry and participation in search of a world worthy of human aspiration. In P. Reason & H. Bradbury (Eds.), *Handbook of action research* (pp. 1–45). London: Sage. doi:10.1016/B978-075065398-5/50002-9

Senanayake, S. G. J. N. (2006). Indigenous knowledge as a key to sustainable development. *The Journal of Agricultural Science*, *2*(1), 87–94. doi:10.4038/jas.v2i1.8117

Shackelford, J. L., & Maxwell, M. (2012). Sense of community in graduate online education: Contribution of learner to learner interaction. *International Review of Research in Open and Distance Learning*, *13*(14), 228–250. doi:10.19173/irrodl.v13i4.1339

Somekh, B. (2006). *Action research: A methodology for change and development*. Berkshire, UK: Open University Press.

Subban, P. (2006). Differentiated instruction: A research basis. *International Education Journal*, *7*(7), 935–947.

Swan, K. (2003). A constructivist model for thinking about learning online. In J. Bourne & J. C. Moore (Eds.), *Elements of quality online education, practice and direction* (pp. 13–30). Needham, MA: Sloan Center for Online Education.

Taylor, J. A. (2011). *Introduction to criminal justice for high school students: A curriculum plan* (Unpublished masters thesis). California State University, Sacramento, CA.

Warner, S. A., & Morford, L. L. (2004). The status of design in Technology teacher education in United States. *Journal of Technology Education, 15*(2), 33–45. doi:10.21061/jte.v15i2.a.3

Watson, H., & Chambers, D. W. (1989). *Singing the land, signing the land.* Geelong: Deakin University Press.

Williams, P. J. (2011). Engineering: Good for Technology Education?. In M. J. De Vries (Ed.), Positioning Technology Education in the curriculum (pp. 87-100). Rotterdam, The Netherlands: Sense.

KEY TERMS AND DEFINITIONS

Action Research: In the context of a school, action research means diverse research methods of evaluation, investigation and analysis applied in diagnosing problems to help educators (and learners) develop practical solutions to address them quickly and efficiently.

Constructivist: A theory of learning which claims that people construct their own understanding and knowledge of the world through experiencing things and reflecting on those experiences.

Design Process: A problem-solving process in a project context which consists of cyclical steps; it involves graphic representations of designed solutions for the purpose of constructing or manufacturing such solutions.

Educational Technology: The technological processes and resources created, used and managed to facilitate learning.

Indigenous: Also native; it means originating in and characteristic of a particular region or country, for example indigenous people of South Africa.

Pedagogical: Deviant from its original meaning, in this chapter this term refers to the work or skill of teaching learners while considering their potential to construct knowledge, thus treating them as active participants in learning.

Synergising: Causing a cooperative interaction between two or more agents, groups or forces so that their combined effect is greater than the sum of their individual effects.

Technology Education: The study of Technology. Learners learn about the intricacies of the techniques, processes and knowledge that relate to technology.

Technology: The use of material and non-material techniques and processes to practically address people's needs, wants and problems by manipulating the environment.

Chapter 5
An Action Research Study Towards Enhancing Community Engagement Partnerships Between ODL Institutions and Schools

Tome' Awshar Mapotse
University of South Africa, South Africa

Sizakele Mirriam Matlabe
University of South Africa, South Africa

Elias E. R. Mathipa
University of South Africa, South Africa

Soane Joyce Mohapi
University of South Africa, South Africa

Magano Meahabo Dinah
University of South Africa, South Africa

ABSTRACT

The South African government has mandated national universities to emancipate and capacitate the teachers of the selected schools in science and technology through an Open Distance Learning (ODL) mode of which the University of South Africa form part of such as Higher Educational Institutions cohort. This chapter reports on the observations, field notes and interviews that were conducted with a group of four teachers and eight learners as a case study at Lovemore Primary School. This Community Engagement chapter focuses on teaching science and technology. In this chapter, researchers argue that focus group participants were able to give enough ideas, thoughts and points to enable the team to design an intervention programme through an Action Research approach. A developmental theory was used to underpin the study while the researcher and participants were engaged in a Community of Practice paradigm. More sessions will be planned to jointly structure the way forward and cultivate the nature of the intervention strategies.

DOI: 10.4018/978-1-5225-2642-1.ch005

INTRODUCTION

The curriculum in postcolonial Africa is still to a large extent confronted by the legacy of colonial education that remained in place decades after political decolonisation, reiterated Higgs (2016) hence is logical to state that Africa education systems still mirror colonial education inherited from former colonial education system. Whilst Africa has its dual mode of education delivery (colonial education and postcolonial operation), Higgs (2016:1) stress that education system in Africa during the 21st century has to operate in both post-colonial and globalising context. It is disturbing to learn that when South African learners are subjected to assessment on a global scale they score lower than most African countries.

Robinsons (2013:3) reports that, "South Africa's performance on international benchmarks tests is low. We need to ask hard questions as to why this is so, bearing in mind that the reasons might go beyond what takes place in the classroom. Are teachers not able to explain the content properly? Are learners not motivated? Is language an issue? Is the culture of the school not conducive to learning? Are the days organised in an orderly fashion? Are the classrooms cold? Are the learners hungry?" To respond to these hard questions, some research teams of academics under the College of Education (CEDU) from the University of South Africa (Unisa), as an Institute for Open Distance Learning (ODL) have embarked on a Community Engagement (CE) project to be part of the solution to the educational problem that is ravaging South Africa (SA). CEDU is running a national 500 schools CE research project. The 500 schools are then divided into 100 schools per five identified provinces that are performing poorly.

The 500 schools research project proposes to investigate the underlying causes of under-performance in Grades 3 and 6, and to develop an intervention approach to assist the sample schools in five South African provinces: Limpopo, Mpumalanga, Eastern Cape, KwaZulu-Natal and North West (Bettman 2012; Smit & Romm 2013). This team of researchers, who were allocated to Lovemore Primary School (a pseudonym is used to conceal the identity of the school), were urged by the words of Nelson (2008) who declares that, "It is the teacher who makes the difference in the classroom. By far the most important factor to school learning is the ability of the teacher. The more capable the teacher, the more successful will be the students. There is an essential body of knowledge about teaching that must be known by all teachers." The 500 schools project intends to ascertain whether teachers are grounded with the essential body of knowledge and what the gaps are that lead to poor learners' performance and meagre teachers' achievement. This project intends to develop the guidelines that may be useful in addressing this underperformance and poor achievement problem.

This study has focused on the subject, natural science and technology, as it is termed in the General Education and Training band of Intermediate Phase in the South African context. In South Africa, natural science and technology is a subject taught in Grade 6 after the Minister of Basic Education merged science and technology subjects from 2012. The term 'natural science and technology' is still applicable in the revised, reviewed and revisited Curriculum and Assessment Policy Statement (CAPS) as adopted in 2011 by SA.

Technology as a subject has different connotations among countries. Mapotse (2015:213) goes on to comment that some call it 'science and technology' (Malawi/Bangladesh), while others label it 'design and technology' (United Kingdom [UK]/Botswana); in SA for instance, it is called 'natural science and technology' in the GET band of the Intermediate Phase and 'technology subject' in the GET band of the Senior Phase, but 'technology field' in the Further Education and Training (FET) band encompasses Grade 10 to 12, while yet others term it 'technology education' (United States of America [USA] and New Zealand [NZ]). In this study it will be referred to as 'natural science and technology', shortened to 'science and technology' education, or just 'S & T'. This study was located in an interpretative orientation to knowledge production and obtained data through focus group interviews with teachers at Lovemore Primary School in the Marapyane Region in Mpumalanga Province of South Africa. The study intends to highlight the challenges of incapacitated teachers handling science and technology education. South Africa has for many years struggled to deliver acceptable results in science, mathematics and technology education at primary and high schools (Gauteng Department of Education [GDE], 2010).

BACKGROUND

International teacher development practices emphasise that professional development for teachers will boost learners' achievement only if teachers engage in long-term in-depth learning opportunities that address the content and pedagogy relevant to the curriculum taught in their classes (Department of Basic Education [DBE], 2011 - Technical Report). Teachers are indispensable in bringing lasting change to future generations (Stars in Education, 2011); hence, their professional development should be continuous. Teachers need to acknowledge the significance of a place – where people live, work and learn – as this is a key to effective school leadership and classroom management (The Teacher, 2013). This is because they spend eight hours a day, five days a week with children of a society shaped by poverty, the effects of apartheid, HIV/AIDS, unemployment and many other social ills (Stars in Education, 2011), especially in most villages in Africa such as Marapyane in SA. These community social ills, in one way or the other, had an impact on both teachers' performance and learners' achievement.

Poor learners' performance in SA has been attributed to different issues such as learners' social and economic circumstances, overcrowding, lack of teacher and learner discipline and commitment, a lack of learner support materials and related resources, and un- and under-qualified teachers to mention but a few (Mahomed 2004; Mapotse 2012). This article reports on the last mentioned attribute of teachers' lack of specialised qualifications as a contributing factor to learners' poor performance. Some education stakeholders hold assumptions that learners perform better when their teachers have received quality education and training in the subjects they teach and at the levels/phases in which they are placed. This on-going study intends to shed light on the engagement progress made thus far by Unisa's research team within CEDU. The Unisa team of researchers had conducted the focus group in their field of specialisation and I on teachers of S & T education in Grades 6 from Lovemore Primary School. I therefore intend to share my findings in that regard.

Quite a number of scholars in the field of ODL, science and/or technology in the GET band have engaged in research targeting a variety of aspects of professional teacher development amid teachers' qualifications and subject knowledge. For instance, the following serve as such scholars: Chetty and Lubben (2010); Mokhele (2011); Williams and Gumbo (2011); Coetzee (2012); Williams, Eames, Hume and Lockley (2012); McLain, Bell and Pratt (2013); Yager (2013); Quasthoff and Wild (2014), Braimoh (2015); McLain, Barlex, Bell & Hardy (2015); and Doyle, Seery and Canty (2016). The aforementioned scholars belonging to both national and global communities have used some common instruments or similar approaches to gather their data, and little has been done in using the Action Research (AR) approach within CE to emancipate incapacitated science and technology education teachers. With this study, the researchers want to attempt to fill that gap by sharing experiences gained through research dialogue with S & T education teachers from Lovemore Primary School.

COMMUNITY ENGAGEMENT WITH LOVEMORE PRIMARY SCHOOL

ODL Institutions' Involvement in Community Engagement

In 1997, the South African government set out a new agenda for transformation in higher education, which included CE as one of the pillars of the system (McKay 2014). ODL Institutions have three pillars of focus, which are tuition, research and community engagement. However, by 2010, studies found that CE was ad hoc and fragmented at many ODL Institutions (Kamal 2014). Of late, CE has become a common course of interactions by ODL Institutions with their communities. The White Paper on Post-School Education and Training recently reassessed CE and acknowledges a lack of conscious definition (McKay 2014). Different scholars (Fawcett, Paine-Andrews & Francisco 1995; Minkler & Wallerstein 2003; Moini Fackler-Lowrie & Jones 2005; Jones & Wells 2007) define CE differently and the above listed scholars describe CE as follows:

- Community engagement in research is a process of inclusive participation that supports mutual respect of values, strategies and actions for authentic partnership of people affiliated with or self-identified by geographic proximity, special interest, or similar situations to address issues affecting the well-being of the community of focus.
- Community engagement is a core element of any research effort involving communities. It requires academic members to become part of the community and community members to become part of the research team, creating a unique working and learning environment before, during and after the research.
- Community engagement is a process that requires power sharing, maintenance of equity, flexibility in pursuing goals, methods, and time frames to fit the priorities, needs, and capacities within the cultural context of communities. Community engagement in research is often operationalised in the form of partnerships, collaborations and coalitions that do the following: help mobilise resources and influence systems, change relationships among partners, and serve as catalysts for changing policies, programmes and practices.

The researchers respectfully offer the following definition of CE from Wikipedia (2014), which stresses that community engagement, refers to the process by which community benefit organisations and individuals build on-going, permanent relationships for applying a collective vision for the benefit of a community. While community organising involves the process of building a grassroots movement-involving communities, community engagement primarily deals with the practice of moving said communities towards change, usually from a stalled or otherwise similarly suspended position. Since this study focuses on teachers, the researchers will therefore share the same sentiments with Mapotse (2017) in defining CE in the following manner:

Community engagement is a regular interaction process with incapacitated teachers as a community of co-researchers led by an action research practitioner for emancipation purposes in pursuit of some new teaching skills, methods and approaches within their cultural context so as to enhance learners' performance and teachers' achievement.

The research teams from Unisa have been distributed countrywide through the CE process to engage with the teachers in low performing provinces. These research teams of academics have been targeting 500 primary schools since 2012. My team was allocated Lovemore Primary School in Marapyane. As a team member, I oversee the teachers coping level in science and technology. Unisa, through CEDU, has extended its hand to work in partnership with the DBE to respond to some hard questions raised by Robinsons. As far as CE is concerned, Unisa would like to utilise all its resources.

Unisa's Resourcefulness Towards Community Engagement

Unisa wants to ensure that all its resources – capital and human – are used optimally for the advancement of both scholarship and CE. Unisa further desires that its allocated resources be aligned to the strategic directives of Unisa and must achieve the social mandate of the university – "towards the African university in the service of humanity". Unisa's Senate Community Engagement Committee annually allocates resources to all Unisa colleges and regions based on project plan submissions. The Senate Community Engagement Committee then recommends the community engagement policy for the university and monitors the progress and impact of CE on scholarship (Unisa Senate Community Engagement Policy, 2013).

After South Africa's democratic government came into power in 1994, policymakers and experts decided to add social effort to the higher education agenda, (Kamal, 2014). As custodians of knowledge and knowledge production, ODL Institutions should play their part in empowering communities and serving humanities. Unisa took a step further with CE and incorporated these communities into teaching and research projects hence; a team of researchers from the CEDU has been dispatched throughout the country. The researchers belong to the Marapyane team, which visited Lovemore Primary School in Mpumalanga to establish an intervention strategy through Action Research (AR) to emancipate S & T education teachers. I will share those science and technology teachers' experiences using the Community of Practice (CoP) paradigm and underpin the study through developmental theory.

COMMUNITY ENGAGEMENT THEORETICAL FRAMEWORKS

CE can trace its roots to the concept of 'community benefit', a term that grew out of an English common law concept, articulated in an 1891 legal decision, that defined three types of charitable organisations: trusts for the advancement of education; trusts for the advancement of religion; and trusts for other purposes beneficial to the community (Pemsel 1891).

Community of Practice at the Heart of Community Engagement

CoP is formed by people who engage in a process of collective learning in a shared domain of human endeavour: a tribe learning to survive; a band of artists seeking new forms of expression; a group of engineers working on similar problems; a clique of learners seeking to define their identity in the school; a network of surgeons exploring novel techniques; a gathering of first-time managers helping each other to cope; and teachers interacting and sharing ways of presenting a new body of knowledge to their learners. Briefly, CoP comprises groups of people who share a concern or a passion for something they do and learn how to do it better as they interact regularly (Lave & Wenger 2014). Back in 2008, the Minister of Education took a swipe at factors that contribute to the mismatch between ODL Institutions and schools. It is against this background that CoP between Unisa and the DBE on underperforming primary schools had been established.

Pandor (2008) contends that several factors contributed to the mismatch between success at school and success at university. She further alluded to the fact that schools did not help pupils become competent in the language of learning at university; first-year students often struggle with the high demand for independent research and self-directed learning and many experience a deep culture shock on arriving at university. Unisa's academic teams of researchers intend to engage teachers who handle these pupils to confront these challenges raised by Pandor at the learner's early learning stage. I will engross these S & T education teachers with the CoP paradigm.

The concept of CoP is often traced back to the work of Lave and Wenger in 1991 who first used the term to explore situated learning in the context of the five apprenticeships they were studying (Roberts 2006; Jita & Mokhele 2013). According to Wenger (2007), three elements/characteristics are crucial in distinguishing CoP from other groups and communities:

- **The Domain:** A CoP is something more than a club of friends or a network of connections between people. 'It has an identity defined by a shared domain of interest. Membership therefore implies a commitment to the domain, and therefore a shared competence that distinguishes members from other people.'
- **The Community:** 'In pursuing their interest in their domain, members engage in joint activities and discussions, help each other, and share information. They build relationships that enable them to learn from each other.'
- **The Practice:** 'Members of a CoP are practitioners. They develop a shared repertoire of resources: experiences, stories, tools, and ways of addressing recurring problems – in short, a shared practice. This takes time and sustained interaction.'

This study therefore focuses on the classroom *practice* of rural schoolteachers teaching science and technology. The interaction between the Unisa team and Lovemore Primary School is expected to be built on activities that reinforce cooperation, bind people together for a common goal, and facilitate a relationship of trust through developmental theory.

Developmental Theoretical Framework Within Action Research

By reflecting critically on the S & T teachers' situation, intervention strategies could be devised to help teachers to think about how they could take action in response to the challenges they face (Mapotse, 2017). The kind of intervention that AR suggests is facilitative because of the co-participatory principles of teachers within AR. Therefore, the intervention would be much more meaningful and emancipatory because it would not be instructive to the teachers; instead, they would be helped to think out the solutions to their challenges and take action. Hence, an assumption is made that engaging teachers in developmental theory had the potential to improve their understanding about the knowledge and pedagogy of S & T in their context.

Developmental theory is the 20th century model used by industrialised nations to represent the most advanced form of society, and how other nations are categorised in terms of their approximation to this model. Hence, these nations can be judged as either more or less developed, and their efforts should be directed towards achieving the 'developed' model in their political, social, *educational* and economic institutions (Scruton in Bothamley 2004:146). SA is listed among developing countries. It is also a member of BRICS, an acronym for the developing countries of Brazil, Russia, India, China and SA. The unemployment rate in SA is directly influenced by its struggle to produce acceptable results in science, mathematics and technology. One of the biggest contributing factors, among other things, is unqualified and under qualified teachers to teach this subject - science and technology.

The Main Purpose of the Focus Group With S&T Teachers

The purpose of these focus group sessions is to gather information about the teaching and learning of S & T. The idea is that, together, Unisa's research team and Lovemore S & T teachers can explore issues connected with teaching and learning more fully. This will help Unisa's research team to design an intervention strategy for improvement of schooling. Facilitators treated all information in the interviews confidentially. Participants were assured anonymity – "If the researchers' want to mention anyone's ideas, researchers' will use pseudonyms to record this so that no-one will recognise any participant in our report." Participants were guaranteed that their answers were going to be used only by Unisa's team to try to find ways, based on participants' input, to make schools better. Participating teachers were also promised that, although Unisa's team of researchers has permission from the DBE to do the research, to the researchers would not give any confidential information to the school principals or the DBE without the teachers' consent.

Interviews started with teachers, followed by learners, school management team (SMT) and parents. These interviews resulted in interesting conversations that, among other things, highlighted areas that need urgent intervention. The purpose of focus groups is for a discussion to take place between various participants on the topic. After the initial phase of building rapport, I moved on to obtain informed consent by explaining the purpose of the project, the activity that the participants will be involved in, and that the idea is to hold a good group discussion on the topic where people can exchange views; all participants were requested to sign their informed consent forms. At this point, the researchers referred the participants to the relevant focus group schedule designed by Unisa to serve as a guide – each of the schedules within a guide explains to the participants what the project is about, how we see their role in it, and it explains how the researchers will report with no reference to their names.

Problem Statement and Research Question

The South African education system has been through a number of challenges since the 1994 democratic elections. It is now the 22nd year since SA democracy and still there are arduous conditions in a number of rural and urban schools. The challenges range from infrastructure to a lack of textbooks, and a lack of teachers who specialise in science and/or technology education, especially in rural areas. This shows that South African education is experiencing serious challenges that need immediate attention from stakeholders. In trying to address the challenges facing Basic Education and infrastructure, there was a drive to build new schools. In October 2012, new schools were opened at villages at Libode and Lusikisiki in the Eastern Cape Province, and textbooks and workbooks were issued throughout the country except in Limpopo Province, which experienced some turmoil around the textbooks. There were efforts to improve the quality of education by introducing the Annual National Assessment (ANA) in grades 3 and 6, and in Grade 6 in Literacy and Numeracy (South African Government Information, 2012/2013). These assessments were mainly introduced to try to improve the quality of Basic Education.

Though these ANAs were implemented in 2010, there are challenges associated with the whole process of assessment. Findings of these ANA results indicated that learners could not spell, count or understand what they read (Masondo, City Press 2013). If the results clearly indicate that learners in South African schools cannot spell and count to the blame usually lies with the teachers within the grades that learners were assessed. Teachers who were not experts in subjects like mathematics, science and technology, and languages were expected to teach those subjects and that was a major challenge in primary schools (Masondo, City Press 2013). Masondo (City Press, 21 July 2013) puts the blame for learners' poor performance on the teachers who lack the skill to deliver the content. The results of the ANAs urged the researchers to explore what teachers' challenges may be in remote rural areas, in particular Grade 3 and Grade 6, hence the research question for this study states: *'How can the ODL Institutions augment the teaching of science and technology education at rural primary schools level through community engagement?'*

RESEARCH METHODOLOGY

Observations, interviews and field notes were employed in gathering data from the teachers of Lovemore Primary School. This chapter report will be from the structured interviews conducted with the participants (teachers and leaners). In total four teachers were used as a sample from this school and eight learners were interviewed as well. Teachers were from different grades: two were teaching Grade 3s and the other two, Grade 6s. This chapter will also be reporting on both observations and field notes. The teacher empowerment centre within the college has grouped under performing schools and allocates such unilaterally to researchers of Unisa teams. Researchers also formed a number of teams to handle all 500 primary schools within the five identified provinces. It is against this background that Lovemore Primary School was allocated to a team of researchers from CEDU in their specialised field to conduct focus groups with teachers.

Questions That Guided the Focus Group

The following series of questions served as a guide for the research dialogue between the participating (S & T) teachers and learners of Lovemore with the Unisa research team. These questions kept the interview facilitation focused as teachers voiced their responses.

1. What challenges do you face in teaching your science and technology subject?
2. What type of assistance do teachers need in order to improve the quality of teaching and learning of S & T?
3. What are the types of interventions that do *not* help in improving the teaching and learning of S & T as you see it?
4. What types of intervention would help in improving S & T teaching and learning as you see it?

Whereas the Grade 6 learners also were accorded a platform to raise their concerns during technology didactic situation. Conversations with the learners are outlined in subsection that outlined "aspects that contribute towards poor learners' performance". The following questions were asked to the learners:

1. Since IDMEC (Investigate; Design; Make; Evaluate, and Communicate) is the backbone of teaching technology how do conduct you investigation?
2. Electrical Technology is the theme of technology subject; do you conduct any demonstration for the theme?
3. How your teachers are teaching about life and living strand?

Science and Technology Teachers' Responses

Many of the teachers' responses are highlighted per question in the following section.

- What challenges do you face in teaching your science and technology subject?
    - Teaching without resources makes my lesson boring for my learners;
    - Teaching in overcrowded classes makes it difficult to give individual attention;
    - Multi-grade teaching is also a challenge as it is time consuming ;
    - Shortage of classrooms makes me work extremely hard and this is tiring;
- What type of assistance do teachers need in order to improve the quality of teaching and learning of S & T?
    - Intensive subject based workshops with follow-ups;
    - To upgrade themselves with higher institutions of learning;
    - Classroom based support would be welcome.
- What are the types of interventions that do *not* help in improving the teaching and learning of S & T as you see it?
    - Workshops organised by the department as the facilitators do not seem to be knowledgeable;
    - Even union organised workshops do not seem to be working as they also do not have any follow-up strategies;
    - Clusters do not seem to be vibrant and active anymore.
- What types of intervention *would help* in improving S & T teaching and learning as you see it?
    - I think class based support could make a difference;
    - If we can do experiments with learners they will understand better;
    - We need a library and more books for our learners to read and increase their vocabulary;
    - Dictionaries are also a problem; we need enough for class use.

Teachers' responses gave the researchers an opportunity to plan for the next visit as well as analyse the collected data.

DATA ANALYSIS

In analysing the data, the researchers followed Creswell's (2009) steps in analysing data. The initial step was to transcribe the focus group discussions onto word processor program. AR practitioner studied the transcripts so that the researchers could make general sense of the data and verify what could have been missed during the discussion process. Of importance was to get the overall depth and credibility of the data (Creswell 2009). I coded the data manually, identified patterns and categorised data. The researchers read the entire list of codes and categories that the researchers formed and tried to get some sense of whether there was coherence (Henning, Van Ransburg & Smit 2004). The researchers then grouped categories with similar meanings and from the clustered categories, the following themes with regard to ODL Institutions and schools partnerships emerged:

- ODL Institutions partnership towards augmenting S & T teachers' content knowledge
- Sleuthing best practices that S & T teachers do, which enhances learning
- Didactical and pedagogical factors that cause barriers to learners' learning S & T education.
- Aspects that contribute toward poor learners' performance in an S & T class.

ODL Institutions Partnership: Augmenting Content Knowledge

Pedagogical Content Knowledge (PCK) has been embraced by many of the recent education practitioners as a way of describing the knowledge possessed by expert teachers. Science PCK is a new form of teacher knowledge transformed from, and embodying the other forms of teacher knowledge (Veal & Makinster, 1999). Pedagogical content knowledge is regarded as an academic construct that represents specialist knowledge of practice (Appleton 2003 & 2006; Bradbury & Koballa 2007; Hashweh 2005; Nilsson 2008). PCK is a way of understanding the complex relationship between teaching and content with specific teaching approaches (Van Driel, Verloop and De Vos, 1998). It should be known that PCK can really supersede ordinary subject matter knowledge but also incorporate the pedagogy and didactic of the subject in diverse forms that enhance learners to comprehend the subject content knowledge. Hence, learners' understanding of the subject depends on how the teacher teaches. In this chapter, an exploration on learners' experiences touches on how they experience the teaching style of different teachers and how these learners experience learning.

Sleuthing Best Practices to Enhance Learning

Learners' experiences with regard to teachers' way of engaging with the learning content evolved around the use of creative teaching using humour in explaining concepts and the use of a dictionary. It was clear from the focus groups that they were able to grasp the content from the teachers who were in a position to detail their explanations and those who went the extra mile.

Didactical and Pedagogical Factors That Cause Barriers to Learning

From the discussions, it was clear that teachers who were shouting at learners were causing a barrier in the learning process. Learners disliked standing in front of other learners in class and being ridiculed. Learners expressed disgust about how they felt when teachers scolded them for not being able to provide correct answers.

Aspects That Contribute Towards Poor Learners' Performance

During technology periods, teachers fail to contextualise their teaching. It was difficult for learners to grasp the content without some misconceptions. There was a lack of detailed explanation for technology processes from the teacher. Theory teaching in technology classes hinders Grade 6 learners' understanding of some technological concepts, as is evident when two learners said:

We don't know how to investigate; our teachers' just send us home to do investigation of technology things.

We wish to connect an electric circuit in class with our teachers and see how it works; we only read about this from the chalkboard.

In science classes, terminology was difficult and there was a lack of detailed explanations for scientific processes from the teacher. The confusion was evident when two Grade 6 learners said:

When they teach about gas in the books it looks like it is flowing, but we are not able to draw it in our books.

We don't understand how the nutrients are absorbed by plants; what are nutrients?

Another difficulty experienced by learners was apparent when concepts are taught only in English without switching to the vocabulary level of the Grade 6 learners. Learners attributed the problem to a lack of dictionaries and teachers who accuse them of not writing at the level of a Grade 6 learner exacerbated this. Learners who did not speak Setswana at home experienced challenges regarding understanding during lesson presentation in both English and/or vernacular.

RESEARCH FINDINGS

The findings revealed teachers' incapacity, in terms of their knowledge and teaching of S & T in the areas of contextualising their teaching, need special attention. In responding to the findings, the results prompted the development of the recommendations to be adopted by the provincial district circuits, ODL institutes of higher learning and the national ministries of education as support tools to empower S & T teachers to confidently teach technology processes to their learners. Cosby (2012), who stated that, "If you want to know who really moulds our children's future, it's not the politicians, movie stars, or corporate leaders, but it's our teachers", confirms this declaration.

Teacher's Classroom Practice Challenges

The teachers highlighted their teaching challenges of Grade 6 natural science and technology in the intermediate phase of GET band. These challenges were uncovered during focus group interviews and are listed as follows:

- **In Science and Technology Classes:** A lack of resources, for example, iodine solution, is eminent;
- **S & T:** Learners do not all speak the same language, hence teachers are bound to switch between languages;
- Teachers change subjects annually. They cannot build experience in subjects;
- Shortage of learner workbooks;
- Learners do not have books and teachers photocopy some work from their textbooks for the learners;

- There is overcrowding due to a lack of classrooms, which that makes individual attention and group work activities impossible.

What type of assistance do teachers need to improve the quality of teaching and learning?

- Enough time for workshops; cluster teaching to improve;
- Science laboratories are needed from lower classes whereas technology classes needs more resources

Types of interventions:

- **Successful Strategy:** Group learners for after-school intervention. This process should prioritise the slow learners before the new content is introduced. It will be fruitful if parents can be involved. The problem is that grandparents do not understand the new curriculum; their understanding varies as some (who are over the age of 60) cannot read or write.

The Unisa team identified a lack of resources such as furniture, classrooms, libraries and science laboratories as urgent needs. The relationship forged with Lovemore Primary School needs to be nurtured in order to improve the performance of the learners. The CEDU at Unisa can intervene in supporting teachers teaching S & T. The Unisa team is looking forward to the second visit to ensure that CEDU leaves a huge footprint in Marapyane.

RECOMMENDATIONS AND WAY FORWARD

Practical community engagement is used as an active method of implementing change. While most current standards employ a more static means through standard marketing techniques, CE involves actively implementing a specific process toward activism such as the 8-step guideline listed below – developed by Hildy Gottlieb of Creating the Future. The process may have similarities to a controversial form of 'friend raising', the emphasis in CE is that of honest relationship building for the sake of community, and not for the sole purpose of money-making/fundraising (Gottlieb, 2007). The steps are:

1. Determine the goals of thc plan
2. Plan who to engage
3. Develop engagement strategies for those individuals you already know
4. Develop engagement strategies for those individuals you do not already know
5. Prioritise those activities
6. Create an implementation plan
7. Monitor your progress
8. Maintain those relationships

The above listed steps were applied at Lovemore Primary School to serve as a reconnaissance study within this AR investigation. These steps will definitely lead the Unisa research team to develop engagement strategies of AR activities through observation, planning, action and reflection.

Since these S & T teachers find it hard to practically teach this subject due to: 1) a lack of content knowledge – AR interaction will serve as a means to bridge that gap; 2) a lack of resources for practical demonstrations – the school needs to purchase some mobile and portable S & T kits owing to the hands-on nature of the subject; and 3) a lack of relevant books – additional resources in classes such as library corners to support learners need to be in place. If possible, both the Lovemore School's S & T teachers and the Unisa research team need to solicit some sponsorship for science and/or technology materials, hand tools and other relevant resources. This will help in maintaining the relationship between Lovemore Primary School and the Unisa research team.

After my first contact session with the Grade 6, S & T education teachers, a rapport that levelled the playing field for future engagement was established with the community of Lovemore Primary School. As a member of the Unisa research team, the researcher have successfully moved from an emic standpoint to an epic position since the teachers were more willing to share their science and technology classroom practice challenges and more open to being engaged in any level of planning. More sessions will be planned to jointly structure the way forward and cultivate the nature of the intervention strategies. The researchers trust that this partnership of mutual understanding will ultimately improve learners' performance and teachers' achievement.

Measuring any construct on PCK is a complex process and any teacher evaluation based on a singular data point should be interpreted with caution. When attempting to comprehensively deconstruct, analyse and measure any complex concept in technology education, including PCK, a more robust system must be in place including content assessment, multiple observations and interviews over time. Few studies have addressed the contention that teachers with strong PCK are more likely to increase student achievement (Abell, 2008; Borowski et al., 2011). It is intended that future work will examine teachers at different career stages, inclusive of student achievement results to further unpack the relationship between PCK.

CONCLUSION

Because of these overlaying problems (poor learners' achievement and teachers' performance), it is noted that community members (natural science and technology teachers) themselves are also directly influenced and therefore similarly hindered with regard to change within their specific regions of development (S & T education didactics and pedagogical development) avers Gottlieb (2001). In the same breath, the Policies web of 31 August 2012, in its report, argues that the performance of South African primary schools as viewed in an international and regional context is seen as a national shame. It is against this background that the Unisa team of researchers saw the necessity of embarking on the current study of exploring challenges that are experienced by teachers in rural primary schools.

Teacher education is a huge task and cannot offer all the knowledge and skills that teachers require for effective teaching of natural science and technology in the GET band. However, as an AR practitioner from Unisa the practitioner should, among other things, attempt to prepare S & T teachers such that: (i) they are aware of the many different forms of knowledge they need for teaching and why; (ii) they can acquire the additional knowledge in practice; (iii) they can adapt to different contexts and situations; and (iv) they can reflect and learn from their own practice and from research. There are further implications that should be considered by the Unisa research team including the following: (i) what can be included in teacher education courses? (ii) what can be learnt in practice? and (iii) how do we know the difference?

The purpose of identifying the problems before making a determination of what the solution could be was successful. This was more like a fact-finding (reconnaissance study in AR) mission before a full roll-out intervention is designed. Knowing the nature of the problem, it is always easy to come up with a programme that is geared toward correcting the situation. A developmental theory was used to underpin the study while the AR practitioner and co-researchers are engaged in a CoP paradigm. For Wenger (1998) CoP is important with regard to negotiation, learning and identity. I view CoP as a crucial avenue for interaction, sharing and establishing mutual trust as was evident with this CE project. In conclusion, the researchers argue that all focus group participants were able to give enough ideas, thoughts and points to enable me as an AR practitioner to design an intervening programme tailored at addressing partnership between Unisa as an ODL Institution and Lovemore Primary School, which is rurally based in Mpumalanga Province of South Africa.

REFERENCES

Appleton, K. (2003). *Pathways in professional development in primary science: extending science pedagogical content knowledge*. Paper presented at the annual conference of the Australian Science Education Research Association, Melbourne, Australia.

Appleton, K. (2006). Science pedagogical content knowledge and elementary school teachers. In K. Appleton (Ed.), Elementary science teacher education: International perspectives on contemporary issues and practice (pp. 31–54). Lawrence Erlbaum in association with the Association for Science Teacher Education (ASTE).

Bettman, M. (2012). *500 School Project – making schools better*. Pretoria: UNISA Press.

Bothamley, J. (2004). *Dictionary of Theories*. New York: Barnes & Noble Books.

Bradbury, L. U., & Koballa, T. R. Jr. (2007). Borders to cross: Identifying sources of tension in mentor-intern relationships. *Teaching and Teacher Education*, *24*(8), 2132–2145. doi:10.1016/j.tate.2008.03.002

Braimoh, D. (2015). Open and Distance Learning in Sub-Saharan Africa: promises and threats. In A. Ravhudzulo & S. Ndlangamandla (Eds.), *Practical and Critical Issues in Open Distance Learning: 2016* (pp. 69–78). Pretoria: Unisa Press.

Chetty, R., & Lubben, F. (2010). The scholarship of research in teacher education in a higher institution in transition: Issues of identity. *Teaching and Teacher Education*, *26*(4), 813–820. doi:10.1016/j.tate.2009.10.018

Coetzee, E. (2012). Community Engagement by Higher Education Institutions – a practical model and guidelines. *Africa Education Review*, *9*(3), 501–517. doi:10.1080/18146627.2012.742669

College of Education. (2012). *College of Education brochure Unisa*. Retrieved from www.unisa.ac.za

Cosby, B. (2012). *About Technology Education*. Academic Press.

Creswell, J. W. (2009). *Research Design: Qualitative, Quantitative, and Mixed Methods Approaches* (3rd ed.). Thousand Oaks, CA: Sage.

Department of Basic Education (DBE). (2011). *Integrated Strategic Planning Framework for Teacher Education and Development in South Africa, 2011–2025: Technical Report*. Pretoria: Government Printers.

DFE. (2016). *Educational Excellence Everywhere*. Available at https://www.gov.uk/government/publications/educational-excellence-everywhere

Doyle, A., Seery, N., & Canty, D. (2016). *Exploring the Relationship between Technology Teachers Orientations towards Teaching and their Associated Professional Life Phases*. Paper presented at PATT–32, Technology Education for 21st Century Skills, Netherlands.

Fawcett, S.B., Paine-Andrews, A., & Francisco, V.T. (1995). Using empowerment theory in collaborative partnerships for community health and development. *Community Psychology*, *23*(5), 677–697.

Fullan, M. (1993). *Change forces: probing the depth of education reform*. Bristol, PA: Falmer Press.

Gauteng Department of Education (GDE). (2010). *Gauteng Mathematics, Science and Technology Education improvement strategy: 2009–2014*. Johannesburg: Government Printers.

Gottlieb, H. (2007). Community Engagement Step-by-Step Action Kit. Tucson, AZ: Renaissance Press.

Hashweh, M. Z. (2005). Teacher pedagogical constructions: a reconfiguration of pedagogical content knowledge. Teachers and Teaching: Theory and Practice, 11(3), 273–292. doi:10.1080/13450600500105502

Henning, E., Van Ransburg, W., & Smit, B. (2004). *Finding your way in a Qualitative Research*. Pretoria: Van Schaik.

Higgs, P. (2016). The African Renaissance and the decolonisation of the curriculum. In V. Msila & M. T. Gumbo (Eds.), *Africanising the Curriculum: Indigenous Perspective and Theories*. African Sun Media Press.

Jita, L. C., & Mokhele, M. L. (2013). Shifting research identities through communities of practice: Reflections on a faculty development programme for teacher education in South Africa. *Progressio. South African Journal for Open and Distance Learning Practice*, *35*(2), 126–142.

Jones, L., & Wells, K. (2007). Strategies for academic and clinician engagement in community-participatory partnered research. *Journal of the American Medical Association*, *297*(4), 407–410. doi:10.1001/jama.297.4.407 PMID:17244838

Kamal, R. (2014). The ethical conundrums of Community Engagement in South Arica. *Unisa e-news*. Available online: https://staff.unisa.ac.za/e-news/2014/07/09/the-ethical-conundrums-of-ce-in-south-africa

Lave, L., & Wenger, E. (2014). Jean Lave, Etienne Wenger and communities of practice, Conference: innovation in Youth Work – Creative practice in challenging times. YMCA George Williams College.

Mahomed, H. (2004). *Challenges in Curriculum Transformation in South Africa*. Paper presented at the 5th Annual Speaking Conference, Gauteng, South Africa.

Mapotse, T. A. (2012). *The teaching practice of senior phase Technology Education teachers in selected schools of Limpopo Province: An Action Research study* (DEd thesis). University of South Africa.

Mapotse, T. A. (2015). An emancipation framework for technology education teachers: an action research study. *International Journal of Technology and Design Education, 25*(2), 213–225. Available at http://link.springer.com/article/10.1007/s10798-014-9275-y/fulltext.html

Mapotse, T. A. (2017). Development of a Technology Education Cascading Theory through community engagement site-based support. *International Journal of Technology and Design Education.* Retrieved from http://download.springer.com

McKay, V. (2014). The ethical conundrums of Community Engagement in South Arica. *Unisa e-news.* Available online: https://staff.unisa.ac.za/e-news/2014/07/09/the-ethical-conundrums-of-ce-in-south-africa

McLain, M., Barlex, D., Bell, D., & Hardy, A. (2015). Teacher perspectives on pedagogical modelling and explaining in design & technology: A Q Methodology study. In Plurality and Complementarity of Approaches in Design and Technology Education. Available at: https://hal.archives-ouvertes.fr/hal-01161553

McLain, M., Bell, D., & Pratt, A. (2013). *Show-how know-how (Part 1): Theory and practice for demonstrating in Design and Technology.* Welles Bourne, UK: Design and Technology Association.

Minkler, M., & Wallerstein, N. (2003). *Community-Based Participatory Research for Health.* San Francisco, CA: Jossey-Bass.

Moini, M., Fackler-Lowrie, N., & Jones, L. (2005). Community Engagement: Moving From Community Involvement to Community Engagement – A Paradigm Shift. Santa Monica, CA: PHP Consulting.

Mokhele, M. L. (2011). *Teachers' perspectives on continuing professional development: A case study of the Mpumalanga Secondary Science Initiative (MSSI) project* (DEd thesis). University of South Africa.

Nilsson, P. (2008). Recognising the needs – Student teachers' learning to teach. *NORDINA, 4*(1), 92–106.

Pandor, N. (2008). *Pandor slams SA's failing education system.* University of Witwatersrand. Available online: http://www.mg.co.za/article/2008-08-05-pandor-slams-sas-failing-education-system

Pemsel, A. C. (1891). Commissioners for Special Purposes of Income Tax v. Academic Press.

Quasthoff, U., & Wild, E. (2014). Learning in context from an interdisciplinary perspective. *Learning. Cultures and Social Interaction, 3*(2), 69–76. doi:10.1016/j.lcsi.2013.12.004

Roberts, J. (2006). Limits to communities of practice. *Journal of Management Studies, 43*(3), 623–639. doi:10.1111/j.1467-6486.2006.00618.x

Robinsons, M. (2013). A call for teachers. In *The Teacher newspaper. October 2013.* M & G Media Ltd.

Smit, B., & Romm, N. (2013). *Focus group field guide, 500 schools' project: making Schools Better. Unisa, College of Education.* Pretoria: Unisa Press.

Stars in Education. (2011). The Teacher and the Teaching Career. In *Education Handbook: Connecting to make a difference.* Stellenbosch: Argo.

Stevens, A. (2006). Technology teacher education in South Africa. In International Handbook of Technology Education: Reviewing the past twenty years. Rotterdam: Sense.

The Teacher. (2013). Understanding the school's location. Principal report by Phumzile Nokuthula Langa. In *The Teacher, Your Guide to Education, October 2013.* M & G Media Ltd.

Unisa Senate Community Engagement Policy. (2013). *Community Engagement Budgets 2014.* Pretoria: Unisa.

Van Driel, J. H., Verloop, N., & De Vos, W. (1998). Developing science teachers pedagogical content knowledge. *Journal of Research in Science Teaching, 35*(6), 673–695. doi:10.1002/(SICI)1098-2736(199808)35:6<673::AID-TEA5>3.0.CO;2-J

Veal, W. R., & Makinster, J. G. (1999). Pedagogical content knowledge taxonomies. *Electronic Journal of Science Education, 3*(4). Available at: http://unr.edu/homepage/crowther/ejse/ejsev3n4.html

Wenger, E. (1998). *Communities of practices: Learning, Meaning, and Identity.* Cambridge University Press. doi:10.1017/CBO9780511803932

Wenger, E. (2007). *Communities of practices: A brief introduction.* Available online: http://www.ewenger.com/theory/

Whitehead, J., & McNiff, J. (2006). *Action Research Living Theory.* London: Sage. doi:10.4135/9781849208536

Wikipedia. (2016). *Community Engagement.* Available online: www.wikipedia.org

Williams, J., Eames, C., Hume, A., & Lockley, J. (2012). Promoting pedagogical content knowledge development for early career secondary teachers in science and technology using content representations. *Research in Science & Technological Education, 30*(3), 327–343. doi:10.1080/02635143.2012.740005

Williams, P. J., & Gumbo, M. T. (2011). New Zealand Technology pedagogical content knowledge. In Pupils' Attitude Towards Technology (PATT) 25 and Centre for Research in Primary Technology (CRIPT) 8 on "Perspective on Learning in Design & Technology Education". Goldsmiths, University of London.

Yager, R. E. (2013). Successes and continuing challenges: Implementing the changes in professional development for teachers in the U.S. *International Journal of Education in Mathematics, Science and Technology, 1*(3), 184–189.

Chapter 6
Trans-Disciplinary Approaches to Action Research for e-Schools, Community Engagement, and ICT4D

Leila Goosen
University of South Africa, South Africa

ABSTRACT

This chapter will highlight how practitioners use trans-disciplinary approaches to conduct Action Research (AR) in the context of e-schools, Community Engagement (CE) and Information and Communication Technologies for Development (ICT4D). The objectives provide details on how: AR is used to align e-education with life-enhancing values and in situations dealing with partnerships towards CE and/or ICT4D matters at local, national and international levels; to promote e-learning and development among those previously excluded from formal education; AR is contributing to transformation and equity in the context of e-schools, CE and ICT4D; AR CE is contributing to improving the quality of life for all; changes and/or growth in the way in which AR is utilized and interpreted can be explored when such questions are raised; AR can stay relevant in light of the ever-increasing speed of change in terms of technological innovations; and examples of ground-breaking AR work can thus be achieved.

INTRODUCTION

As early as the end of the previous millennium, Avison, Lau, Myers and Nielsen (1999) were presenting the case in favor of Action Research (AR): Researchers ought to try out the theories they propose together with practitioners in real-world situations and with real organizations, towards making academic research relevant. According to these same authors, there is therefore a need for a book on Action Research, such as this one, and this chapter thus responds to a challenge to Action Research practitioners, especially when working in socio-economically challenged contexts, to explain how they are contributing to conducting Action Research in specific fields relating to e-schools and Community Engagement (CE) in the context of Information and Communication Technologies (ICTs) for Development (ICT4D).

DOI: 10.4018/978-1-5225-2642-1.ch006

In terms of describing the general perspective of the chapter, similar to that of Blake, Tuckery and Glaser (2014), this chapter uses the point of view of Action Research as approach. This perspective encourages a holistic understanding of applicable contexts by offering an actionable framework for helping to choose appropriate interventions (Marshak & Heracleous, 2005).

Both Blake, Tuckery and Glaser (2014) and Van Biljon and Lotriet (2014) regard the nature of research in the field with regard to ICT4D as inherently tending towards being multi-, inter- and/or trans-disciplinary. In light of the 'Cross-Disciplinary Approaches to Action Research' theme of this book, the purpose of this chapter will therefore be to highlight how academics, researchers, teachers and/or other practitioners, who are considered to be included in the audience that this chapter wishes to address, use trans-disciplinary approaches to conduct Action Research for empowerment in the context of e-schools, Community Engagement and Information and Communication Technologies for Development (Avison, Lau, Myers, & Nielsen, 1999).

In terms of indicating under which areas of specialization this chapter falls, the areas covered in this chapter will include, but are not limited to, e-education and e-schools, community engagement and Information and Communication Technologies for Development. This chapter will therefore include Action Research work being done on local/global scales in both formal and informal education contexts.

With some similarities to what was done in an article by Marshak and Heracleous (2005), this chapter will present a conceptualization of Action Research, which is then illustrated through analyses of development interventions. The conceptualization and application of the approaches to action research are discussed.

Weber (2009) identified challenges to building useful ICT4D models of reality, in that the phenomena of interest are not clearly defined, the theories that underpin the research are not articulated carefully, and the research methods and data-analysis methods used do not conform to norms for conducting high-quality empirical research. Steyn (2011) agreed that no clear theories and/or models are available, while also pointing out that ICT4D often depends on globalization models, even when well-meaning social activists implement it as a tool for social change. Similarly, Heeks (2010) indicated that until quite recently, obtaining evidence, e.g. on the digital provide and of ICTs' contribution to development, and especially evidence from research, which is well-conceptualized and conducted rigorously, was still limited. Evidence of such a contribution has only now started to emerge and even more is needed.

The problem that the major areas of this chapter wishes to address and will therefore focus on include exploring topics that would enable a more profound understanding of the formulation and/or design of theory, conceptual frameworks or typologies, the testing of theories (including the validation of existing typologies, theories and/or frameworks) and methodologies (including novel methods or approaches to collect or analyze data) that have received less attention. Problems related to these issues will be addressed in order to identify the standards applicable to trans-disciplinary and methodological distinctions for ensuring and supporting high-quality research in ICT4D fields (Gomez & Pather, 2011).

This introductory section will now end by specifically stating the objectives of the chapter, which include, but is not limited to, providing details on how the Action Research reported on in this chapter promote learning in e-schools and Information and Communication Technologies for Development among those previously excluded from, and/or disadvantaged in terms of, especially education. The chapter will further show how AR is contributing to transformation and equity in the context of e-schools, community engagement and ICT4D, and is being used in situations dealing with partnerships towards community engagement and/or ICT for Development matters at local, national and international levels. AR community engagement is also contributing to improving the quality of life for all, while changes and/or growth in the way in which AR is utilized and interpreted, are explored in light of questions raised. Finally, the chapter will provide examples of how AR stays relevant, despite the ever-increasing speed of change in terms of technological innovations, as well as how ground-breaking AR work can be achieved.

BACKGROUND

This first section of the chapter after the introduction will discuss key term topics as they relate to Action Research, incorporating the views of other researchers, in the form of a literature review, into the discussion to support, refute, and/or demonstrate the position of the author on these and related topics.

In the section on Action Research in the 'Encyclopedia of Research Design', Adams (2010) argued that Action Research not only contributes to social change, but that this is, in fact, the primary goal. From work that Fourie and Krauss (2010) undertook in a developing South African context, it would seem as if ICT4D might offer a research focus, which could help in deepening the understanding of, and direction for extending, research into the use of ICTs in support of economic and social development.

Action Research is increasingly being utilized in health-related fields (MacColl, Cooper, Rittenbruch, & Viller, 2005) – just one example of some evidence from literature of the latter would be e.g. the article by Vaismoradi, Turunen and Bondas (2013), on the implications of content and thematic analysis for conducting a qualitative descriptive study in nursing and health sciences.

According to Adams (2010), Action Research has also been widely used in a discipline such as education, for example: Smith and Sela (2005) implemented Action Research as a bridge between pre-service teacher education and in-service professional development for students and teacher educators, Auger and Wideman (2000) used Action Research to open doors to life-long professional learning, and Brown and Macatangay (2002) studied the impact of Action Research for professional development through case studies in two Manchester schools. Ponte, Beijard and Ax (2004) also investigated Action Research and initial teacher education in three different countries, whilst Price (2001) looked at the interplay between pedagogy and change in terms of the transformative potential of Action Research in teacher education.

Other literature include Baker Collins (2005), who used Action Research to develop an understanding of poverty from those who are poor, while Sharma and Sturges (2007) implemented an action research program on making use of ICT to facilitate poor people's access to public services. Cecchini and Scott (2003) used their experiences in rural India and lessons learnt there and elsewhere in the developing world in an attempt to answer questions around whether Information and Communications Technology applications can be utilized to support human development and poverty reduction strategies.

According to Blake, Tuckery and Glaser (2014), Action Research is a research method, which was established on a pragmatist epistemology, needing to be viewed in the milieu of experimental ICT4D. In this same vein, Adams (2010), however, pointed out that although there are plenty of ontological and epistemological groupings of qualitative and quantitative research approaches, predominantly when debating the authority of action research studies, especially quantitative research is often condemned when conclusions are drawn, which are pragmatically irrelevant.

Although Avison, Lau, Myers and Nielsen (1999) *categorized* four distinctive types of action research, this chapter will only be discussing three of these that are actively applicable:

1. The Action Research strategy that is strongly *focused* on action, change and reflection; on doing something in a real situation and then seeing what the results would be after that. Research such as case studies often details what practitioners say that they do. What practitioners actually do is more important in action research.
2. Action Science, which tries resolving inconsistencies between the ways in which theories are advocated and applied; and
3. Participatory Action Research, emphasizing substantial *collaboration* and participation. According to Van Biljon and Lotriet (2014), methodologically, the discourses in the special issue of the South African Computer Journal (SACJ) on ICT4D related largely to such a participatory design, with the article by Blake, Tuckery and Glaser (2014) reflecting on the use of community-based co-design methods in terms of ICT4D towards communication and information access for Deaf people. The latter authors specifically urged that while working cooperatively in this way, it should be born in mind that community members are co-designers on an equal footing with researchers. Similar to what Baker Collins (2005) described, participatory research projects such as those described in this chapter can begin committing to the inclusion of the experiences of community members.

The *characteristics* of Action Research as set out by Adams (2010) cover aspects related to reflection, iterancy, collaboration and the role of the researcher.

Reflection have already been mentioned in the first of the types of action research above, and further details will also be provided - see the first of the steps of Action Research supplied on the next page.

The second characteristic of Action Research that Adams (2010) listed include the idea of it being repetitive and/or involving an iterative approach, comprising of the engagement in successive cycles in the collection of data, to make sense of it, or, stated slightly differently, while carrying out an action research study, the activities recur related to planning, acting, observation and reflection in a spiraling fashion. The application of Action Research thus involves an iterative cycle of plan-act-reflect (Oates, 2006), or, according to Blake, Tuckery and Glaser (2014), leads to a design-implement-reflect cycle. This process can be done repeatedly to see what the results would be after several tries.

Collaboration have also already been mentioned in the third of the types of action research, while further details will also be provided - see the third of the steps of Action Research supplied on the next page.

In terms of the role of the researcher, action researchers are often not only defined as external experts, but are also seen as full participants (Adams, 2010). This could lead to the researcher obtaining more accurate information, since (s)he is a part of the process. In Action Research, there are a number of roles, which can guide both the researcher and participants:

1. The *complete participant* role, where the researcher is not disguised, but identified;
2. The *participant observer* role, encouraging action researchers towards negotiating the levels of approachability and membership in the participating group(s);
3. The *observing participant*, where the researcher does not become part of the group of participants, but asks for permission to make detailed notes; and
4. The *complete observer* sees the researcher adopting passive involvement in activities and/or events.

According to MacColl, Cooper, Rittenbruch and Viller (2005), Action Research is usually seen as operating over a reasonably short time span, with a variety of action research formulations available for implementation. Incorporating contributions from Adams (2010), Avison, Lau, Myers and Nielsen

(1999), Blake, Tuckery and Glaser (2014) and Oates (2006), this chapter suggests the following steps in Action Research:

1. Reflective analysis by the researcher and/or participants, which also overlaps with the first characteristic of Action Research listed by Adams (2010). Mentioned in connection to both Dewey (1910) and Lewin (1946), hypothesizing, or problem diagnosis, are usually indicated as part of this initial phase. Such reflection can take place regarding both the current state of affairs,
2. As well as what the plan of action could be. This then leads to the formulation of theory - according to Adams (2010), citing e.g. Avison et al. (1999), Action Research is often viewed as a process involving linking or combining theory and practice – this step obviously overlaps to some extent with the Action Science category suggested by Avison, Lau, Myers and Nielsen (1999). As indicated in the first sentence provided for this step, this can include a planning or design stage, which could also involve the fact-finding concept of Lewin (1946) referred to above.
3. This is followed by *collaborative* change: As proposed by Dewey (1910), the hypothesi/(e)s would now be tested through the execution and/or implementation of certain actions interventions.
4. Finally, the (observed) results of such an action / intervention could be studied and analyzed through reflection on and/or reflective learning about the whole process.

THEORETICAL FRAMEWORKS

Standal and Rugseth (2014) recently indicated that the educational theory of Dewey (1910) is still relevant, with particularly the concept of reflective thinking providing the framework, which guides the action research process.

In his authoritative original book '*How we think*', Dewey (1910) proposed the following phases of inquiry: inquisitiveness, intellectualization, conjecturing, thinking and trying out hypotheses through action (Adams, 2010). The latter author also referred to another historical antecedent towards developing theory around Action, in the form of Lewin (1946), who was indicated as having conceptualized Action Research to involve hypothesizing, planning, fact-finding, execution and analysis.

A potential problem that manifests as part of the definition of action research is that when researchers and practitioners work together to conduct Action Research, there is a need for sharing an ethical framework that is mutually acceptable (Avison, Lau, Myers, & Nielsen, 1999). It is, however, unlikely that action research will be successful when conflict occurs between the researchers and practitioners or amongst the practitioners themselves.

Towards serving as comprehensive framework and guidance for the larger community and consisting of four dimensions, Avison, Lau, Myers and Nielsen (1999) proposed considering, firstly, the *category* in terms of the types of action research used – this was detailed by these same authors in the previous section of this chapter - as well as the *focus*, as specifically discussed in terms of the first of the three categories proposed. Secondly, contemplate the tradition and beliefs, which the applicable assumptions imply. Next, the research process, which includes the theme, the level of organization involved, the extent of change, and the researcher's role(s) should be thought through – again, see discussion of these in the previous section of this chapter - and finally, the style of presentation adopted.

The framework that Avison, Lau, Myers and Nielsen (1999) referred to is foundational for the Action Research pedagogy in systems development, which could be debated to refine it, and can also perhaps help to establish such a unified framework in systems development. This framework, however, should still be supplemented through the definition of a comprehensive set of explicit criteria in terms of which action research is conceived, designed, processed, conducted, presented, and evaluated, before carrying out the research in such a way that anyone would later be able to judge the outcome(s), as well as ways for managing alterations in these criteria.

Relating more specifically to e-schools and community engagement, Mlitwa and Koranteng (2013) used an Activity Theory perspective in their study on the integration of ICT into the curricula of e-schools in the Western Cape province of South Africa. While the study that the paper by Goosen and Van der Merwe (2015) reported on was located within a framework that clarified issues around ICT and e-schools, the study by Goosen and Gouws (2016) addressed clarified issues around community engagement.

Both a chapter by Steyn (2011) and a paper by Gomez and Pather (2011) motivated needing a fundamental paradigm shift in ICT4D evaluation. According to the latter authors, the value of ICTD evaluation can be found in the production of data substantiating socio-economic outcomes and impact. Convincing evidence of impact is, however, not being provided, as there is a lack of suitable frameworks that can go beyond tangibly and quantifiably indicating the success of ICT4D projects.The paper by Duncombe and Heeks (2005) therefore presented a framework to understand the roles of ICTs across a range of Micro, Small and Medium-scale Enterprise (MSME) applications in developing countries.

In terms of suggestions for a trans-disciplinary planning approach, an article by Fourie and Krauss (2010) argued that communities could benefit from planning that is tailored and influenced by ICT4D literature and takes into account factors like the building of relationships with adaptation of the definitions and standards for information handling/gathering and the selection of applicable methods and theories for supporting on-going research. These authors also suggested selecting appropriate research paradigms and methodological approaches for studying impact and the experiences of participants as some of the guidelines and/or principles influencing insights from ICT4D. According to these same authors, Critical Social Theory (CST), as a research paradigm relevant in the field of ICT4D, seem to hold potential in this regard.

Finally, the paper by Van Zyl and Vannini (2013) explored the possibilities related to Design-Based Research (DBR) methodologies in the broader domain of ICT4D. Specifically, these authors reflected on DBR offering important advantages reminiscent of those related to Action Research, including engagement of participants, iterative design and learning becoming self-directed.

ISSUES, CONTROVERSIES, PROBLEMS

This section of the chapter will present the perspective of the author on issues, controversies, problems, etc., as these relate to AR and arguments supporting the position of the author. This section will also compare and contrast these with what has been, or is currently being, done.

Mlitwa and Koranteng (2013) conducted an action research study in order to understand the issues and problems that deploying ICT could present, especially when working in socio-economically challenged contexts, such as in disadvantaged schools. The latter authors concluded that the e-education goal of the then South African Department of Education (2004) has not been reached, citing financial resources as the most prominent hindrance. They did, however, go on to note that the problem was not only financial, but that a general lack of communication between schools, communities, policy makers and teachers was also prevalent. In addition, an all too pervasive lack of support, both socially and technically, was observed. It was, therefore, somewhat unsurprising that the final hindrance was observed as being a lack of clear guidelines and training.

In their conference paper, MacColl, Cooper, Rittenbruch and Viller (2005) specifically highlighted some of the ethical issues related to ethnographic action research. The latter authors pointed out that although considerable previous work exists on ethical considerations relating to research, there is little on issues, which specifically arise from particular research approaches. Similarly, Avison, Lau, Myers and Nielsen (1999) indicated that while examples are available of action research articles, these still lack detailed guidelines to novice researchers and practitioners for understanding and engaging with action research studies.

According to the latter authors, if researchers do not explicitly follow the principles of action research when they work in real-life situations, such work could be more accurately described as consulting (Avison, Lau, Myers, & Nielsen, 1999). On the other hand, the interview and observation of people in such contexts devoid of the insight related to intervention also do not qualify as action research - the latter should, instead, rather be described as case study research.

One problem associated with the fact that action researchers are often not only defined as external experts, but are also seen as full participants, is that the people involved in the study may also change their attitudes, so that how they behave will be in line with what they believe is the outcome that the researcher wants (Adams, 2010).

In the context of e-schools, Action Research could have been the best most appropriate strategy, if the researcher had the means and authority to act and implement changes at the school being studied. Also it is nearly impossible to convince teachers, school leaders and learners to be active participants in the research. Another problem with regard to using action research in this context is that it is too dependent on some major event or action taking place, in the form of, for example, an e-learning tool implementation, which requires some level of capital and investment.

According to Adams (2010), Action Research can meet determinants with regard to reliability and validity if the key contextual variables consistently remain the same, and researchers are disciplined when they gather, analyze and interpret the evidence from their studies by using triangulation strategies, and in purposefully using participant validation.

In connection to the key contextual variables mentioned in the previous paragraph, and with regard to generalizability, Adams (2010) asked whether any claim of replication was possible in studies that involve human researchers and participants. Further unto this issue, the first of the concerns regarding the methodology used in the Action Research milieu mentioned by Adams (2010) comprise minimizing possibilities for the repeatability of the situation/context and results: it would be out of the question to reproduce settings, given that these are live, actual classroom situations with all their inherent complexities (Marshak & Heracleous, 2005).

Corresponding to researchers' discipline mentioned previously, and due to the second issue mentioned by Adams (2010), i.e. restrictions with regard to the process when gathering and documenting data, it is all the more important to keep a record of as much appropriate data as is feasible, as correctly as is workable within the conditions. In this same vein, Avison, Lau, Myers and Nielsen (1999) agreed on the importance of properly documenting the research process, while the action researcher may participate in experimenting to improve such writing-up by using diaries and concept maps.

Finally, in this regard, it remains essential to employ thoughtful, critical study in reflection upon and analysis of the meaning of data, thereby coming to applicable, well-founded insights that contribute to the way in which the knowledge, skills and values of researchers in the field could potentially affect investigations (Adams, 2010).

SOLUTIONS AND RECOMMENDATIONS

This section will discuss solutions and recommendations in dealing with the issues, controversies, and/or problems presented in the preceding section of the chapter.

A research strategy is the overall approach implemented in order to obtain solutions in terms of answering the question or questions towards dealing with the issues, controversies, and/or problems presented in research, of which Oates (2006) mentioned action research, case studies, design and creation, ethnology, experiments and surveys. As one of the research strategies thus outlined by Oates (2006), Action Research can specifically be used in the fields of Information Systems and Computing as applicable to ICT4D. Huxham and Vangen (2003) used case studies as part of their design choices while researching organizational practice through Action Research.

According to Davison (1998), the following align with positivism and could possibly be employed as research strategies during studies, such as those described in this chapter: action research, case studies laboratory and field experiments, and surveys. This same author used an Action Research perspective of group support systems to find a solution towards improving meetings in Hong Kong.

Goosen (2016a) reported on learners, teachers and managers wanting e-learning in the South African Basic Education (school) sector, while Goosen (2016b) looked at the policy issues relating to ICT education towards effective teaching and meaningful learning. The latter author provided demographic details with regard to a sample of seventeen participants from different African countries, as well as details around the extent to which the schools in her sample qualified to be labelled as being e-schools and engaged with their communities.

The purpose of the study by Goosen and Gouws (2016) was to explore research questions around a community engagement activity, while the topic of the study that the paper by Goosen and Van der Merwe (2015) reported on was investigating research questions around learning in e-schools in a specific South African (SA) district. The authors specifically provided details around the extent to which the schools in their sample were characterized as institutions exhibiting e-learner characteristics regarding the utilization of ICTs for enhancing social and collaborative e-learning, qualified and competent teachers adopting e-learning and using ICTs for enhancing their teaching, as well as qualified and competent managers, who use ICTs to plan for, manage and administer the integration of e-learning into their educational environments, across South Africa, the rest of Africa and even further afield.

An initial study by Tangkuampien (2013) reacted to challenges related to efforts towards introducing ICTs into a school in a developing community, by combining Ethnographical Action Research techniques with the Process approach. This combination of the Process approach's encouragement of participants, together with the researcher passively taking on the role of complete observer, were contributing factors to the positive outcome achieved.

Finally, Kalanda and De Villiers (2013) employed a variant of Action Research (VoAR) in their research design, in order to conduct a longitudinal study investigating the progress made in terms of the integration of ICTs in e-learning and the teaching of the educational Science curriculum in selected high schools, in Lesotho. In addition to surveying the state of affairs, the aim of their work centered on establishing possible solutions to the issues and problems identified regarding ICT integration. With regards to sampling, Kalanda and De Villiers (2013) conducted their research in nine schools and involved forty participating teachers. Qualitative data was gathered by way of field notes taken during observation, open-ended questions during face-to-face interviews, as well as questionnaires. The qualitative data was analyzed using a thematic analysis and a simplified Grounded Theory approach. Vaismoradi et al. (2013, p. 400) defined thematic analysis by juxtaposing between Grounded Theory and content analysis: "a systematic coding and categorizing approach used for exploring large amounts of textual information unobtrusively to determine trends and patterns of words used, their frequency, their relationships, and the structures and discourses of communication". Moreover, Vaismoradi et al. (2013) explicated thematic analysis as being a method to identify, analyze and report on patterns and/or themes in the data.

Recommendations

In addition to what was just mentioned, Vaismoradi et al. (2013) recommended thematic analysis as being able to provide a rich and detailed account of the data in question.

MacColl, Cooper, Rittenbruch and Viller (2005) refer to work by Williamson and Prosser, who recommended researchers and participants negotiating around the issues, controversies, problems, etc., as described in the previous section of this chapter, before starting on a new project.

According to Avison, Lau, Myers and Nielsen (1999), the importance of action researchers clarifying their research aim and theory and being explicit about explaining their research approach and method from the outset, as well as all the way through to application, and at the time of publication is true for action research, just like for any other research approaches, keeping in mind that research will need to be evaluated in terms of the ability it has to explain practice.

FUTURE RESEARCH DIRECTIONS

This section will discuss emerging and future trends, as well as provide insight about the possible future of AR from the perspective of e-schools, community engagement and ICT4D. The viability of Action Research as a paradigm and/or implementation issues around the contexts mentioned will also be included in this section. Future research opportunities within these domains will be suggested.

Referring to Dewey (1910), Adams (2010) pointed out that rather than representing a single research methodology, Action Research is characterized by an *attitude* of inquiry. Similarly, Blake, Tuckery and Glaser (2014) indicated that the approach they implemented had been one where the application of Action Research was viewed mostly as a paradigm, as opposed to a specific research methodology. The broad aim of the article by Huxham and Vangen (2003) was making contributions to the growth of understanding relating to the qualitative phenomenological methodological paradigm known as Action Research.

In their work, Kalanda and De Villiers (2013) discovered that there exists a number of issues and problems, resulting in a great need for hands-on interaction, from the perspective of the learners, and that teachers require on-going development and support (in terms of ICT) at the same time. From the latter perspective, their research, therefore, coincides with the findings of Goosen (2015), on the use of educational technologies for growing innovative e-schools in the 21st century, in a community engagement project. Moreover, it reinforces the case for more future research to be done in an effort towards understanding the greater Southern African context.

As a possible avenue for future research, Mlitwa and Koranteng (2013) suggested that a similar study to theirs, on the integration of ICT onto curricula in the schools of the Western Cape province of South Africa, be performed in South African provinces other than the Western Cape.

Jensen, Iipito, Onwordi and Mukumbira (2012) indicated that there were many more aspects to investigate in future before they could conclude that the solution they described in their paper would work: other aspects, including organizational, legal, social, cultural and ethical ones, should also be factored in.

CONCLUSION

This section will provide a discussion of the overall coverage of the chapter.

In terms of concluding remarks, the objectives addressed in this chapter specifically included, but were not limited to, providing details on how the Action Research reported on in this chapter promoted learning in e-schools and Information and Communication Technologies for Development among those previously excluded from, and/or disadvantaged in terms of, education.

Similarly to what was suggested by Steyn (2011), this chapter attempted to critically contemplate and review especially the second dimension of a comprehensive framework, proposed by Avison, Lau, Myers and Nielsen (1999), towards serving as guidance for the larger community, as these relate to traditions, beliefs and implications of the applicable assumptions.

The chapter has also shown examples of contributions in situations dealing with partnerships towards transformation and equity in the context of e-schools, community engagement and/or ICT for Development matters at local, national and international levels. AR community engagement is thus contributing to improving the quality of life for all. Changes and/or growth in the way in which AR is utilized and interpreted, were explored in light of questions raised.

The expectations that development contexts set for ICT, together with the impact that national, societal and ethics cultures have on ICT4D implementation and usage in terms of social development (Takavarasha, Hapanyengwi, Chimanikire, & Kabanda, 2013), serve as important reminders of AR and ICT4D practitioners' ethical tasks (Sutinen & Tedre, 2010).

AR therefore stays relevant, despite the ever-increasing speed of change in terms of technological innovations. Although examples of ground-breaking AR work thus being achieved were provided, Heeks (2010) warned that due to issues related to working in an area where there is such fast-moving change, any conclusions may, however, be time-contingent.

REFERENCES

Adams, P. (2010). Action Research. In N. J. Salkind (Ed.), Encyclopedia of Research Design (Vol. 1, pp. 4 - 9). Thousand Oaks, CA: SAGE Publications, Inc. doi:10.4135/9781412961288.n3

Auger, W., & Wideman, R. (2000). Using action research to open the door to life-long professional learning. *Education*, *12*(1), 120–127.

Avison, D., Lau, F., Myers, M., & Nielsen, P. A. (1999, January). Action Research. *Communications of the ACM*, *42*(1), 94–97. doi:10.1145/291469.291479

Baker Collins, S. (2005). An understanding of poverty from those who are poor. *Action Research*, *3*(1), 9–31. doi:10.1177/1476750305047983

Blake, E., Tuckery, W., & Glaser, M. (2014, October). Towards communication and information access for Deaf people. *South African Computer Journal*, *54*, 10–19. doi:10.18489/sacj.v54i0.236

Brown, M., & Macatangay, A. (2002). The Impact of Action Research for Professional Development: Case studies in two Manchester schools. *Westminster Studies in Education*, *25*(1), 35–45. doi:10.1080/0140672020250104

Cecchini, S., & Scott, C. (2003). Can information and communications technology applications contribute to poverty reduction? Lessons from rural India. *Information Technology for Development*, *10*(2), 73–84. doi:10.1002/itdj.1590100203

Davison, R. M. (1998). Research Methodology. In *An Action Research Perspective of Group Support Systems: How to Improve Meetings in Hong Kong. PhD thesis* (pp. 3-1 - 3-20). Hong Kong: City University of Hong Kong. Retrieved September 13, 2016, from http://www.is.cityu.edu.hk/staff/isrobert/phd/ch3.pdf

Department of Education. (2004, September 2). White Paper on e-Education: Transforming Learning and Teaching through Information and Communication Technologies (ICTs). *Government Gazette*, (26734), 3 - 46.

Dewey, J. (1910). *How we think*. Boston: D. C. Heath. doi:10.1037/10903-000

Duncombe, R., & Heeks, R. (2005). *Information & Communication Technologies (ICTs), Poverty Reduction and Micro, Small & Medium-Scale Enterprises (MSMEs)*. Vienna: United Nations Industrial Development Organization. Retrieved from http://www.unido.org/file-storage/download/?file_id=83230

Fourie, I., & Krauss, K. (2010, December). Information literacy training for teachers in a developing South African context: Suggestions for a multi-disciplinary planning approach. *Innovation (Abingdon)*, *41*, 107–122.

Gomez, R., & Pather, S. (2011). ICT Evaluation: Are we asking the right questions?. *The Electronic Journal of Information Systems in Developing Countries, 50*(5), 1 - 14. Retrieved January 20, 2014, from http://www.ejisdc.org

Goosen, L. (2015). Educational Technologies for Growing Innovative e-Schools in the 21st Century: A Community Engagement Project. In D. Nwaozuzu, & S. Mnisi (Ed.), *Proceedings of the South Africa International Conference on Educational Technologies* (pp. 49 - 61). Pretoria: African Academic Research Forum.

Goosen, L. (2016a, February 18). *"We don't need no education"? Yes, they DO want e-learning in Basic and Higher Education!*. Retrieved from http://uir.unisa.ac.za/handle/10500/20999

Goosen, L. (2016b). Policy Issues in ICT Education towards Effective Teaching and Meaningful Learning. In J. Kriek, B. Bantwini, C. Ochonogor, J. J. Dhlamini, & L. Goosen (Ed.), *Proceedings of the Institute of Science and Technology Education (ISTE) Conference On Mathematics, Science And Technology Education* (pp. 667 - 679). Mopani Camp, Kruger National Park, South Africa: UNISA Press.

Goosen, L., & Gouws, P. (2016). Inspiring and Sustaining Learners' and Their Communities' Interest in Science, Engineering and Technology. In W. Mwakapenda, T. Sedumedi, & M. Makgato (Ed.), *Proceedings of the 24th Annual Conference of the Southern African Association for Research in Mathematics, Science and Technology Education (SAARMSTE)* (pp. 530 - 541). Pretoria: Tshwane University of Technology.

Goosen, L., & Van der Merwe, R. (2015). e-Learners, Teachers and Managers at e-Schools in South Africa. In C. Watson (Ed.), *Proceedings of the 10th International Conference on e-Learning (ICEL)* (pp. 127 - 134). Nassau: Academic Conferences and Publishing International.

Heeks, R. (2010). Do information and communication technologies (ICTs) contribute to development? *Journal of International Development*, *22*(5), 625–640. doi:10.1002/jid.1716

Huxham, C., & Vangen, S. (2003, July). Researching organizational practice through action research: Case studies and Design Choices. *Organizational Research Methods*, *6*(3), 383–403. doi:10.1177/1094428103254454

Jensen, K. L., Iipito, H. N., Onwordi, M. U., & Mukumbira, S. (2012). Toward an mPolicing solution for Namibia: leveraging emerging mobile platforms and crime mapping. *Proceedings of the South African Institute for Computer Scientists and Information Technologists Conference* (pp. 196 - 205). New York, NY: ACM. doi:10.1145/2389836.2389860

Kalanda, K., & De Villiers, M. R. (2013). E-Learning in the Science curriculum: A study in selected high schools in Lesotho. In J. Herrington (Ed.), *Proceedings of World Conference on Educational Media and Technology* (pp. 1664-1673). Victoria, Canada: Association for the Advancement of Computing in Education (AACE).

Lewin, K. (1946). Action research and minority problems. *The Journal of Social Issues*, *2*(4), 34–46. doi:10.1111/j.1540-4560.1946.tb02295.x

MacColl, I., Cooper, R., Rittenbruch, M., & Viller, S. (2005). Watching Ourselves Watching: Ethical Issues in Ethnographic Action Research. *Proceedings of the Annual Conference of the Computer-Human Interaction Special Interest Group (CHISIG) of the Human Factors and Ergonomics Society of Australia (HFESA) (OZCHI)* (pp. 1 - 4). Canberra, Australia: ACM.

Marshak, R. J., & Heracleous, L. (2005). A discursive approach to organization development. *Action Research*, *3*(1), 69–88. doi:10.1177/1476750305049966

Mlitwa, N. W., & Koranteng, K. (2013). Integration of ICT onto Curricula in the Western Cape Schools: The Activity Theory Perspective. *The Journal of Community Informatics*, *9*(4). Retrieved from http://ci-journal.net/index.php/ciej/article/view/957/1055

Oates, B. J. (2006). *Researching Information Systems and Computing*. London: Sage.

Ponte, P., Beijard, D., & Ax, J. (2004). Don't wait till the cows come home: Action research and initial teacher education in three different countries. *Teachers and Teaching*, *10*(6), 591–621. doi:10.1080/1354060042000304792

Price, J. N. (2001). Action research, pedagogy and change: The transformative potential of action research in pre-service teacher education. *Journal of Curriculum Studies*, *33*(1), 43–74. doi:10.1080/00220270118039

Sharma, G., & Sturges, P. (2007). Making use of ICT to facilitate poor peoples access to public services: An action research programme. *Information Development*, *23*(1), 15–23. doi:10.1177/0266666907075623

Smith, K., & Sela, O. (2005). Action research as a bridge between pre-service teacher education and in-service professional development for students and teacher educators. *European Journal of Teacher Education*, *28*(3), 293–310. doi:10.1080/02619760500269418

Standal, O. F., & Rugseth, G. (2014). Practicum in adapted physical activity: A Dewey-inspired action research project. *Adapted Physical Activity Quarterly; APAQ*, *31*(3), 219–239. doi:10.1123/apaq.2013-0105 PMID:25028475

Steyn, J. (2011). Paradigm Shift Required for ICT4D. In J. Steyn & G. Johanson (Eds.), *ICTs and Sustainable Solutions for the Digital Divide: Theory and Perspectives* (pp. 19–44). Hershey, PA: IGI Global; doi:10.4018/978-1-61520-799-2.ch002

Sutinen, E., & Tedre, M. (2010). ICT4D: A Computer Science Perspective. In T. Elomaa, H. Mannila, & P. Orponen (Eds.), *Algorithms and Applications* (Vol. 6060, pp. 221–231). Berlin: Springer-Verlag. doi:10.1007/978-3-642-12476-1_16

Takavarasha, S., Hapanyengwi, G., Chimanikire, D., & Kabanda, G. (2013, October-December). An IT Project Management Framework for Assessing the Dynamism of Culture under Globalization: Evidence from Zimbabwe. *International Journal of Information Technology Project Management*, *4*(4), 66–81. doi:10.4018/ijitpm.2013100104

Tangkuampien, J. (2013, February). *A Community-based Approach to New Medium Integration in South African Education: A Combination of ICT4D Process Approach and Ethnographic Action Research Techniques* (PhD Thesis). Cape Town, South Africa: University of Cape Town. Retrieved January 28, 2017, from http://open.uct.ac.za/bitstream/handle/11427/10671/thesis_sci_2013_phdthesis_ttangkuampien_jakkaphan.pdf?sequence=1

Vaismoradi, M., Turunen, H., & Bondas, T. (2013). Content analysis and thematic analysis: Implications for conducting a qualitative descriptive study. *Nursing & Health Sciences*, *15*(3), 398–405. doi:10.1111/nhs.12048 PMID:23480423

Van Biljon, J., & Lotriet, H. (2014, October). ICT for Development in Southern Africa: Guest editorial. *South African Computer Journal*, *54*, v–vi. doi:10.18489/sacj.v54i0.282

Van Zyl, I., & Vannini, S. (2013). Participatory re-action: reflecting on a Design-Based Research approach in ICT4D. In J. Steyn, & A. G. Van der Vyver (Ed.), *Public and private access to ICTs in developing regions: Proceedings of the 7th International Development Informatics Conference* (pp. 206 - 223). Bangkok: Development Informatics Association.

Weber, R. (2009). Research on ICT for development: Some reflections on rhetoric, rigor, reality, and relevance. In J. Steyn, P. Plantinga, B. Van der Vyver, & J. Meyer (Eds.), *Proceedings of the 3rd IDIA Conference* (pp. 2–27). Berg-en-Dal, Kruger National Park: School of Information Technology, Monash University (South Africa). Retrieved from http://arrow.monash.edu.au/hdl/1959.1/514447

Chapter 7

Anecdotes of Action Learning:
Initiations of Sustainability Lessons by an Indian B-School

Saravan Krishnamurthy
Symbiosis International University, India

Vishal Pradhan
Symbiosis International University, India

ABSTRACT

The objective of this chapter is to narrate the initiations of sustainability lessons for inclusion within an MBA-IT Business Management Curriculum. This process is elucidated in a narrative style. Initiation of AL began in the year 2013 and improvised in 2014. 'Trial and Error' experimentation in the year 2013, with 'out of the box' lessons, achieved initial success with highly positive student feedback. AL was improvised in 2014 with industry visits, reflecting on corporate social responsibility and collaborations with civil society organizations. For easy comprehension of the thought process of AL iterations, it is recommended to understand Figure 1. Each AL activity is marked with year-wise sub-headings and AL reflections. For the stakeholders interested in modifying in higher education with AL, this institutional choice of sustainability and emphasis on sustainability AL is useful. Unlearning of previous social responsibility action patterns, usage of multiple intelligence, and experiential learning of social issues by IT Business students are the highlights of this chapter.

INTRODUCTION

In India, at master's level business education programs, students are learning profit orientation on a daily basis. Business lessons aimed for the firm's profitability usually discount the significance of triple bottom line of people-planet-profit, as a synchronized development for society, environment and economy (Elkington, 2004; Thomas, 2009). Business curricula could explore learning opportunities better (Corey, 1949; Harrison, Leitch, & Chia, 2007; Wu, Shen, & Kuo, 2015). Business students were forming habits to be 'only-profiteers' (Brook, Pedler, Abbott, & Burgoyne, 2016; Bowles & Gintis,

DOI: 10.4018/978-1-5225-2642-1.ch007

1976), unconsciously molding their learning methods and behavioral patterns. The attitude of 'engage in something only if there is a personal benefit for me' rather than 'I am a student willing to explore the unknown' was observed. Thus, the issue of critical concern was the apathy developing within students towards the social ailments in their society. Such apathy demands perspectival changes in education, to prevent gradual degradation towards long-term harms of an insensitive society. This chapter narrates the changes attempted in an Indian IT Business School, where sustainability lessons within an MBA-IT Business Management program were incrementally included into the curriculum. Two AL iterations in two years are presented. For ease of comprehension, the developmental thought process from cohort to cohort is presented in Figure 1. Each activity in the chapter is marked with year-wise sub-headings. These experiential learning of stakeholders are shared for the benefit of AL practitioners in higher education.

- **(Year 2013) 'TRIAL and ERROR':** Out of the box lessons, within a course curriculum on globalization were experimented. The first AL cycle was evaluated and received high student participation and appreciations. This initial success and reflections motivated to transition to next year to include augmented social interactions.
- **(Year 2014) 'DEVELOPING NEW ACTIONS':** The significance of learning sustainability was comprehended better by reflections and faculty discussions. Upgrading AL with industry visits and collaborations with civil society organizations were attempted. Shortcomings and corrections were noted, followed by reflections and evaluations. These follow-through exercises helped to develop an internal capacity building environment for the AL team.

Figure 1.

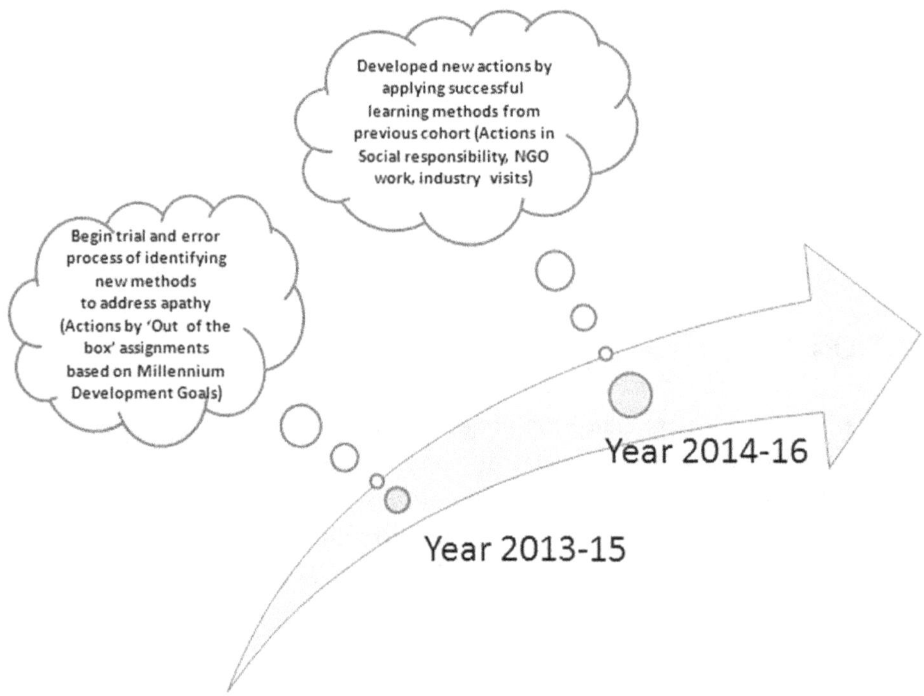

BACKGROUND

Sustainability Awareness in India: Where Are We Now?

During India's independence in 1947, the country was filled with landless laborers, unemployment, and lack of resources for industrial and agricultural growth (Kapila, 2009). Poverty, famine, malnutrition and a lack of morale to develop one's country were observed. Later decades of development began with the reorganization of efforts in the 1950's. During the 1970's when political ideologies were strongly formed, the nation had arrived at a developmental economy status, but many of the needs of a growing nation were unfulfilled. The 80s was a period of stabilization and maturity in science and technology (Kohli, 2006). The decade of the 1990s took off with a much-awaited breath of fresh air, with stronger industrialization. India had adopted 'Liberalization, Privatization and Globalization' policies to free the economy from former autocratic methods (Goyal, 2006). Private universities modernized their learning approaches, and nimble institutions attempted renewing collegiate education (Agarwal, 2015; Altbach & Knight, 2007; Yeravdekar & Tiwari, 2014). The 2000s was the decade of changes. Together with the liberalization of the economy, a social change had crept up in the country (Mitra & Singh, 1999). Neo-middle classes with new consumption patterns had begun to command a high demand for hitherto unexplored products in the market. The disparity in wealth between the neo middle classes and lower classes became an alarming phenomenon (Jayadev et al., 2007). It was observed that, as an indirect consequence, current students in colleges today are a generation who seemed to have grown up with little or no exposure to poverty. The youth of Generation Z had not learned enough about the dark pages of the history of post-independence India. Civic sense had changed markedly, with modern youth demanding more attention and fast results. Although many developed nations have seen such changes before, educators in India are surprised to find such apathy among its youth.

Whereas Indian youth in the 1980s would have participated in civic duty with patriotism and pride, the neo youth posed questions such as 'Why should I do this? What do I get if I care? Where is my benefit in this?' Educators, who were primarily of the previous generation, Gen Y, are surprised at this change in their youth's attitude and often wonder how to improve responses to this attitudinal change. According to a Harvard study on millennials, selfishness and opportunistic behavior have been rated higher than previous generations (Alsop, 2008) and is detrimental to sustainability principles. The youth of India may be growing with inadequate competency in foresighted thinking (De Haan, 2006), unintentionally recommitting blunders previously associated with developed economies. Meanwhile, the neo-generations of educators are at crossroads: Should I continue 'business as usual' (all is well while the economy is good) or should we generate modifications (i.e. include sustainability lessons in our education system)?

Based on earlier studies on rectification of systemic lacunae in higher education in India, India's noteworthy prior engagement with sustainable development education, the momentum towards sustainability education and community outreach indicates the time for change is now (Banga Chhokar, 2010; Bantanur, Mukherjee, & Shankar, 2015; Yeravdekar & Tiwari, 2012). If India does not conduct meaningful educational research to attempt ontological process modifications (Mack, 2010), towards sustainability leanings now, its education sector may stumble into a sea of post-fault-corrections and adjustments for a whole generation. Citizens are already wondering 'when is this country going to change? In spite of making many things affordable, why are we still stuck in the same societal ills?' In a country where technology rules well, a lack of education to understand fellow human beings has prompted well-known engineering and technology institutions to develop their individual social sciences programs. When young engineers were encouraged to serve the weaker socio-economic sections, it resulted in AL students' improved emotional intelligence and development into a far better citizen (Basu, 2015).

Sustainability Research for India's Educational Institutions

The three pillars of sustainability are economic feasibility, social equality and environmental soundness (United Nations [UN], 1992). In India, typical rural populations have declined in number, with many migrating to urban areas for improving chances of survival. The occupations of rural people are overcast with worries about being dominated by white-collared jobs. Struggles of sustaining the rural life are assuaged by governmental and civil society interventions. Such intricate connections within sustainability studies of India context generates academic debates. Sustainability is a widely-discoursed subject, but also one of the most challenging concepts to develop concrete action plans.

A seminal study in corporate sustainability points out the lack of descriptive research on specific target groups, sectors and countries (Salzman, 2005). Contradicting Salzman's disregard to the significance of anecdotal data, another study advises acknowledging a story teller's approach for good qualitative research (Sandelowski, 1986). Given the dearth of AL narratives about Indian academic environments, we posit that a narrative approach would provide ample background for further empirical research (Bryman, 2003) encourage active debate and result in a useful commentary. It is surmised that such a qualitative approach would be helpful lessons for adoption at Indian institutions and comparable socio-cultures and economies. Traditionally, AL and AR studies in schools are limited to one cohort. In contrast, the current study aims to demonstrate incremental development over two cohorts of 2013-15 and 2014-16. In the future, by gathering such multiple cohorts and institutional data, could enable validation for AL.

A study advocates education sector's pivotal role in creating profound changes to pave the way for sustainability mindset in societies (United Nations Educational Scientific and Cultural Organization [UNESCO], 1997). In affirmation of the need for sustainability mindset, the decade of 2010 has seen further changes, dividing entire macro-societies with deep fissures evidenced in the global financial crisis (Rajan, 2011). Sustainability issues relate to overall social well-being, being concepts that require a higher level of mental abstraction by a learning audience. The urgent need of educating sustainability has been further strengthened in this decade, confirming earlier premises. For society to overcome challenges of divergent problems posed by multiple and often conflicting objectives in the well-being of society, education systems need transitions towards sustainability education (Georgescu-Roegen, 2002). The call to the persuasion of the individual towards sustainable development and to act in the common interest needs to be answered by educational institutions strengthening regulatory frameworks (Brundtland et al., 1987; Rao et al., 2013). Envisioned as one of the four pillars of learning (UNESCO, 1996), we posited that it is time to revive 'Learn by Doing' i.e. AL to educate sustainability principles in the year 2013. The transitions in AL narrates one Indian B-School's efforts to initiate sustainability education, tabulated briefly in Table 1 to outline the chapter contents and then described in subsequent sections.

ACTION LEARNING IN YEAR 2013: TRIAL AND ERROR

Comparison to Business Case Discussions

In business education, the case discussion method offers advantages over other learning methods (Rippin et al., 2002). The business case story usually describes a well-designed scenario for students to prepare

Table 1. AL components for learning sustainability

Year and Theme	Whole Action	AL Components for Learning Sustainability
Year 2013 – TRIAL and ERROR	Initiated AL for Sustainability Lessons, feedback data gathered	• Comparing AL to Business Case Discussions • Initiating 'Learning by doing.' • Designing out-of-the-box learning experiences • Following select Millennium Development Goals (MDGs) • Addressing urbanization and globalization issue • AL Reflections and Improvement for Next Cohort
Year 2014 – DEVELOPING NEW ACTIONS	Attempted Initiation for Behavior Change of Students, deeper reflections	• Comprehending pluralist approach to express learning • Enabling survey in a remote rural area • Observing Praxis and social transformation of AL group • Reflecting on cynicism and emotional support • Linking sustainability lessons to Industrial Visits • Developing holistic views on sustainability as a • AL Reflections and Improvement for Next Cohort

and discuss on chosen topics (Hammond, 1980: Harvard, 2016). This well-established learning method successfully puts the student in the shoes of the proponent for a business decision. In business scenarios simulated in a classroom, decisions by the students are to be challenged to demonstrate thoroughness in analysis, practice leadership skills, to develop realistic views about society and people (Fullan, 2004; Timperley, 2005). However, during in-class practice, a challenge arose. The responses of students regarding human and social issues often followed shortcut methods to address business problems. Frequently, unethical solutions were acceptable to students. The students' viewpoint to adopt only-business-and-only-profiteering reduced into myopic reasoning, tending towards hasty conclusions. Real life business scenarios might not allow corrections if less than fastidious decisions were made. Indian businesses often admit that business is devoid of emotion or social feeling, reaffirming the power of experiential learning (Cope & Watts, 2000). The tolerance for human flaws was missing. A significant apathy in treating people of the business case indicated a lack of compassion, and need to educate humane perspectives. Thus, a priori conditions provided doxic certainty that students' decision-making skills needed improvement. The time was ripe for training students in sustainability.

Contemplation to Teach Sustainability Lessons

The trial to teach sustainability began in November 2013, with the cohort of 2013-15. A moral rearmaments training program (Initiatives of Change, 2013) inspired a fresh look at values of unselfishness and honesty. With moral value sensitizations, the student's groups were primed for approaching sustainability lessons society's ailments. The syllabus for a course on globalization was set to include sustainability assignments by AL. Together with traditional lectures and case discussions, the practical AL approach of 'learning by doing' was adopted. Narrative learning outcomes were planned, to develop analytical sensitivities like a socially aware critic (Carlson, 2010; Sandelowski, 1991;).

Initiating Sustainability AL

We learn 10% of what we read, 30% of what we see, 70% of what is discussed with others, 80% of what is experienced personally. – William Glasser (Adapted)

To learn global issues such as lack of education, poverty, difficulties of the underprivileged required the student to experience life together with them briefly and empathize. Out-of-the-box learning experiences were designed to address Millennium Development Goals (MDGs), to enable experiencing the actual emotions and feelings of fellow human beings (End Poverty, UN, 2015). The AL objectives of year 2013 were to: gather efforts of students inspired towards 'giving back to society,' practice AL aimed at maturing views on sustainability, develop learning habits of 'reflections on sustainability.' More than 90 students participated in the exercise, with each small team selecting one MDG. Deliberations were carefully designed to help learners to travel from initial curiosity to full awareness of the scope of the assignment. Teacher and students co-created learning objectives for each team, applied extensive discussion time, supplemental readings and videos to develop robust comprehension of their assignment. Some examples are:

Assignment on Campus Sustainability Practices

Learners were assigned to conduct an exploratory study on sustainability practices of urban educational campuses. Then to follow up with an extended study of sustainability issues, solutions implemented in corporate IT campuses in India, and prepare a comparison study report. The AL objective was to study, interview, observe and develop initial learning on sustainability concepts by relating to nearby campuses. Learners applied networking skills, developed basic assessment skills to review campus sustainability practices, learned the importance of sustainability practices by close examination of one or two campuses and comprehended scale-up of worst case scenarios and solutions.

Assignment on Urbanization Patterns

Learners conducted a pilot study on urbanization patterns in the world, to understand its importance in Globalization. Then they followed up with an exploratory study of the water facilities problems of urban India, developed neighborhood networking, conducted interviews and gathered data, discussed and presented their reports in class. The AL objective was to interview fellow citizens, observe lacunae in the local water supply system, grasp the importance of water problems, develop learning and linking solutions. Learners experienced by putting themselves in the place of water-poor citizens and realized the helplessness of their situations. Comprehending the harms of uncontrolled urbanization and its effects on sustainability, they provided recommendations to reduce unpredictability of urban water problems.

Reflections of Students

Students were pleasantly surprised with the co-creation method of goal setting and scope setting of assignments. Many students were delighted at the opportunity of out-of-the-classroom activity and encouragement for creating one's individual views. While participating in the field work, there was a range of feelings: number of unresolved issues (incredulity), coping skills of the underprivileged (surprise), inability to resolve the issues (helplessness). During the conclusion of the semester, a globalization seminar was organized to disseminate new learning for peers in the university. Appreciations for AL were expressed in student feedback survey:

This project gave us the freedom to assign our own way of performing on the chosen objective. This project gave us chances to learn from field experience. This assignment was a great and think-out-of-box project!

Negative comments were accepted as useful feedback towards corrections for next cohort.

Assessing the groups and their experiences and trying to reflect thereon was not easy. It could have been smooth if other teams were cooperative. Also, there have to be some means to see if the person given responsibility is doing his job!

Reflections of AL Practitioner

Post-fieldwork, in-class discussions became a lively environment. Students appreciated the opportunity to participate in debates, to reflect on sustainability issues that were naturally transcending across multiple disciplines. For example, student teams studying 'sustainability in campuses' initiated a discussion on water supply problems, followed by another student group expressing complementary viewpoints from their study on 'water-poverty of citizens in the vicinity of campuses.' This debate was further enriched by the 'urbanization pattern' team who explained the larger picture of water woes across many cities in India. Therefore, the whole class received a comprehensive discussion on water issues. Thus, class debates were designed to link multiple aspects of a sustainability issue, to develop a higher order of thinking, to demonstrate the value of intrapersonal learning and to practice new skills such as distributed leadership (Krishnamurthy et al., 2014a; Mapotse, 2014; Timperley, 2005). Active critical thinking skills and deliberations for critical unlearning were well applied (Brook et al., 2016). Students learning complex global sustainability issues demonstrated maturity in utilizing a scientific qualitative approach, not merely artistic (Becker, 1958). Further contemplation on inter-disciplinary research in the future, may be considered for improvements in AL. Overall, the new teaching experience was highly satisfying, and a pleasure to observe students' evolved thinking, expressing their feelings and emotions in class. As improvisation, explorations of realms of industry and civil society organizations were considered for the next year.

ACTION LEARNING IN YEAR 2014: DEVELOPING NEW ACTIONS

Initiating Behavior Change for Students

Teachers contemplated on 'What can augment the effectiveness of education?' A cohort survey informed that 95% of students were motivated towards job placement, to the exclusion of all other goals. While students' aspirations were well acknowledged, the teaching faculty's deliberation remained about 'are we raising future citizens who would become successful in career but socially unresponsive'? It was deliberated that student teams need to study wealth generation and disparity of wealth distribution in India. It was felt that additional AL experiences about sustainability solutions implemented in the industry sectors are necessary. For 2014 cohort's AL, sensitization happened early within the first year by way of field work in remote rural areas (refer Million SoULs project on p.7).

Planning for AL

"The career goals of business students can be achieved without compromising their learning towards society" (internal communication). This pivotal revelation led to AL by sensitizing towards the 'other half of society.' Engaging in naturalistic activities such as tree planting, visits to underprivileged homes, rural survey, etc., were designed to resemble 'the other three billion' (Williamson, 2009). The intelligence of students involves six types (Gardner, 1995; Silver et al., 2000). The intelligence exercised habitually in IT-business learning are:

1. **Logical-Mathematical (Quantifying in IT Courses in Business Management):** 'The ability to calculate, quantify, consider propositions and hypotheses, and carry out complex mathematical operations' (Tai, 2014). Logical-mathematical is an essential ability in a student of IT business management, given the predominance of technology learning components within most of the courses in the program.
2. **Visual-Spatial (Problem-Solving in Business):** 'The ability to generate, retain, retrieve, and transform well-structured visual images' (Ackerman & Lohman, 2006). A highly desirable ability in students training for the techno-managerial role, given the skills requirement for mapping entities in business systems, designing complex procedures for implementation in corporate business processes.
3. **Interpersonal (Sharing, Interacting and Collaborating During Team Assignments, Case Study Method):** Includes measures such as 'analysis, judgment, planning, risk-taking, oral communication, business sense, self-management' (Dulewicz & Higgs, 1998). Emphasis on business expertise alone, at the expense of ignoring their emotional intelligence, would be incomplete education and later leads to poor decision-making skills at a workplace (Goleman, 2001).

Spirals of reciprocity within the institution can be achieved if other intelligences of students are applied with explicit intentions of making learning symbiotic (Salomon & Perkins, 1998). Utilizing other forms of intelligences enable a pluralist approach to express their learning (Domask, 2007; Gardner, 1995) in aspects of:

4. **Intrapersonal (Connecting to the Self, Reflecting, Making Authentic Choices):** spending alone time, writing in one's native language to express with the 'native' feeling, self-prepared notes on authentic choices made by the learner. Upon self-contemplation, this approach leads to transformative learning (Thomas, 2009). The emancipatory process of transforming has been attempted through community-building occupations (Barth, 2003; Townsend, 1993)
5. **Naturalist (Experiencing Nature, Care for Living Things, Explorer's Approach to Life):** 'Ability to distinguish and categorize objects or phenomena in nature' such as recognizing ecological footprinting (Gardner, 1995; Tainter, 2006). Tree plantation activities open up one's connection to nature when student ventures into rural landscapes. Naturalism provides an intellectual undercurrent to qualitative research, prompting learners to spontaneously act in positive ways in an emergent coordination (Bryman, 2003; Knoblich Butterfill, & Sebanz, 2011).
6. **Bodily-Kinesthetic (Act, Build, Touch and Feel Things, Sing Together):** Preparing and enact street plays; practicing listening to issues of people while living with the underprivileged and participating in their community activities such as village group discussions. Actions devoid of

urban comforts provide the essential experiential learning environment of undergoing the same inconveniences as fellow citizens who are being observed. Caring for the underprivileged opens up one's compassion. Community service would enable elements of social constructivism (Nguyen & Larson, 2015).

Rural Survey 2014 and Learners' Reflections

A project designed by IIT Bombay aimed to distribute a million SoULs (Solar Urja Lamps) exclusively to impoverished school children in remote rural locations in India (Sawal et al., 2015; Solanki et al., 2016). The rural survey, in collaboration with IIT Bombay, was the follow-up action of the project to study the impacts in recipient households. Eight junior students were field workers for three days, surveying and living among the rural poor. This rural immersion experience was an eye opener. One student surveyor had offered a chocolate bar to a poor child in a remote village. He considered it as a casual interaction since the cost of the chocolate was negligible for him. He was shocked to hear that the child had never seen chocolate with a gold colored wrapper. This single small incident taught the student surveyor lessons much beyond words. After an extended reflection, the action learner remarked:

We have every modern comfort, but we usually turn a blind eye to the poor person we pass on the street. For them, even small things become difficult to buy. How can this be? Why is society providing so well to a few, but is so malevolent to others?

In spite of the poverty, the rural people's readiness to share their scarce meal with student visitors was seen as an act of mighty benevolence. Such atypical instances of simultaneously enacting a humanity lesson and practicing a skill led to raised critical consciousness by 'Praxis' (Hofman, 2006; Mack, 2010). Praxis paved the way to social transformations in the AL team. The experiential learning was expressed in blogs co-created by co-researchers (Household Survey, 2015), to sustain the AL process. Member checking was implemented to iterate checking by the self, participants, and mentor to avoid errors in this set of learner's narratives (Carlson, 2010; Creswell & Miller, 2000). Messages of appreciation and acknowledgments from faculty and other student teams were observed to produce positive reinforcement effects for the AL team. The behavior of student peers who joined social responsibility activities after this field survey was seen to be heartfelt and earnest than before.

A few early comments of cynicism were noted from peers. Cynicism denies a wealth of learning opportunities by denying the value of out-of-the-box thinking (Dean et al., 1998; Reynolds, 2009). A positive and patient attitude was mentored, to moderate. It was observed that students overcame reactions to cynicism and helped themselves further in intrapersonal learning. Allowing extra freedom of expression to students significantly improved self-belief of AL team members (Goel et al., 2014). The reflection exercises of such students indicated an improved morale and high self-esteem. As one action learner remarked: *'This is what I wish to do for my country. I feel this is my duty to be completed; ignoring cynicism is part of the game'.*

Rural Survey 2014, AL Practitioner's Reflections

Society still equates emotional maturity with the control or repression of feelings and continues to use the word "emotional" in a derogatory sense. (Smith & Sharma, 2002)

University Education is to encourage openness, emotional exposure, vulnerability, and involvement. (Harrison et al., 2007)

As experienced in rural survey AL, having spent the emotional cost for learning, students had learned to be emotionally detached (Harrison et al., 2007), but actively encouraged novice learners to invest emotionally (Dreyfus, 2008). They helped novice learners to participate, reduce anxiety and offered emotional support after the learning action. Students in mature age groups can adopt spiral curriculum more effectively (Bruner, 2009; Corey, 1949). Thus, given its capacity building nature the sustainability visions of co-curricular activities were easier to adopt (Dickens & Watkins, 1999; Fullan, 2004; Krishnamurthy et al., 2014a). If we liberate from conventional views on community outreach being temporary service, it can be seen as substantially contemplative in learning, pragmatic for learning sustainability and emancipatory for the learner (Townsend, 1997). Some students explained that they would rather maintain focus on their core professional goals instead of sustainability lessons. Such disagreements were accepted, and a 'no coercion' approach was clarified. Some students who felt that their educational goals may become confused had the freedom 'to discontinue without consequences.' Initial disagreements were resolved with authentic feedback. Teachers from technology backgrounds accepted the novel approach and adherence to national development visions. AL Improvements achieved in year 2014:

- Selection of AL goals to study industry-community participation.
- The inclusion of industry-based social responsibility live-cases.
- Same teams were visiting industry follow up with AL activities.

Linking to Industrial Visits

Visits to agro-manufacturing industries was organized to learn the supply chain of raw material to the retail shelf. Throughout the supply chain, sustainability practices of the organizations were studied. Within the processes of manufacturing, eco-friendliness was the focus of learning. Industry employees' social responsibility activities were examined to comprehend its suitability to the local community improvement. Business organizations' corporate social responsibility (CSR) initiatives were studied to assess its national importance and employees' perceptions (Pradhan, Sharma, Krishnamurthy, 2016). Examples were: Rural sanitation campaign to build toilets, Relief trust development for rebuilding schools damaged in an earthquake disaster. The differences in experiences of organizations in entirely different markets i.e. NGO and industry were discussed, resembling a learning laboratory. By contrasting mental models of student team members (Romme & Dillen, 1997), previous archetypes of social responsibility and sustainability were discovered. Thus views on sustainability as a 'greater whole' were explored, leading to new ideas for on theme of 'building bridges between NGO, industry, education.' Thus, the institutional internal community's capacity building for future field research was learned as 'to generalize cautiously and tentatively from the evidence collected' (Corey, 1960; Dickens & Watkins, 1999).

AL Practitioner's Reflections

Reflective knowledge is a form of cognition which instills conviction in the knower, and the courage to go with it, and commits him or her to action. (Park, 1999)

When learners are invited to stretch themselves, to grapple with challenges (Barnett, 2004) a sincere involvement of AL practitioner and conviction to AL is essential. To not ignore effects on community stakeholders, facilitating required robust preparation in moral, emotional, and social content by the AL practitioner (Cohen, 2006; Rosing, 2015). For beginners in AL practice, learning theory backgrounds and counseling requires arduous preparation (Cope & Watts, 2000; Lackritz, 2004). Together with students' feelings and occasionally overwhelmed emotions, the AL practitioner needs to undergo the same. Such experiences can be exhausting if there are many students, given the individual needs observed during one on one conversations. Free flow discussions pave the way towards counseling. A helplessness to solve societal problems was often expressed and observed. But, the same helplessness may tend to militate and initiate action, as a catalyst for the community to come together (Dickens & Watkins, 1999). In spite of well-explained commitments, cynicism and exploitation of the credits by the uninvolved stakeholders were observed. The AL practitioner in coordination with the head of the institution needs to develop clear codes ethical conduct for the future (Cohen, 2006; Fullan, 2004). In India, as in many countries, intentions of social work sometimes border on political activism. Therefore, peaceful learning intentions of the practitioner-learner group were often clarified. If engaging in service and activism is necessary, a stronger integration of service learning into the curriculum is required, as adopted by renowned universities (Rosing, Reed, Ferrari, & Bothne, 2010).

Reflections on Future Sustainability Lessons by AL

Taking up a known social issue and a known solution was an appropriate means for learning national development. Selection of the right learning device i.e. field survey method was understood well. '*Too much time in reflection is just armchair theorizing; too much time in action is mere activism*' (Dickens & Watkins, 1999). Collaborating with a renowned institution such as IIT Bombay, aided development of further partnerships. Follow-through disseminations by blogs, social media and sharing lessons with the university sustainability committee were well appreciated. Synergistic frameworks to develop interdisciplinary approaches for sustainability lessons need to be explored further to sustain action learning method (Krishnamurthy et al., 2014b; Rao & Patil, 2015). While appreciations were excellent for student morale, it did not adequately reflect in rewards of grade points (feedback communications of AL students). The shortcoming of connected academic rewards was noted as needed to be corrected for next cohort. These exercises had set an inspiring example for a future acceptance of more students into AL. AL had indeed augmented students' maturity, and empathy later observed during business case discussions.

ISSUES, CONTROVERSIES, AND RECOMMENDATIONS

This section summarizes the learning of AL practitioners and the students as co-researchers. Observations upon the learner group's issues, role perceptions, dilemmas, and group dynamics, helped in recollected during AL group reflections and further discussed with faculty. Collectively, recommendations were developed to improvise Al for future. These are tabulated in Table 2.

Table 2. Summary of issues and controversies (during AL initiations in years 2013 and 2014); relevant recommendations developed

	Issue/Controversy	Recommendations
1	**Competition:** Student AL teams occasionally faced inter-teams competition from other AL teams, in the customary spirit of business learning. It was observed that this leads to a lack of empathy for the social issue. Such an unreasonable competitiveness makes the assignment a competition for grades and achievements, instead of learning sustainability.	**Collaboration, not Competition:** Students were used to following a competitive spirit while engaging in assignments of the course. This competitiveness demonstrates individuality and enhances the quality of assignment outputs. However, for learning sustainability, a co-operative environment is far more suitable. Democracy in dialogues and debates with fair play are vital characteristics to be molded in sustainability learners. Hence, AL practitioners need to encourage collaboration, not competition among their learner groups.
2	**Unawareness about Sustainability:** During the initial trials, students were found to associate sustainability with eco-friendliness. Such partial awareness or total unawareness about the complexity of sustainability goals hinder learning processes.	**Sensitization, Experiential Learning, and Awareness Campaigns:** An awareness building workshop was useful to overcome unawareness on several sustainability issues. Observations, face to face interviews with the people affected by sustainability issue are vital steps to be embedded for relevant experiential learning. In the end, adequate connections between the sub-topics of sustainability need to be completed, to illustrate the interconnected nature of sustainability issues. Mere exposure to issues would not suffice. A maturity to campaign for the affected people would demonstrate completeness in the students' development.
3	**Helplessness:** Students frequently faced helplessness feelings. The inability to change a situation leads the student to relinquish learning sustainability, rather than face the uphill task of action for the issue. Feeling their action is inconsequential may lead to dropouts, a disheartening experience for all.	**Small but Meaningful Changes:** As awareness about the problems and the difficulties increase, students who face helplessness in solving a sustainability issue may conclude their assignment with increased feelings of discouragement and disempowerment (Domask, 2007). It is essential to demonstrate by AL, that even small changes can make a difference, and that these changes count as valid lessons. Ultimately, in the larger picture of sustainability, such small changes invigorate learning for surrounding micro-community members. Small drops do make an ocean, a valuable lesson that counters helplessness and meaningful insight into sustainability practices.
4	**Emotional Involvement:** It was observed that business students have a tendency to avoid emotional involvement with the social issue. They felt this approach facilitates precise decision making, whereas expressions of their emotions were often received by peers in a derogatory sense. The complete avoidance of feelings may lead to just rational human beings, devoid of feeling.	**Mindfulness During Interventions:** Affective learning involves the learner emotionally (Shephard, 2008). In contrast, customary business learning methods expect students to be mostly in an unemotional state. For holistic improvement of managerial skills, emotional intelligence training is essential (Goleman, 2002). Society's general disparages about emotional involvement can be changed during AL. Given AL's prospects for upheavals of emotions, such emotional moments can become ideal learning opportunities for both AL learner and practitioner. AL practitioners need to develop maturity in their interventions, to participate with empathy, to help students learn and to observe. Instances of high emotions are not weaknesses, but rather a chance to practice emotional intelligence. Additionally, facing cynicism with fewer emotional upsets was noted as a successful learning outcome.

FUTURE RESEARCH DIRECTIONS

Learning more about the emotional involvement of students in ALs can be a useful exploration for further study. The different feelings about a particular social cause are worthwhile studying to comprehend the current sentiments of learners. It would be valuable to adopt an empirical dependence relationships study among AL participation measures such as motivation to the social cause, outlook about the reflection activity, etc. (Pradhan et al., 2016) Collaborative practices for the industry with educational sector are suitable explorations. A study of effects of an external review on sustainability AL activities, evaluations of teachers' external and internal factors may be informative to redesign academic rewards system (Krishnamurthy et al., 2014a; Mapotse, 2015).

CONCLUSION

Unlearning students' previous patterns in relating social responsibility were the core of AL in year 2013, while the experiential learning in emotional intelligence was the core of AL in year 2014. These activities did not provide personal benefits but changed their perspectives about learning social issues and consequently liberated their learning from habitual patterns. This break away i.e. unlearning was encouraging for the AL team, and was noted as the most successful outcome of both years. Senior students validating such encouraging outcomes were important because feedback is passed on to junior students and therefore clarifies perceptions for newly admitted students towards learning sustainability lessons. Accepting corrections, adopting feedback from seniors' reflections is equally important as a validation of the democratic decision-making process in the sustainability arena. In India, the revived sense of patriotism by youthful ways such as sports, music, intellectual means such as debating community issues, and political issues was observed as a commendable trait among youth. Such a youthful quality could be directed to a nation that needs to learn about sustainability in the right time. The more the attention paid to economic growth to the exclusion of social issues, the more seems to be the inattention paid to sustainability. Such a pathway with unawareness about sustainability ultimately degrades the social fabric of the nation, with visible harms to society's well-being already showing, such as severe income disparities. Over-concentration on economic growth or job creations or infrastructural growth promotes eco-unfriendliness, a grave long-term damage to our living environment and the destruction of future generations' resources. The sustainable development pathway for India may be a complex process, demanding clarity in current practices (Yaduvanshi, et al., 2016). But the key participants to a turnaround in the society are its education systems.

Modern education, particularly business instruction needs to a turnaround in its heels to re-examine itself and modernize its curriculum structures. To suit the needs of current students, but ignore the future generations' well-being is partially short-sighted. While a few institutions in India are practicing significant curriculum upgrades to include sustainability lessons, it is argued that such an important agenda needs nation-wide adoption. For educational institutions in India and comparable economies, it is urged to find space for sustainability lessons as an essential curricula modification strategy. There may be numerous ways to educate students on sustainability. Given its constructivist methodology, AL was found to be an expedient means to educate for sustainability. The complex nature of sustainability stipulates carefully prepared lessons, so as to craft meaningful conceptual linkages within the cohort group. Such linkages were the chief advantage of AL over other conventional instruction methods. Complex phenomena linked with socio-eco concepts can be learned well by AL methods. Exploring intricately connected socio-economic-eco issues were achievable by AL. The needs of sustainability lessons were well-served by AL taught methods, with its' prominence of reflection exercises leading students from apathy to empathy. Dissemination of successful learning techniques and issues faced in this chapter are meant to serve as a learning support for institutions attempting AL routines for sustainability lessons. While the AL experiences can be adopted in many curricula with different social sciences emphases, business schools would benefit especially well with AL methods, as a new path away from the beaten path.

ACKNOWLEDGMENT

This book chapter is dedicated to the student batches of the year 2013-15, 2014-16 and 2015-17, in the of SCIT MBA ITBM program. Special thanks to the students who participated enthusiastically and supported the action learning trials, experiments, and reflections within the ecosystem of an IT and business-oriented institution. Their support enabled this attempt to document and share learning with all stakeholders in higher education. The authors wish to express sincere thanks to Dr. Urvashi Rathod and Dr. Ajey Kumar for their support and encouragement in reviewing SCIT's Institutional Social Responsibility. This chapter validates scholastic benefits of inclusion of sustainability lessons within IT business curriculum.

REFERENCES

Ackerman, P. L., & Lohman, D. F. (2006). Individual differences in cognitive functions. In P. R. Alexander & P. H. Winne (Eds.), *Handbook of educational psychology* (2nd ed.; pp. 139–161). Mahwah, NJ: Lawrence Erlbaum Associates.

Agarwal, P. (2015). Private deemed universities in India. *International Higher Education*, (49).

Alsop, R. (2008). *The trophy kids grow up: How the millennial generation is shaking up the workplace*. John Wiley & Sons.

Altbach, P. G., & Knight, J. (2007). The internationalization of higher education: Motivations and realities. *Journal of Studies in International Education*, *11*(3-4), 290–305. doi:10.1177/1028315307303542

Banga Chhokar, K. (2010). Higher education and curriculum innovation for sustainable development in India. *International Journal of Sustainability in Higher Education*, *11*(2), 141–152. doi:10.1108/14676371011031865

Bantanur, S., Mukherjee, M., & Shankar, R. (2015). Emerging dimensions of sustainability in institutes of higher education in India. *International Journal of Sustainable Built Environment*, *4*(2), 323–329. doi:10.1016/j.ijsbe.2015.03.004

Basu, M. (2015). A class apart at IIT-Bombay - Some students of IIT Bombay takes an unconventional career decision. *Indian Express*. Retrieved Sept 8, 2016 from http://indianexpress.com/article/cities/mumbai/a-class-apart-at-iit-bombay/

Becker, H. S. (1958). Problems of inference and proof in participant observation. *American Sociological Review*, *23*(6), 652–660. doi:10.2307/2089053

Bowles, S., & Gintis, H. (1976). *Schooling in capitalist America* (Vol. 57). New York: Basic Books.

Brook, C., Pedler, M., Abbott, C., & Burgoyne, J. (2016). On stopping doing those things that are not getting us to where we want to be: Unlearning, wicked problems and critical action learning. *Human Relations*, *69*(2), 369–389. doi:10.1177/0018726715586243

Brundtland, G., Khalid, M., Agnelli, S., Al-Athel, S., Chidzero, B., Fadika, L.,… & Singh, M., (1987). *Our common future*. Brundtland Report.

Bruner, J. S. (2009). *The process of education.* Harvard University Press.

Bryman, A. (2003). *Quantity and quality in social research.* Routledge.

Carlson, J. A. (2010). Avoiding traps in member checking. *Qualitative Report, 15*(5), 1102.

Cope, J., & Watts, G. (2000). Learning by doing-An exploration of experience, critical incidents and reflection in entrepreneurial learning. *International Journal of Entrepreneurial Behavior & Research, 6*(3), 104–124. doi:10.1108/13552550010346208

Corey, S. M. (1949). Curriculum development through action research. *Educational Leadership, 7*(3), 147–153.

Corey, S. M. (1960). Action Research in Education. *Guidance Readings for Counselors*, 603.

Creswell, J. W., & Miller, D. L. (2000). Determining validity in qualitative inquiry. *Theory into Practice, 39*(3), 124–130. doi:10.1207/s15430421tip3903_2

De Haan, G. (2006). The BLK 21programme in Germany: A Gestaltungskompetenz-based model for Education for Sustainable Development. *Environmental Education Research, 12*(1), 19–32. doi:10.1080/13504620500526362

Dean, J. W., Brandes, P., & Dharwadkar, R. (1998). Organizational cynicism. *Academy of Management Review, 23*(2), 341–352.

Dickens, L., & Watkins, K. (1999). Action research: Rethinking Lewin. *Management Learning, 30*(2), 127–140. doi:10.1177/1350507699302002

Domask, J. J. (2007). Achieving goals in higher education: An experiential approach to sustainability studies. *International Journal of Sustainability in Higher Education, 8*(1), 53–68. doi:10.1108/14676370710717599

Dreyfus, H. L. (2008). *On the internet.* Routledge.

Dulewicz, S. V., & Higgs, M. J. (1998). *Emotional intelligence: can it be measured reliably and validly using competency data?* Competency Journal.

Elkington, J. (2004). Enter the triple bottom line. *The Triple Bottom Line: Does It All Add Up,* 1-16.

Fullan, M. (2004). *Leadership & sustainability: System thinkers in action.* Corwin Press.

Gardner, H. (1995). Reflections on multiple intelligences: Myths and messages. *Phi Delta Kappan, 77*(3), 200.

Georgescu-Roegen, N. (2002). Four challenges of sustainability. *Conservation Biology, 16*(6), 1457–1460. doi:10.1046/j.1523-1739.2002.01668.x

Glasser, W. (1986). *Control theory in the classroom.* Perennial Library/Harper & Row Publishers.

Goel, A., Gupta, A., & Bande-Vilela, B. (2014). It's all in interpersonal interactions: Empowerment, organizational cynicism, and perception of organizational politics. In Values in Shock - The role of contrasting management, economic, and religious paradigms in the workplace. Academic Press.

Goleman, D. (2001). Emotional intelligence: Issues in paradigm building. *The Emotionally Intelligent Workplace, 13*, 26.

Goleman, D. (2002). *The New Leaders: Transforming the art of leadership into the science of results.* London: Little Brown.

Goleman, D., Boyatzis, R. E., & McKee, A. (2002). *The new leaders: Transforming the art of leadership into the science of results*. London: Little, Brown.

Goyal, K. A. (2006). Impact of Globalization on Developing Countries (with special reference to India). *International Research Journal of Finance and Economics, 5*, 166–171.

Hammond, J. S. (1980). *Learning by the case method.* Boston: Harvard Business School.

Harrison, R. T., Leitch, C. M., & Chia, R. (2007). Developing paradigmatic awareness in university business schools: The challenge for executive education. *Academy of Management Learning & Education, 6*(3), 332–343. doi:10.5465/AMLE.2007.26361624

Harvard. (2016). *The Harvard business school case method.* Retrieved from www.hbs.edu/mba/academic-experience/Pages/the-hbs-case-method.aspx

Hofman, N. G., & Rosing, H. (2006). *Pedagogies of Praxis: Course-based action research in the social sciences*. Anker Publishing Company, Incorporated.

Household Survey. (2015). Household Survey by student volunteers, Household Survey conducted by the 'One Million SoUL' project in Palghar District of Maharashtra from 15th to 17th of November 2014, with help of Symbiosis Centre for Information Technology [SCIT]. Retrieved Sep 15, 2016, from: http://www.millionsoul.iitb.ac.in/outreach_blogs.html

Initiatives of Change. (2013). *Friends of Moral Re-Armament (India), an educational public charitable trust established in 1961 and registered under the Bombay Public Trust Act, 1950*. Panchgani, Maharashtra, India. Retrieved Sep 15, 2016, from: http://in.iofc.org/

Jayadev, A., Motiram, S., & Vakulabharanam, V. (2007). Patterns of wealth disparities in India during the liberalisation era. *Economic and Political Weekly*, 3853–3863.

Kapila, U. (2009). *Indian Economy since independence*. Academic Foundation.

Knoblich, G., Butterfill, S., & Sebanz, N. (2011). 3 Psychological research on joint action: Theory and data. *Psychology of Learning and Motivation-Advances in Research and Theory, 54*, 59–101. doi:10.1016/B978-0-12-385527-5.00003-6

Kohli, A. (2006). Politics of economic growth in India, 1980-2005: Part I: The 1980s. *Economic and Political Weekly*, 1251–1259.

Krishnamurthy, S., Joseph, S., & Bharathi, V. (2014b). Creating environment friendly projects in rural India–A synergy framework for sustainable renewable energy. *International Journal of Applied Engineering Research, 9*(24), 26719–26738.

Krishnamurthy, S., Joseph, S., Bharathi, V., & Pradhan, V. (2014a). Inclusion of sustainability education in business schools – An Indian B-school case study of MBA-ITBM curriculum. *International Journal of Applied Engineering Research*, *9*(23), 22703–22725.

Lackritz, J. R. (2004). Exploring burnout among university faculty: Incidence, performance, and demographic issues. *Teaching and Teacher Education*, *20*(7), 713–729. doi:10.1016/j.tate.2004.07.002

Mack, L. (2010). *The philosophical underpinnings of educational research*. Academic Press.

Mapotse, T. A. (2015). An emancipation framework for technology education teachers: An action research study. *International Journal of Technology and Design Education*, *25*(2), 213–225. doi:10.1007/s10798-014-9275-y

Mitra, S. K., & Singh, V. B. (1999). *Democracy and social change in India: a cross-sectional analysis of the national electorate*. Sage Publications Pvt. Ltd.

Nguyen, D. J., & Larson, J. B. (2015). Dont forget about the body: Exploring the curricular possibilities of embodied pedagogy. *Innovative Higher Education*, *40*(4), 331–344. doi:10.1007/s10755-015-9319-6

Park, P. (1999). People, knowledge, and change in participatory research. *Management Learning*, *30*(2), 141–157. doi:10.1177/1350507699302003

Poverty, E. (2015). *Millennium development goals*. United Nations. Retrieved August 20, 2016, from: http://www.un.org/millenniumgoals/

Pradhan, V. S., Sharma, R., & Krishnamurthy, S. (2016), (in Press). Analyses of factors affecting employee perception on organizational Corporate Social Responsibility, SCIT Journal 2016, Volume XVI, August 2016 (ISSN 0974-5076)

Prince, M. J., & Felder, R. M. (2006). Inductive teaching and learning methods: Definitions, comparisons, and research bases. *The Journal of Engineering Education*, *95*(2), 123–138. doi:10.1002/j.2168-9830.2006.tb00884.x

Rajan, R. G. (2011). *Fault lines: How hidden fractures still threaten the world economy*. Princeton University Press. doi:10.1515/9781400839803

Rao, P., & Patil, Y. (2015). Integrating energy and environment in postgraduate management education: A case study from Symbiosis institute of international business, Pune, India.. In *Implementing Campus Greening Initiatives* (pp. 265–276). Springer International Publishing. doi:10.1007/978-3-319-11961-8_20

Rao, P., Patil, Y., & Gupte, R. (2013). Education for sustainable development: trends in Indian business schools and universities in a post liberalization era. In *Sustainability Assessment Tools in Higher Education Institutions* (pp. 417–432). Springer International Publishing. doi:10.1007/978-3-319-02375-5_24

Reynolds, M. (2009). Environmental ethics. In M. Reynolds, C. Blackmore, & M. J. Smith (Eds.), *The Environmental Responsibility Reader* (pp. 40–51). London: Zed Books and the Open University.

Rippin, A., Booth, C., Bowie, S., & Jordan, J. (2002). A complex case: Using the case study method to explore uncertainty and ambiguity in undergraduate business education. *Teaching in Higher Education*, *7*(4), 429–441. RetrievedSep152016 doi:10.1080/135625102760553928

Rosing, H. (2015). Tracking Culture: The Meanings of Community Engagement Data Collection in Higher Education. *Metropolitan Universities*, *26*(2), 147–164.

Rosing, H., Reed, S., Ferrari, J. R., & Bothne, N. J. (2010). Understanding student complaints in the service learning pedagogy. *American Journal of Community Psychology*, *46*(3-4), 472–481. doi:10.1007/s10464-010-9338-5 PMID:20859675

Salomon, G., & Perkins, D. N. (1998). Individual and social aspects of learning. *Review of Research in Education*, *23*(1), 1–24. doi:10.3102/0091732X023001001

Salzmann, O., Ionescu-Somers, A., & Steger, U. (2005). The business case for corporate sustainability: Literature review and research options. *European Management Journal*, *23*(1), 27–36. doi:10.1016/j.emj.2004.12.007

Sandelowski, M. (1986). The problem of rigor in qualitative research. *ANS. Advances in Nursing Science*, *8*(3), 27–37. doi:10.1097/00012272-198604000-00005 PMID:3083765

Sandelowski, M. (1991). Telling stories: Narrative approaches in qualitative research. *Journal of Nursing Scholarship*, *23*(3), 161–166. doi:10.1111/j.1547-5069.1991.tb00662.x PMID:1916857

Sawal, N., Venkateswaran, J., Solanki, C. S., & Narayanan, N. C. (2015, December). Million Solar Urja Lamp Programme: A supply chain experience. In *Industrial Engineering and Engineering Management (IEEM), 2015 IEEE International Conference on* (pp. 342-346). IEEE. doi:10.1109/IEEM.2015.7385665

Shephard, K. (2008). Higher education for sustainability: Seeking affective learning outcomes. *International Journal of Sustainability in Higher Education*, *9*(1), 87–98. doi:10.1108/14676370810842201

Silver, H. F., Strong, R. W., & Perini, M. J. (2000). So each may learn: Integrating learning styles and multiple intelligences. Association for Supervision and Curriculum Development.

Smith, P. A., & Sharma, M. (2002). Rationalizing the promotion of non-rational behaviors in organizations. *The Learning Organization*, *9*(5), 197–201. doi:10.1108/09696470210442132

Solanki, C. S., Narayanan, N. C., Venkateswaran, J., Joshi, L., Arora, N., & Rajagopalan, S. (2016). *Concurrent Evaluation Report of Million SoUL Program in India*. IIT Bombay. Retrieved from http://www.millionsoul.iitb.ac.in/Infographics/Four%20States%20Final.pdf

Tai, F. M. (2014). Exploring multiple intelligences. *The Journal of Human Resource and Adult Learning*, *10*(1), 11.

Thomas, I. (2009). Critical thinking, transformative learning, sustainable education, and problem-based learning in universities. *Journal of Transformative Education*, *7*(3), 245–264. doi:10.1177/1541344610385753

Timperley, H. S. (2005). Distributed leadership: Developing theory from practice. *Journal of Curriculum Studies*, *37*(4), 395–420. doi:10.1080/00220270500038545

Townsend, E. (1997). Occupation: Potential for personal and social transformation. *Journal of Occupational Science*, *4*(1), 18–26. doi:10.1080/14427591.1997.9686417

(1993). Townsend (1993) EA. Muriel Driver Memorial Lecture: Occupational therapys social vision. *Canadian Journal of Occupational Therapy*, *60*, 174–184. doi:10.1177/000841749306000403

UNESCO. (1996). *Learning, the Treasure Within: Report to UNESCO of the International Commission on Education for the Twenty-first Century: Highlights*. UNESCO Publication.

UNESCO. (1997). *Educating for a Sustainable Future*. Retrieved from http://www.unesco.org/education/tlsf/TLSF/theme_a/mod01/uncom01t05s01.htm#edu

United Nations. (1992). *The Rio declaration on environment and development*. Rio de Janeiro: UN.

Williamson, M. (2009). Connecting the other three billion. *Engineering & Technology*, *4*(4), 70–73. doi:10.1049/et.2009.0417

Wu, Y. C. J., Shen, J. P., & Kuo, T. (2015). An overview of management education for sustainability in Asia. *International Journal of Sustainability in Higher Education*, *16*(3), 341–353. doi:10.1108/IJSHE-10-2013-0136

Yaduvanshi, N. R., Myana, R., & Krishnamurthy, S. (2016). Circular Economy for Sustainable Development in India. *Indian Journal of Science and Technology*, *9*(46). doi:10.17485/ijst/2016/v9i46/107325

Yeravdekar, V. R., & Tiwari, G. (2012). The higher education system in India and its impact on the economy. doi:10.2139/ssrn.2139894

Yeravdekar, V. R., & Tiwari, G. (2014). Chinas Lead in Higher Education: Much to Learn for India. *Procedia: Social and Behavioral Sciences*, *157*, 369–372. doi:10.1016/j.sbspro.2014.11.041

KEY TERMS AND DEFINITIONS

Affective Outcomes: Psycho-physiological reparations of an activity.

Concept of Self: A collection of beliefs about oneself. It is also called self-concept, self-construction, self-identity, and self-perspective or self-structure.

Ecological Foot Printing: Measures the amount of biologically productive land and water area an individual, a city, a country, a region, or all of humanity uses to produce the resources it consumes and to absorb the waste it generates with today's technology and resource management practices.

Explicit Intention: A mental state that represents a commitment to carrying out an action or actions in the future.

Institutional Social Responsibility (ISR): Institutional Social responsibility pertains tp Institutional conscience, Institutional citizenship, social responsibly or responsible business. ISR is one of the most important business ethics quality for an organization to become successful.

Pluralist Approach: The pluralist approach is to assess a situation or circumstance while using multiple sciences or philosophies to find a solution or explanation. When studying or observing the behavior of people on an international level, the pluralist approach is often useful.

Positivist: A philosophical theory stating that positive knowledge is based on natural phenomena and their properties and relations. Thus, information derived from sensory experience, interpreted through reason and logic, forms the exclusive source of all authoritative knowledge.

Praxis: An integration of action and thought. Simultaneously enacting a humanity lesson and practicing a skill is a desirable Praxis. Praxis may be used for developing integration between theory and practice, research and development.

Reflexive Approach: Means 'reflecting' and specifically, as part of the social research, reflexivity is the process by which the researcher reflects upon the data collection and interpretation process.

Social Constructivism: A sociological theory of knowledge according to which human development is socially situated and knowledge is constructed through interaction with others. The phrase was coined by Peter L. Berger and Thomas Luckmann in The Social Construction of Reality.

Social Transformation of AL Group: The process by which an AL group alters the socially ascribed social status of their parents into a socially achieved status for themselves or it refers to large scale social change as in cultural reforms or transformations.

Transformative Group Learning: The process of perspective transformation in a cohesive group. It has three dimensions: psychological (changes in understanding of the self), convictional (revision of belief systems), and behavioral (changes in lifestyle).

Chapter 8
Anecdotes of Action Learning:
Realizations of an Indian B-School's Sustainability Lessons

Saravan Krishnamurthy
Symbiosis International University, India

Vishal Pradhan
Symbiosis International University, India

ABSTRACT

The objective of this chapter is to narrate the realizations while developing sustainability lessons by Action Learning (AL) for inclusion within an MBA-IT Business Management curriculum. This process is elucidated in a narrative style. The premise was set by ALs of earlier cohorts' trials, errors, improvisations, and reflections. In the year 2015, the Institutional Social Responsibility (ISR) team attempted scaling up 'Learning by Doing.' In 2016, a 'Consolidate and Optimize' approach was adopted. Each activity used a reflexive approach during AL evaluations. Together, AL practitioner and learners developed incrementally maturing sustainability lessons for the subsequent cohort. The chapter aims to share experiential learning with all stakeholders in higher education. An overall shift from apathy to empathy by virtue of AL to implement sustainability lessons is the highlight of this chapter.

INTRODUCTION

Indian Business school curricula need to include sustainability lessons to reshape into new patterns of learning. A disturbing trend was observed i.e. business students focusing on solely on profiteering rather than inclusive growth. Personal benefits had gained more importance than citizen's participation, in the growth of business and country together. Such a mentality may have promoted apathy, to the exclusion of empathy. Taking cue from the three pillars of sustainability of economic feasibility, social equality and environmental soundness (United Nations [UN], 1992), sustainability lessons were deliberated upon by faculty. As a solution, incremental inclusions sustainability lessons were implemented, in the year 2013. These experimented 'Trial and Error' methods received high student participation and ap-

DOI: 10.4018/978-1-5225-2642-1.ch008

preciations. After reflections and evaluations, the inclusions of sustainability lessons continued in the year 2014 by developing new actions for AL sustainability Lessons. Upgrading AL with industry visits and collaborations with civil society organizations were attempted. Shortcomings and corrections were noted throughout, followed by reflections and evaluations. These follow-through exercises helped to develop an internal capacity building environment for the AL team. Based on these initiations of AL for sustainability lessons in earlier cohorts, the current chapter narrates further AL developments in the years 2015 and 2016. Beginning with a re-organization in 2015 of the Institutional Social Responsibility (ISR) team to adopt sustainability lessons, maturing in delivering sustainability lessons within traditional business courses and culmination in interdisciplinary research in 2016 are described in this chapter. For easy comprehension of the whole process, it is recommended first to understand Figure 1 that illustrates the flow of AL spirals in a detailed 'Plan-Act-Observe-Reflect' diagram. The iterations of an Indian B-School's attempts at AL are listed below.

- **(Year 2015) SCALING UP 'LEARNING by DOING':** The objective was to address nationally critical issues, and to simultaneously align with University vision. The Institutional Social Responsibility (ISR) team was revamped, and a scaled up bustling environment ensued with multiple AL actions, gathering more students' participation. ALs on a selected cluster of sustainability issues, relevant evaluations of each activity with reflexive approaches yielded useful augmentation for next cohort. Shortcomings, reflections, and corrections were recorded.
- **(Year 2016) 'CONSOLIDATE and OPTIMIZE':** The objective was to consolidate into lesser number of activities than before and optimize with better quality of AL reflections ensued. The cohort that began AL in 2015 received carefully designed interventions during their courses in each subsequent semester. Student teams were empowered to try self-governing vision setting and self-organizing themselves towards one social cause per team. After receiving the full cycle of sustainability lessons, this cohort has concluded well with student research publications on sustainability.

BACKGROUND

India's education on sustainability had transition through changes since her independence in 1947. Progressing from poverty ridden history to a formation of political ideologies in the 1970s, and then to rapid development in the 1990s, education systems had undergone adaptive changes. Among the neo youth of the 2010s, a generation that had grown up without much exposure to difficulties of the previous generations, a marked variation in the civic sense of citizens was observed by the AL practitioners. In the observation of authors who are also AL practitioners, the modern youth are demanding more attention and fast results, and educators in India are surprised to find apathy among its youth. Taking caution from a Harvard business study, the millennials' selfish and opportunistic behavior had been rated higher than previous generations (Alsop, 2008). Taking clues from experts in sustainability education, to develop adequate competencies in foresighted thinking (De Haan, 2006), and avoid unintentional blunders, educators are advised to include sustainability lessons within their curricula. According to an empirical study based on evidence from India, quality education components with Universal values of sustainability is advocated (Bhattacharya & Bhattacharya, 2014).

According to the Brundtland et al., (1987), sustainability development "meets the needs of the present without compromising the ability of future generations to meet their own needs." A landmark reference for society's social health expounds income equality thoroughly (Wilkinson & Pickett, 2009), arguing for greater income equality, to achieve better outcomes for all. India is one of the countries where the predominantly agro-economy dependent people resort to urban non-agricultural employment, a typical survival strategy of the rural poor to overcome high inequality of income (Hussein & Nelson, 2016). ALs to learn sustainability need to incorporate qualitative methods. Recent qualitative research supports the view of changing teaching methods for better acquisition of sustainability-social responsibility competences (Albareda Tiana, 2016). A seminal study in corporate sustainability points out the lack of descriptive research on specific target groups, sectors and countries (Salzman, 2005). While Salzman disregards the value of anecdotal data, in contradiction, another study advises acknowledging a story teller's approach for good qualitative research (Sandelowski, 1991). We posit that a narrative approach would provide ample background for further empirical research (Bryman, 2003), would encourage active debate and result in a useful commentary. It is surmised that such a qualitative approach would generate beneficial experiential lessons for adoption at Indian institutions and comparable socio-cultures.

Holistic Inclusion of AL for Sustainability

Having ascertained the need for including sustainability lessons, we now proceed to discuss the method of including it. The sustainability phenomenon is by nature, a complex one. Wood (2012), argues for addressing the complex interrelationships between biodiversity and human societies. Decrying the reductionist practices of twentieth century sciences, he deems science and sustainability are historical enemies. As he calls for a humble, non-quantitative respect for nature's complexity, to address the metaphorical "gulf" between data and meaningful human behavior towards environment, he advises allowing the accumulation of data to give way to social and narrative analysis. Similarly, discussing the complex nature of sustainability, deliberations on urban systems call for a combined anthropogenic and technological approach to cross disciplinary boundaries (Trantopoulos, Schläpfer & Helbing, 2011). Inferring from these studies, modern complex urban systems seem to remain a challenge for researchers, due to the inadequate attention paid to the interdependencies between the socio-technological modernizations and sustainability.

Taking a historical view, Tainter (2006), theorizes that societies have become increasingly information-dependent, placing higher demands on the education of its members. When institutions are unmindful, complexity becomes more insidious. Hence, institutions need to imbibe and address sustainability issues as effective problem-solvers of their society. Initiating a study for a long-term vision for Indian society's sustainability agenda, we adopted the notion of BSTE (Bounded Socio-Technical Experiment) to address multiple characteristics for a geographically bound small community of students and faculty of an institution (Elzen, Geels, & Green, 2004). Our collective endeavor design needs actors from diverse backgrounds such as civil society organizations, education experts, University policy makers and the democratic governmental vision for sustainability.

Well-known comprehensive approaches to evaluating sustainability, advice significance of participatory learning (Astier, Masera, & Lopez-Riduara, 1999; Masera, & Lopez-Riduara, 2000). Participation in sustainability and social learning emphasize the need to develop reflection-evaluation processes that discern more than mere factual knowledge, or individual actors aided inferences on the capabilities of participatory methods for fostering social learning (Garmendia & Stagl, 2010). Systems thinking underpinned in sustainability principles advice against sterile debates on social issues or mere compliance to a national syllabus, instead allow the open-ended enquiry, encourage dialogue and action learning

(Martin, Brannigan, & Hall, 2005). Knowing the effects of complexities and uncertainties of education for sustainability, we concluded that teaching to perceive oneself as part of the whole is an important lesson for future generations (Garmendia & Stagl, 2010; Shephard, 2008; Steiner & Posch, 2006). To address sustainability holistically in a developing country, transcending departmental silos, gaining complementary senior authority's support, developing evaluation at every stage through participatory methods (Ngyun et al., 2012; Rosing et al., 2010) were noted as the building blocks for a systems approach in AL for sustainability lessons.

Propelled by significances, now decision making became a quandary. Will we make good decisions? How should we go about it? Philosophers ponder what our decisions say about our values (Buchanan & O Connell, 2006). We decided upon a pragmatic approach to achieving learning outcomes, scaling up or diagonal, developing a nuanced understanding of AL practitioner and learner's behavior. Thus, a combination of reason and intuition were employed in decision-making in this process. With time and practice, initiated educational priorities to reorient students' worldview needs to become resilient in its pathway to "true sustainability" (Smith, Lopes, & Carrejo, 2011). Domask (2007) elaborates on the struggles of academicians attempting to carry the entire burden of experiential learning while managing customary pressures of teaching and publishing (Kezar & Rhoads, 2001). If not well-supported by administration, funding and reward systems, the result could be a fragile future for AL.

Traditionally, AL studies in institutions are limited to one cohort. In contrast, the current study aims to demonstrate incremental development in two cohorts. In the future, by gathering such multiple cohorts and institutional data, could enable validation for AL. A study by United Nations Educational Scientific and Cultural Organization [UNESCO] (1997), advocates education sector's pivotal role in creating profound changes to pave the way for sustainability mindset in societies. In affirmation of the need for sustainability mindset, the decade of 2010 has seen further changes, dividing entire societies with deep fissures evidenced in the global financial crisis (Rajan, 2011). Sustainability issues relate to overall social well-being, being concepts that require a higher level of mental abstraction by a learning audience. The urgent need of educating sustainability has been further strengthened in this decade, confirming earlier premises. For society to overcome challenges of divergent problems posed by multiple and often conflicting objectives in the well-being of society, education systems need transitions towards sustainability education (Georgescu-Roegen, 2002). The call to persuasion of the individual towards sustainable development and to act in the common interest needs to be answered by educational institutions (Brundtland et al., 1987; Krishnamurthy, Joseph, & Bharathi, 2014b).

In year 2013, we had posited to revive 'Learn by Doing' i.e. AL to educate sustainability principles by following select Millennium Development Goals (MDGs). Later, in 2014, we developed holistic views on sustainability by AL in a remote rural survey, industrial visits and deeper reflections. In 2015, the Institutional Social Responsibility (ISR) team adopted a cluster of India's nationally critical Issues, included sustainability lessons within business courses leading to culmination in interdisciplinary research in 2016. In this chapter, we posit that transdisciplinary collaborations encourage changes in teaching methods. Maturation of reflections within the AL team helping the formation of incrementally mature sustainability lessons indicates development of successful AL practice. Collaborations of faculty and students were reiterated by valuable learning outcomes such as the expansion of a shared language (Dmochowski et al., 2016). Closing with students' research applied to current social issues of the nation, illustrate careful intervention choices of the AL team. AL developments in the years 2015 and 2016 are tabulated in Table 1.

Table 1. Components of AL transitions: efforts towards sustainability education

Year and Theme	Whole Action	AL Components for Learning Sustainability
Year 2015: SCALING UP 'LEARNING by DOING'	Attempted ALs scaling up, inclusions of sustainability lessons within business curriculum	• Revamping Institutional Social Responsibility (ISR) • Aligning to a cluster of India's nationally critical Issues • Aligning with University's Community Outreach approach • Reflecting on 'Philosophy of AL for Change' • AL on Social Awareness (Female infanticide) • AL on Education for the underprivileged • AL in community engagement (visits to old age homes) • AL Reflections and improvement for next cohort
Year 2016: CONSOLIDATE and OPTIMIZE	Adopted inter-disciplinary research, collective critical reflections	• Adapting Single-loop learning towards double-loop learning • Training for inter-disciplinary research • Selecting Sustainable Development Goals (SDGs) • Developing IT-enabled solutions by sensitized cohort • Publishing student research papers on Socio-technical issues • AL Reflections, conclusion and future research directions

Figure 1.

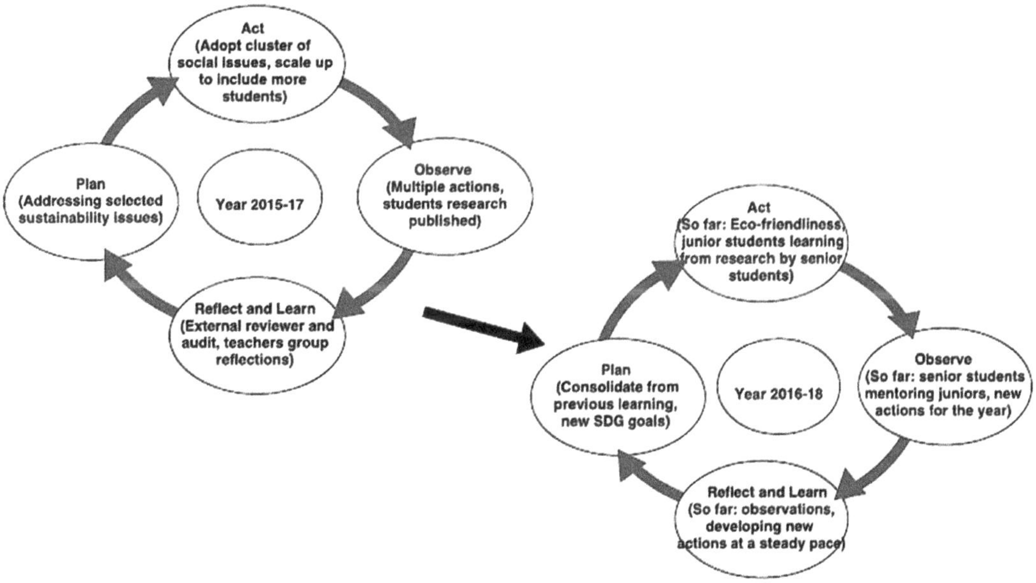

ACTION LEARNING IN YEAR 2015: SCALING UP "LEARNING BY DOING"

India has had a 5-year development plan from the 1950s (GOI, 1956). The immersion into rural areas or afflicted regions was a successful traditional method to teach social responsibility (Sekhar, 2002). In current times, the challenge was not a lack of enthusiasm or willingness, but a lack of time as per students interests survey. Thus, objectives of AR was set in the year 2015: to synthesize views of stakeholders in the institute for curriculum development, including key inputs from faculty on AL for learning sustainability; to continue with AL practices and reflections aimed to comprehend the positive trend of learning sustainability. Based on outcomes of previous year AL reflection, the set agenda for the year 2015, was to:

- Revamp the mission of the ISR Committee to encompass critical national development priorities and university vision. Develop standard operating procedures for ISR committee (internal communications)
- Attempt distributed leadership to develop transformational leadership qualities for students with elaboration on student leadership, code of ethical conduct, etc. (Nolen & Vander Putten, 2007; Timperley, 2005)
- Develop clarity for inclusions of AL Sustainability Lessons within MBA ITBM curriculum, synchronizing with successful redesign for sustainability based curricula by University stakeholders (Rao & Patil, 2015).

Revamping ISR Committee

The Planning Commission of India was being replaced by NITI *Aayog* (National Institution for Transforming India [NITI], 2015) in the year 2015. NITI *Aayog* was envisioned as a think-tank to reduce the centre-to-state-one-way flow of policy and to strengthen intentions to evolve shared vision of national development priorities. Participation across a spectrum of stakeholders was invited, including state governments, domain experts, and relevant institutions. Extensive brainstorming with students and faculty teams were required to align with national development vision. The appropriate set of nationally critical issues selected as action areas for social responsibility activities for the year 2015 were: Social Awareness, Education, Community Engagement, Environment, Healthcare (Internal Communications). Simultaneously aligning with the University's mission statements (Symbiosis, 2015) the ISR initiatives for the year 2015 were:

- To foster the spirit of national development (sensitizing to a relevant set of nationally critical issues, applying ISR activities thereof)
- To instill sensitivity amongst the youth towards the community and environment (initiating sensitivity in students towards community and eco-friendliness)
- To promote ethical and value-based learning ("learning by doing" instead of passive learning, cognizance of ethical values)
- To contribute towards knowledge generation and dissemination (sharing best practices, motivating inter-disciplinary research, application of relevant teaching)

Symbiosis Community Outreach Programme and Extension (SCOPE) Activities

Symbiosis International University (SIU) launched SCOPE activities in July 2012 to positively impact the community around SIU. Earlier services initiatives for underprivileged such as healthcare, finance, banking, computer literacy, legal counsel were consolidated under SCOPE activities (SIU, 2012). It formalized Institutional Social Responsibility (ISR), student committees of various constituents of SIU who continued to work with various NGOs on projects of relevance for them. Initiatives for undergraduate students to earn floating credits by service learning, a structured approach for faculty-in-charge for each constituent to mentor and guide social responsibility work, were formalized. Mentoring faculty undertook reviewed actions periodically, with annual reviews to envision and plan activities for the upcoming year. Symbiosis as a welfare organization had inculcated values to take timely correct action.

The Philosophy of AL for Change

Everyone thinks of changing the world, but no one thinks of changing himself. – Leo Tolstoy

Be the change that you wish to see in the world. – Mahatma Gandhi

Mahatma Gandhi's profound reflection on Leo Tolstoy teaches today's action learners first to change one's self if they wish to see a change in the world. Thus, '*Be the change*' became the motto of ISR. To avoid reproducing unsustainable actions and the consequent social ailments, a new approach to life with transformations in teaching for sustainability are essential (Sipos, Battisti, & Grimm, 2008; Timperley, 2005). All the actions and reflective exercises were aimed at aiding the willing change of the learner, to wait for the positive change to happen. If the change occurred, the action learner usually underwent emotional upheavals. More than emotional linking, a personal responsibility to the act is essential (Bennet & Bennet, 2008). A peculiarity of the emotionally involved approach is to be empathetic to the action learner and simultaneously remain in scientific objectivity (Tedlock, 1991). Ethnographically, such an approach entails follow up unaided reflections by the AL practitioner so as to produce sufficient narrations. Post-comprehension of the social nature of learning, the AL practioner needs to encourage personal journal keeping habits to develop reflection skills (Barth, Godemann, Rieckmann, & Stoltenberg 2007; Bolhuis, 2003). The shared thoughts in meetings and shared responsibility of documenting activities aids a gradual peer-learning and the critical construction of choices (Barth, et al., 2007; Reeve, J., Bolt & Cai, 1999; De Haan, 2006). By monitoring and documenting each AL activity's plan-do-act-check cycle for quality improvement, a systematic method was developed for students shared journal keeping. This detailed approach recorded minutes of meetings, budget plans, workload matrices, and reflections. This systematic approach to transition from philosophy to implementation was instrumental in accomplishing thirty activities during the year. These methods were audited by University authorities, noted as a best practice for the University (Internal Communications), and acknowledged as a best practice during an accreditation committee audit of the entire University. The achievements by AL students were appreciated with constructive criticisms by faculty and student peers.

AL: Social Awareness

To develop social awareness on Female infanticide and foeticide, AL students organized a peace march. They connected to common Indian causal issues were dowry avoidance, the patriarchal dominance of high-caste communities, and desire to keep wealth within the family. This injustice was observed and first suppressed during the British rule in the 1800s, then partly attributed to the patriarchal nature of Indian society (Nath, 1973). Today, the selective abortion of girls in the womb is a social issue (Agarwal, 1988; Sudha & Rajan, 1999). The peace-march action recognized the senseless killing of female fetuses, value of female gender and interviewed citizens to voice their support for the social cause. Affirmative banner messages such as 'daughters are blessings,' a sardonic query in a poster asking the patriarch 'you want a mother, you want a sister, you want a wife, but you *don't* want a daughter?' were bold expressions of AL students' creativity. Evaluating, this AL encouraged learners to voice their serious social protests strongly yet peacefully, with appropriate sensitization for themselves. Many students felt liberated about their first experience in marching out on the street for a social cause. The larger evil was noted as 'we have done something about it,' but felt helplessness in solving it was reflected during AL meeting. From the AL practitioner's point of view, it was a liberating experience to mentor boldness in peaceful action against a social evil. The interactions with many common people, to remember a lesson and to participate in the social process of a country were the key learning outcomes of the activity. It broke the ice for the next activity.

AL: Education

AL team observed a lack of quality education for the underprivileged (Kremer Brannen, & Glennerster, 2013). This critical need was attended to with extra tutorial classes were organized for children of the staff of the Institute. The supplemental tuition classes in English, math, reading and writing for children in the age group of 5 to 15 were organized in weekly schedules to arrange appropriate child-lesson-tutor combination. Parents of tutees appreciated AL students' team. While evaluating, the difficulty noted was that during examinations both tutors and tutees were stressed about their relevant exams, stipulating a reorganization of pre-exam tutorials. The progress of children was recorded with simple measures, with a resulting observation that much more training is required to develop children's skills consistently. It was also noted that children learn much faster from their college level 'big brothers' and 'big sisters' role models. Reflection of AL Practitioner was felt as an emancipating experience to observe current college students teaching their younger generation, ready to take responsibility for a nationally critical issue. The personal satisfaction felt was noted as if the future was in good hands, despite current appalling primary school standards in India.

AL: Community Engagement

Visits to an old age home was organized to sensitize action learners about the underprivileged. When AL students saw the elderly cast-off of as unwanted, the erosion of human values was a delicate lesson. The fragility of human lives is not an easy lesson to digest for the youth. After sharing youthful energy by games and street play entertainment, AL students commented that '*at the end of each hour, the smile we left on them was precious*.' When learners visited another poorer old age home with squalid facilities, where residents were suffering from depression and loneliness, all AL students felt emotionally overwhelmed. Evaluating, the pity they felt had induced complex emotions of helplessness, guilt, deep sympathy towards the elders. It became an urgent need to re-bond with their familial elders. Their feelings were reconciled during reflection session, by reminding them to hold the compassion in their heart, not guilt. The best lesson reflected and philosophized by action learners was surmised by quoting the Dalai Lama: "Compassion is the wish to see others free from suffering" – Dalai Lama (Miller & Rollnick, 2012)

Reflections of Action Learners

The quiet times practiced by each action learner, enhanced clarity in thinking and mature reflection comments. Intuition aids to concretize AL, thus leading to resolutions to allocate well planned quiet time in the future. Student's ISR committee operated democratically as a team with free flow debates and disagreements. They gathered several times during the year to evaluate individual activity performances and concluded with the annual evaluation of all activities. Substantiating democratic mechanisms lead to adeptness in dealing with uncertainty, changes and inherent multiple perspectives (Garmendia & Stagl, 2010). Students' reflexive approach was demonstrated in their unaided recommendations for improvement:

- Constitute a dedicated junior students team from 10% of the cohort, mentored by five senior AL students. (Faculty observed that this was a positive outcome, an improvement in taking self-responsibility for one's team and its leadership).
- The dedicated AL need to observe for quality improvement of activities and reflections. Evolve a steady flow of ALs throughout the year, instead of intense activities during a selected month (Faculty recognized the positive outcome of maturing teamwork, where students anticipated difficulties, plan to overcome challenges, and allocating more time for reflection within actions and post-actions).
- Distributed leadership worked only occasionally in this year. Change this back to hierarchical leadership in student AL team. (Faculty observed backsliding towards rigidity. Both students and faculty accepted a decision for more liberty to explore other roles within the team)

Reflections of AL Practitioner

Providing support during AL student's intense sensitization and emotional upheavals was an important practice for the AL mentor (Cohen, 2006; Lackritz, 2004). Waiting and allowing the student to struggle through changes, instead of solving the student's problem, stipulates patience and fortitude (Bolzern-Konrad, 2014). The common urging of a teacher to 'make the student practice organizing skills' was dissimilar to the counseling skills and patience demonstrated during reflection. Consequently, it sometimes left the students confused and wondering '*is the teacher pushing me to practice organizing and management skills or demanding me to change myself as a person*.' It became necessary to announce for each meeting such as '*this session is to critique the team's organizing*' or '*Let us take some quiet time as a group to reflect on the activity*.' Clarifications were essential to improving reflection exercises for AL.

A socially constructed leadership is necessary to mediate perceptions, beliefs, and assumptions, to compose meanings collectively (Harris, 2003). In the current attempt with distributed leadership, AL students occasionally emerged as team leaders, with significant interaction skills and earned the trust of their peers. The typical competitive environments of B-Schools provide fewer opportunities for students to unite, but more to compete. Social causes inspire unity towards the social responsibility cause, resulting in support for the AL team. With a lesser risk of failure due to inter-team competition, every student now has an opportunity to demonstrate leadership in one way or another (Goleman, Boyatzis, McKee, 2002). Upon reflection, a similarly distributed leadership was practiced during the faculty discussions on adopting sustainability lessons, where feelings of other members were given ample space (Corey, 1949). With a view towards the larger picture of sustainability lessons, openness was demonstrated by faculty in mutual sharing of information, aligning to AL goals within their respective courses, provisions for gathering observations and sharing evidence of academic credits.

Quantitatively, a rapid scale up to include a large number of activities was useful to increase participation of students. Nearly all the students of the cohort were involved in at least one of the social responsibility activities during the year. But learners felt pressurized and sometimes complained of coercion, noted as a significant ethical concern to add legitimacy for next cohort (Bolzern-Konrad, 2014; Nolen & Vander Putten, 2007). Qualitatively, a large number of AL activities added stress to the AL mentors' role. Work overload can be emotionally exhausting during social responsibility learning (Lackritz, 2004; Maslach & Leiter, 2005; Smith et al., 2002). Much commitment is required towards the learner, to observe and counsel a young person. Adverse consequences of scaling up too much were discerned as controversial, can lead to skepticism and reduce acceptance of social work. AL team deliberated on substituting scaffolding effects instead of discovery learning (Hmelo-Silver, Duncan, & Chinn, 2007; Pea, 2004) as a stabilizing approach for the upcoming year. Within an optimum number of activities, improvements in reflection methods such as personal journal writing, additional one-on-one counseling were decided. Inclusions of sustainability lessons within business courses were adopted by three more faculty members, indicating positive perceptions, risk-taking, and involvement in training for emotional intelligence (Goleman, 2001). This association has aided further acceptance and added new AL reflections on serving as evidence of new forms of learning. Continued critical reflections of faculty adopting AL is expected to develop deeper exchanges harmonizing with sustainability learning practices.

ACTION LEARNING IN YEAR 2016: 'CONSOLIDATE AND OPTIMIZE'

Transition into the Last Stage of AL

It is essential for AL practitioners to undertake the primary responsibility of decision-making with critically informed actions, and to evaluate implemented strategies (Mapotse, 2015; Petrovic, Kittl, & Teksten, 2001). Some previous AL models were replicated, owing to their success in inspiring students, individualized learning models and ability to embed sustainability learning goals (Yiu Yuen, & Cheong Cheng, 2000). The cohort's first loop of successful learning enabled the formation of new mental models, assimilation into sustainability education, and application in interdisciplinary research. This second loop on socio-technical topics was aimed to conclude with disseminations to University and publish research papers (Corey, 1960; Petrovic et al., 2001; Pradhan, Sharma, Krishnamurthy, 2016). Single-loop learning was adapted to form double-loop learning in the next stage (Argyris & Schön, 1978; Fiol & Lyles, 1985). The following commentary recapitulates AL in year 2016.

Interdisciplinary Research in the University

Symbiosis International University encouraged collaborative interdisciplinary research efforts. Together with experts in AL, the author conducted a workshop for community outreach for AL colleagues of University. Learning from interdisciplinary research workshops aided the transition into last stage of AL. Connecting with the cohort's earlier development from AL in junior year to research on socio-technical topics in their senior year, selected AL teams brainstormed with AL mentor to develop research manuscripts. Tangible academic rewards were designed within the course and intangible rewards of satisfaction in advanced study. When national issues are complex, interdisciplinary research evolves as a useful tool to develop new fundamental understanding (Wu, Shen, & Kuo, 2015; Bolzern-Konrad, 2014; Hmelo-Silver, et al., 2007). Students cultured by their previous AL experiences, combined with IT skills core strength, were now mentored to inquire into social issues and shortcomings in governance. The selection of learner-centric research exercises in higher education was appropriate, considering the average age of students was 25 years, most of them with a bachelor's degree and one or two years of IT work experience. This sensitized cohort was ready for mature study, patiently reflect on critical national issues, attempt to innovate IT-enabled solutions for socio-technical issues (Nguyen, Graham, Ross, Maani, & Bosch, 2012). Reduced inequality, inclusive growth, and national development were selected as the themes of research. When students approached their topics with a business-like efficiency, they were reminded about as learned in previous year's learning about society's sluggishness to adopt solutions and changes. Clarity in their writing was rechecked several times, to outlast resistance of societal systems. When some AL students displayed a scornful attitude in criticizing their government, it was reminded that as advanced students, it is their civic duty to recommend solutions by study, not mere criticism. The following are AL activities of MBA students, aligned to learn selected Sustainable Development Goals (SDGs).

AL Example 1

Student's published Socio-technical Research topic: "Application of Internet of Things in Healthcare Sector for Bottom of Pyramid in India" (Deshpande, Mathur, & Krishnamurthy, 2016). Aligned to SDGs: SDG 9) Innovation and infrastructure SDG 12) Responsible consumerism SDG 17) Partnership for goals. 'Internet of Things' (IoT) is advanced technology but with a potential to solve issues of accessibility to healthcare for the rural and poor citizens. Disparate infrastructures are stumbling blocks for doctors desiring to deliver healthcare services in remote locations, the typical dwelling of rural poor. Lack of time and access to diagnosing instruments are tackled better by IoT and remote monitoring. This research articulates implementation feasibilities of IoT for remote health monitoring of Indian rural masses and anticipation of socio-technical challenges.

AL Example 2

Student's published Socio-technical Research topic: "A Study of Social Media Usage in Rural Development Schemes of Government of India" (Kothari, Mohanty, Yadav, & Sisodia, 2016). Aligned to SDG 8 (Good jobs and economic growth), SDG 9 (Innovation and infrastructure) SDG 10 (Reduced inequalities). Rural citizens of India have become more active in social media and are connecting better to the urban society. This research developed recommendations to develop stronger rural participation in rural-centric infrastructure development. Thus, the student's earlier sensitizations on social responsibility, and responsiveness towards democratic processes helped to address rural people issues with compassion, not disdain.

AL Example 3

Student's published Research topic: "Circular Economy for Sustainable Development in India" (Yaduvanshi, Myana, & Krishnamurthy, 2016). Aligned to SDG 12) Responsible consumerism SDG 17) Partnership for goals SDG 11) Sustainable cities and communities SDG 10) Reduced inequalities. This research explored Circular Economy dedicated to sustainable development in India. Increased affordability of modern Indian middle class has led to considerable increases in consumerism and substantial increments in waste generation. Current urban India's inadequate waste management practices were critiqued, and social challenges were examined. Solutions for urban India were studied, with inferences drawn from other comparable economies. Inducing consumer's responsible behavior towards their environment could simultaneously reduce resources consumption and enable resources availability to those who require them most.

AL Example 4

Student's published Research topic: "Educating Consumers about Digital Wallets" (Nair, Dahiya, Gupta, Yadav, & Mehta, 2016). Aligned to SDG 10 (Reduced inequalities), SDG 11 (Sustainable cities and communities), SDG 17 (Partnership for goals). To understand that digital wallets are an efficient solution for money transactions is easy for an IT student. But it required patient inquiry to discover that most average people were resistant towards this solution due a lack of basic technology skills. Thus, their research focal point became 'educating citizen consumers on digital wallets,' published before the announcement of demonetization of currency by Reserve Bank of India (RBI, 2016). Poor people with small cash transactions needed more protection and accuracy. While they are the ones who can receive the best benefits if they become consumers of digital wallets, they are also the ones with the greatest resistance, and lacking basic technology skills, hence the ignorance.

Reflections of AL Practitioner and Co-Researchers

Reflexive connections enabled students to apply research and IT business skills towards sustainability and national development. During conceptualization stage, action learners need to avoid reinventing the wheel. An application of research skills to solve social issues facilitates higher level professional development (Dickens & Watkins, 1999; O'Hara, Beaty, Lawson, & Bourner, 1997). By disseminated their learning within the University, useful results to learner's society were manifested. While grassroots implementations of their research may take some years, the current AL contributions to the existing body of knowledge provided much personal satisfaction for AL students in researching real-world sustainability issues and earning publishing rewards (Prince & Felder, 2006; Steiner & Posch, 2006). These enhancements of AL i.e. applying learners' intelligence towards solving national issues accomplished last stage AL goals. With growing acceptance, three more teachers adopted AL components for sustainability lessons within their business courses of Marketing Management, Research Methodology, and Human Resources Management. Their critical reflections are summarized in Table 2.

Demonstrating the ethical principle of 'reciprocity,' the research was of some mutual benefit to researchers and participants i.e. members of the society (Bell & Bryman, 2007). Debating among the research teams with other socio-technical topics demonstrated interpersonal learning and empowerment (Goel, Gupta, & Bande-Vilela, 2014). Occasionally, AL co-researchers arrived for counseling sessions with some frustration, with frustrations about the tardiness of common people in accepting smart IT solutions. The affective aspect of love for one's country was inspiration enough to overcome minor frustrations. Practicing restraint, the AL practitioner encouraged student-driven debates during research discussions. By themselves, learners arrived at an acceptance of the democratic society's awkward struggles in technology acceptances. Leaning in too much with self-views of a teacher on societal issue solutions was avoided since it would turn into indoctrination, not facilitation (Prince & Felder, 2006; Shephard, 2008; Sipos, et al., 2008). The natural dynamism of youth to 'correct all at once' even with national issues, was simultaneously amusing and heartening. Tales of caution were used to remind action learners to pace themselves. Occasional disheartening was the youth resigning to helplessness and abandonment. Amusing anecdotes were used to cheer up learners, thus facilitating equilibrium. AL practiced here reinforced the need for advanced business students to apply themselves as best as one can, instead of becoming apathetic. The AL objectives achieved by the 2015-17cohort:

- **Critical Reflections:** Collected data frequently for on each activity. Recorded faculty's collective critical reflections for developing a harmony of AL for sustainability lessons within the curriculum for ITBM (summary of responses shown in Table 2).
- **Continuation in Sustainability AL:** Generated positive trends in including sustainability lessons, developed possibilities for other courses, attempted solving sustainability issues with interdisciplinary research, applied AL team to study real-time sustainability solutions in India.
- **Chief Practices of AL:** AL stakeholders learned leadership lessons, reaffirmed the values of ethical codes of conduct, applied themselves to complex socio-technical issues of national importance, comprehended AL practices well, reflected in a mature manner and mentored junior students.

ISSUES, CONTROVERSIES AND RECOMMENDATIONS

This section summarizes the learning of AL practitioners and the students as co-researchers. Noting by the qualitative nature of AL, the student participants underwent breakoffs in learning. AL group members negotiated for consensus in learning goals, and progressed in developing a common identity. Thus a convergence of values was observed during the year (Meeussen, L., Delvaux, E., & Phalet, K., 2014). Observing and reflecting upon the learner group's issues, role perceptions, dilemmas, and group dynamics, recommendations were developed.

Table 2. Summary of responses (reflexive questions asked to faculty adopting AL in year 2016)

	Query to Faculty (Adopting AL in Year 2016)	Summary of Faculty Responses
1	Noteworthy actions to learn sustainability and their effectiveness.	*During orphanage and old-age home visits, direct impacts to the life of recipients were observed. Students went the extra mile, motivated by grade points as incentives. Planting and maintaining plant facilities within campus was a significant intervention.*
2	Impacts (of sustainability lessons) observed in students	*Students exhibited changes in attitude towards nature. We noted improved humane aspects, useful for building leadership role! This will a create long-term impact rather than immediate results. Improved human elements are evident in alumni, who had learned sustainability and social responsibility with us and are currently active social responsibility volunteers at their workplace.*
3	Did the social responsibility activities result in students learning sustainability lessons?	*It may be too soon to comment. We should discuss if being aware and being responsible are the same. However, sustainability lessons will have positive impacts in the long run.*
4	How can we become better teachers?	*Education is beyond mere IT literacy. To become leaders with vision in the business world, we need to emphasize to 'Learn from life' and to explore beyond the textbook. A good teacher must be able to bring out the human values of students.*
5	As a teacher, your motivation in including sustainability lessons?	*To expose students beyond theories and classrooms and to make them emotionally intelligent. An education that develops rational but unemotional human beings can be dangerous.*
6	Where your course learning goals achieved to satisfaction?	*They took a deep interest in socially relevant assignments. Students' environmental and social intelligence were challenged. Some attempts had limited successes. But, we could make students think and feel.*
7	Has this experience changed you as a teacher?	*It gave us opportunities to take risks. We are empowered to make decisions. We have become more confident in the process.*

Table 3. Summary of realizations: issues, controversies, recommendations

	Issue/Controversy	Recommendations
1	**Coercion:** Student teams may be coerced to participate in social responsibility work or sustainability lessons. Such an approach becomes counterproductive and unnecessarily increases students' resistance to valid lessons.	**'Less risk – No coercion' policy:** Mentors and committees within institutions may appraise initial interests of students in social responsibility work. Interest surveys, group meetings can be organized to select student teams. Allow choice for students to withdraw from social action. A lack of restrictions or over-compromising learning or incomplete tasks is not recommended here. Free will helps to enhance commitment and strengthens self-governance of student committees.
2	**Confusion:** Action learners and facilitators may become confused in their roles occasionally. The roles of experienced learners can sometimes become mere managerial coordination and fail to transfer the learning culture. The purpose of actions sometimes becomes a competition to accumulate high scores instead of learning.	**'Above all, freedom!':** Learners and facilitators need sufficient reflection time for learning to be absorbed. The candor of participants during reflection meetings is essential to avoid malfunction of sustainability curricula. Learning sustainability entails sustaining learning spirit by committed seniors, helpful faculty peers, and an accommodating administration. However, a reality of check during group reflection meetings enables critique. Confusions in roles, trials and course corrections can become interesting processes of learning for business students.
3	**Rapid Scale-up:** Institutions may authorize AL with 'weak actions followed by too many projections of learning,' without actually learning enough. The other imbalance of 'too many actions and scarce reflection' is also unfavorable for AL. The eagerness to scale up actions to include all students may churn out all of the above issues.	**'Slow and Steady' Approach:** Rapid scale-up is ill-advised. Mentors of AL need time to develop maturity and to train another faculty to adopt AL. Adequate AL skills development meeting time is required for departments. A culture of 'collaboration, not competition' sets a constructivist approach to all stakeholders in this realm. This seemingly utopian philosophy is not easy to adopt in competitive environments but is an authentic style that needs to be practiced to be believed.
4	**Lack of empathy:** Seizing social issues AL activities with vested interests, with no empathy for the people or issue of concern. If such studies are allowed, it may open doors for publishing premature or unsound social responsibility practices.	**The ethical code of conduct:** Mentors and committees within institutions need to screen social responsibility research topics. Adequate review procedure by university panels are required to monitor social issues research work and appraise conduct of researcher. Students and faculty commitment are required to keep social responsibility activities aimed at sensitization, experiential learning, reflection and learning sustainability.
5	**'One hit wonders':** Institutions may have inspired social work among teachers, who may achieve great success. Repetition of the same process year after year without improvements, may not be accomplishing sustainability learning goals. This recognizable shortcoming needs to be identified. For some teachers, dealing with too many challenges and issues could lead to burnout, leading to replication of lifeless routines to learners.	**'Trial-error-correction-reflection' approach:** Critical evaluations enable fresh perspectives, an essential requirement for AL course corrections. One method to solve this shortcoming is to organize well-spaced reflection times for AL practitioners, in solo and together with peer advising groups. Among AL peers, interdisciplinary advice, enhances camaraderie and cross-breeding ideas. Another method is to plan progressively for an entirely out-of-campus AL engagement for a small group of AL students, at least once in a semester. Three whole days of action-reflection-recording-further study can become memorable lessons to share. For B-School programs willing to iterate and take bold steps, this can further lead to a two-week social responsibility internship, once a year.
6	**Motivation:** Motivation for AL drop during formal academic evaluations, when the performance pressures are high. Seasonal excitement about major cultural events of the year, internship and job placements cause such reductions in motivation. Here, the drop-in motivation hinders learning and could render AL ineffective.	**Role Clarity, Non-interference:** If students can participate in long-duration AL (a week or month) they could demonstrate commitment. After action-reflection-recording-plan for further study, they can return to routine academia. Core AL team members could be allowed for higher responsibility to organize activities for their non-ISR peers. Without a procedure for non-interference in AL, it may not change learning effectiveness.

FUTURE RESEARCH DIRECTIONS

Authentic human qualities of carefulness, thoughtfulness, humility, criticality, receptiveness, resilience, courage and stillness need to be construed within pedagogical tasks to achieve transformatory curriculum. (Ronald Barnett in 'Learning for an unknown future')

All learning is preparing for a future, where the essence of future itself is that it is unknown. We realized that predictability of business situations is counter-intuitive to social learning. Sustainability lessons are urgently required for current generation of students in India. Synchronically, institutions need time to imbibe curriculum transformation and lead this change process (Fullan, 2004; Timperley, 2005). As learning organizations, the curriculum development processes of institutions call for a responsibility towards developing authentic human qualities (Barnett, 2004). Future research directions could involve assessments of current university programs for their effectiveness in teaching sustainability, to avoid mere requirement tallies and fragmented learning (Romme & Dillen, 1997). As discussed before, inclusions of sustainability lessons need not compromise primary business curriculum, but multiple trials by B-schools can be researched to groom clarity, to assist realizations and to foster sustainability curriculum modifications (Krishnamurthy, Joseph, Bharathi, & Pradhan, 2014a; Prince & Felder, 2006; Rao, Patil & Gupte, 2013). Upon gathering such multiple institutional data, AL could be validated further (Dickens & Watkins, 1999). AL researchers would often confront different versions of a life event and a different account of the same life event after some time. Instead of dismissing the storyteller as an unreliable informant, embracing the discursion and careful consideration of inconsistencies, is necessary to avoid rigor mortis in qualitative research (Sandelowski, 1986).

CONCLUSION

As true replication cannot be attained in social systems, an action researcher has to archive a situation in which the research process is recoverable by interested outsiders. (Petrovic et al., 2001)

Evaluations of sustainability lessons inclusions within a niche business program are archived in this chapter. Based on critical thinking, the chapter narrated self-reflections, year-wise reflections over three years. The changes generated are still in the making. As learned in one business school, AL lives up to the challenge very well. Ideally, the enriched learner groups can and need to apply themselves to improving their society, no matter how small. For institutions initiating sustainability lessons by AL, this chapter provides experiential narrations. During initial attempts, interesting twists and unexpected turns for the AL learners and AL practitioners may provide immense learning. When considering business curricula, seemingly unrelated study areas can be connected by interdisciplinary research and applied to resolve social issues. AL for sustainability demands rigorous evaluations for successful inclusion into business school curricula. If reflected well and included appropriately, AL practitioners may aspire for business students to rise up to learning sustainability with a bottom-up approach instead of a top-down approach, a heartening move from apathy to empathy.

ACKNOWLEDGMENT

This book chapter is dedicated to the student batches of the year 2014-16, 2015-17 and 2016-18, in the of SCIT MBA ITBM program. Special thanks to the students who participated enthusiastically and supported the action learning trials, experiments, and reflections within the ecosystem of an IT and business-oriented institution. Their support enabled this attempt to document and share learning with all stakeholders in higher education. The authors wish to express sincere thanks to Dr. Urvashi Rathod and Dr. Ajey Kumar for their support and encouragement in reviewing SCIT's Institutional Social Responsibility. This chapter validates scholastic benefits of inclusion of sustainability lessons within IT business curriculum.

REFERENCES

Agarwal, B. (1988). Patriarchy and the 'modernizing' state: An introduction. *Structures of Patriarchy*, 1-27.

Albareda Tiana, S., & Alférez Villarreal, A. (2016). A collaborative programme in sustainability and social responsibility. *International Journal of Sustainability in Higher Education, 17*(5), 719–736. doi:10.1108/IJSHE-07-2016-0134

Alsop, R. (2008). *The trophy kids grow up: How the millennial generation is shaking up the workplace*. John Wiley & Sons.

Argyris, C., & Schön, D. A. (1978). *Organizational learning: A theory of action perspective* (Vol. 173). Reading, MA: Addison-Wesley.

Astier, M., Masera, O., & López-Ridaura, S. (1999). *Sustentabilidad y manejo de recursos naturales* (M. P. México, Ed.). SA de CV.

Barnett, R. (2004). Learning for an unknown future. *Higher Education Research & Development, 23*(3), 247–260. doi:10.1080/0729436042000235382

Barth, M., Godemann, J., Rieckmann, M., & Stoltenberg, U. (2007). Developing key competencies for sustainable development in higher education. *International Journal of Sustainability in Higher Education, 8*(4), 416–430. doi:10.1108/14676370710823582

Bell, E., & Bryman, A. (2007). The ethics of management research: An exploratory content analysis. *British Journal of Management, 18*(1), 63–77. doi:10.1111/j.1467-8551.2006.00487.x

Bennet, A., & Bennet, D. (2008). A new change model: Factors for initiating and implementing personal action learning. *Vine, 38*(4), 378–387. doi:10.1108/03055720810917633

Bhattacharya, S., & Bhattacharya, S. (2014). Indian Renaissance in the Twenty-First Century: Management Education, Thoughts and Practices for the Betterment of Business, Nation and Society. *Journal of Human Values, 20*(1), 45–58. doi:10.1177/0971685813515602

Bolhuis, S. (2003). Towards process-oriented teaching for self-directed lifelong learning: A multidimensional perspective. *Learning and Instruction, 13*(3), 327–347. doi:10.1016/S0959-4752(02)00008-7

Brundtland, G., Khalid, M., Agnelli, S., Al-Athel, S., Chidzero, B., Fadika, L., ... Singh, M. (1987). *Our common future*. Brundtland Report.

Bryman, A. (2003). *Quantity and quality in social research*. Routledge.

Buchanan, L., & Connell, O, A. (2006). A brief history of decision making. *Harvard Business Review*, *84*(1), 32. PMID:16447367

Cohen, J. (2006). Social, emotional, ethical, and academic education: Creating a climate for learning, participation in democracy, and well-being. *Harvard Educational Review*, *76*(2), 201–237. doi:10.17763/haer.76.2.j44854x1524644vn

Corey, S. M. (1949). Curriculum development through action research. *Educational Leadership*, *7*(3), 147–153.

Corey, S. M. (1960). Action Research in Education. *Guidance Readings for Counselors*, 603.

De Haan, G. (2006). The BLK 21programme in Germany: A Gestaltungskompetenz-based model for Education for Sustainable Development. *Environmental Education Research*, *12*(1), 19–32. doi:10.1080/13504620500526362

Deshpande, A., Mathur, A., & Krishnamurthy, S. (2016). Application of Internet of Things in Healthcare Sector for Bottom of Pyramid in India. IJEAST. *International Journal of Engineering Applied Sciences and Technology*, *1*(11), 28–33.

Dickens, L., & Watkins, K. (1999). Action research: Rethinking Lewin. *Management Learning*, *30*(2), 127–140. doi:10.1177/1350507699302002

Dmochowski, J. E., Garofalo, D., Fisher, S., Greene, A., & Gambogi, D. (2016). Integrating sustainability across the university curriculum. *International Journal of Sustainability in Higher Education*, *17*(5), 652–670. doi:10.1108/IJSHE-10-2014-0154

Domask, J. J. (2007). Achieving goals in higher education: An experiential approach to sustainability studies. *International Journal of Sustainability in Higher Education*, *8*(1), 53–68. doi:10.1108/14676370710717599

Dreyfus, H. L. (2008). *On the internet*. Routledge.

Elzen, B., Geels, F. W., & Green, K. (Eds.). (2004). *System innovation and the transition to sustainability: theory, evidence and policy*. Edward Elgar Publishing. doi:10.4337/9781845423421

Fiol, C. M., & Lyles, M. A. (1985). Organizational learning. *Academy of Management Review*, *10*(4), 803–813.

Fullan, M. (2005). *Leadership & sustainability: System thinkers in action*. Corwin Press.

Garmendia, E., & Stagl, S. (2010). Public participation for sustainability and social learning: Concepts and lessons from three case studies in Europe. *Ecological Economics*, *69*(8), 1712–1722. doi:10.1016/j.ecolecon.2010.03.027

Georgescu-Roegen, N. (2002). Four challenges of sustainability. *Conservation Biology*, *16*(6), 1457–1460. doi:10.1046/j.1523-1739.2002.01668.x

Goel, A., Gupta, A., & Bande-Vilela, B. (2014). It's all in interpersonal interactions: Empowerment, organizational cynicism, and perception of organizational politics. Values in Shock - The role of contrasting management, economic, and religious paradigms in the workplace, 418.

GOI. (1956). *Government of India, Sixth Five Year Plan*. Planning Commission, Government of India.

Goleman, D. (2001). Emotional intelligence: Issues in paradigm building. *The Emotionally Intelligent Workplace, 13*, 26.

Goleman, D., Boyatzis, R. E., & McKee, A. (2002). *The new leaders: Transforming the art of leadership into the science of results*. London: Little Brown.

Harris, A. (2003). Teacher leadership as distributed leadership: Heresy, fantasy or possibility?. *School Leadership & Management, 23*(3), 313–324. doi:10.1080/1363243032000112801

Hmelo-Silver, C. E., Duncan, R. G., & Chinn, C. A. (2007). Scaffolding and achievement in problem-based and inquiry learning: A response to Kirschner, Sweller, and Clark (2006). *Educational Psychologist, 42*(2), 99–107. doi:10.1080/00461520701263368

Hussein, K., & Nelson, J. (2016). *Sustainable livelihoods and livelihood diversification*. Academic Press.

Kezar, A., & Rhoads, R. A. (2001). The dynamic tensions of service learning in higher education: A philosophical perspective. *The Journal of Higher Education, 72*(2), 148–171. doi:10.2307/2649320

Kothari, A., Mohanty, P., Yadav, R., & Sisodia, T. (2016). A Study of Social Media Usage in Rural Development Schemes of Government of India. *International Journal of Engineering Applied Sciences and Technology, 2*(2), 1–5.

Kremer, M., Brannen, C., & Glennerster, R. (2013). The challenge of education and learning in the developing world. *Science, 340*(6130), 297–300. doi:10.1126/science.1235350 PMID:23599477

Krishnamurthy, S., Joseph, S., & Bharathi, V. (2014b). Creating Environment Friendly Projects in Rural India–A Synergy Framework for Sustainable Renewable Energy. *International Journal of Applied Engineering Research, 9*(24), 26719–26738.

Krishnamurthy, S., Joseph, S., Bharathi, V., & Pradhan, V. (2014a). Inclusion of Sustainability Education in Business Schools–An Indian B-School Case Study of MBA-ITBM Curriculum. *International Journal of Applied Engineering Research, 9*(23), 22703–22725.

Lackritz, J. R. (2004). Exploring burnout among university faculty: Incidence, performance, and demographic issues. *Teaching and Teacher Education, 20*(7), 713–729. doi:10.1016/j.tate.2004.07.002

Mapotse, T. A. (2015). An emancipation framework for technology education teachers: An action research study. *International Journal of Technology and Design Education, 25*(2), 213–225. doi:10.1007/s10798-014-9275-y

Martin, S., Brannigan, J., & Hall, A. (2005). Sustainability, systems thinking and professional practice. *Journal of Geography in Higher Education, 29*(1), 79–89. doi:10.1080/03098260500030389

Masera, O., & López-Ridaura, S. (2000). *Sustentabilidad y Sistemas Campesinos. Cinco experiencias de evaluación en el México rural*. MundiPrensa-GIRA-UNAM.

Maslach, C., & Leiter, M. P. (2005). Reversing burnout. *Stanford Social Innovation*, 43-49.

Meeussen, L., Delvaux, E., & Phalet, K. (2014). Becoming a group: Value convergence and emergent work group identities. *The British Journal of Social Psychology*, *53*(2), 235–248. doi:10.1111/bjso.12021 PMID:23316715

Miller, W. R., & Rollnick, S. (2012). *Motivational interviewing: Helping people change*. Guilford press.

Nair, A., Dahiya, M., Gupta, N., Yadav, R., & Mehta, R. (2016). *Educating Consumers about Digital Wallets* (Vol. 3). IJCSN International Journal of Computer Science and Network.

Nath, V. (1973). Female infanticide and the Lewa Kanbis of Gujarat in the nineteenth century. *The Indian Economic and Social History Review*, *10*(4), 386–404. doi:10.1177/001946467301000404

Nguyen, N. C., Graham, D., Ross, H., Maani, K., & Bosch, O. (2012). Educating systems thinking for sustainability: Experience with a developing country. *Systems Research and Behavioral Science*, *29*(1), 14–29. doi:10.1002/sres.1097

NITI Aayog. (2015). *National Institution for Transforming India*. Retrieved from niti.gov.in

Nolen, A. L., & Vander Putten, J. (2007). Action research in education: Addressing gaps in ethical principles and practices. *Educational Researcher*, *36*(7), 401–440. doi:10.3102/0013189X07309629

O'Hara, S., Beaty, L., Lawson, J., & Bourner, T. (1997). Action learning comes of age-part 2: action learning for whom?. *Education+ Training*, *39*(3), 91-95.10.1108/00400919710164080

Pea, R. D. (2004). The social and technological dimensions of scaffolding and related theoretical concepts for learning, education, and human activity. *Journal of the Learning Sciences*, *13*(3), 423–451. doi:10.1207/s15327809jls1303_6

Petrovic, O., Kittl, C., & Teksten, R. D. (2001). *Developing business models for ebusiness*. Available at SSRN 1658505.

Pradhan, V. S., Sharma, R., & Krishnamurthy, S. (2016). Analyses of factors affecting employee perception on organizational Corporate Social Responsibility, SCIT Journal 2016, Volume XVI, August 2016 (ISSN 0974-5076)

Rajan, R. G. (2011). *Fault lines: How hidden fractures still threaten the world economy*. Princeton University Press. doi:10.1515/9781400839803

Rao, P., & Patil, Y. (2015). Integrating Energy and Environment in Postgraduate Management Education: A Case Study from Symbiosis Institute of International Business, Pune, India. In Implementing Campus Greening Initiatives (pp. 265-276). Springer International Publishing.

Rao, P., Patil, Y., & Gupte, R. (2013). Education for sustainable development: trends in Indian business schools and universities in a post liberalization era. In *Sustainability Assessment Tools in Higher Education Institutions* (pp. 417–432). Springer International Publishing. doi:10.1007/978-3-319-02375-5_24

RBI. (2016). *Withdrawal of Legal Tender Status for ₹ 500 and ₹ 1000 Notes: RBI Notice* (Revised). Reserve Bank of India.

Reeve, J., Bolt, E., & Cai, Y. (1999). Autonomy-supportive teachers: How they teach and motivate students. *Journal of Educational Psychology*, *91*(3), 537–548. doi:10.1037/0022-0663.91.3.537

Romme, G., & Dillen, R. (1997). Mapping the landscape of organizational learning. *European Management Journal*, *15*(1), 68–78. doi:10.1016/S0263-2373(96)00075-8

Sandelowski, M. (1986). The problem of rigor in qualitative research. *ANS. Advances in Nursing Science*, *8*(3), 27–37. doi:10.1097/00012272-198604000-00005 PMID:3083765

Sandelowski, M. (1991). Telling stories: Narrative approaches in qualitative research. *Journal of Nursing Scholarship*, *23*(3), 161–166. doi:10.1111/j.1547-5069.1991.tb00662.x PMID:1916857

Sekhar, K. (2002). Three decades of the National Service Scheme The social work component and development issues. *International Social Work*, *45*(1), 99–114. doi:10.1177/0020872802045001322

Shephard, K. (2008). Higher education for sustainability: Seeking affective learning outcomes. *International Journal of Sustainability in Higher Education*, *9*(1), 87–98. doi:10.1108/14676370810842201

Sipos, Y., Battisti, B., & Grimm, K. (2008). Achieving transformative sustainability learning: Engaging head, hands and heart. *International Journal of Sustainability in Higher Education*, *9*(1), 68–86. doi:10.1108/14676370810842193

Smith, C. L., Lopes, V. L., & Carrejo, F. M. (2011). Recasting paradigm shift: "True" sustainability and complex systems. *Human Ecology Review*, *18*(1), 67–74.

Steiner, G., & Posch, A. (2006). Higher education for sustainability by means of transdisciplinary case studies: An innovative approach for solving complex, real-world problems. *Journal of Cleaner Production*, *14*(9), 877–890. doi:10.1016/j.jclepro.2005.11.054

Sudha, S. S. I. R., & Rajan, S. I. (1999). Female demographic disadvantage in India 1981–1991: Sex selective abortions and female infanticide. *Development and Change*, *30*(3), 585–618.

Symbiosis International University. (2012), Social Activities, http://siu.edu.in/social-initiatives.php, retrieved Aug 2016.

Symbiosis International University. (2015), Vision and Mission, http://siu.edu.in/symbiosis-international-university.php#visi, retrieved Aug 2016.

Tainter, J. A. (2006). Social complexity and sustainability. *Ecological Complexity*, *3*(2), 91–103. doi:10.1016/j.ecocom.2005.07.004

Timperley, H. S. (2005). Distributed leadership: Developing theory from practice. *Journal of Curriculum Studies*, *37*(4), 395–420. doi:10.1080/00220270500038545

Trantopoulos, K., Schlopfer, M., & Helbing, D. (2011). Toward sustainability of complex urban systems through techno-social reality mining. *Environmental Science and Technology-Columbus*, *45*(15), 6231–6232. doi:10.1021/es2020988 PMID:21744879

UNESCO. (1997). United Nations Educational Scientific and Cultural Organization. In *Educating for a Sustainable Future*. Retrieved Oct 3, 2016, from http://www.unesco.org/education/tlsf/TLSF/theme_a/mod01/uncom01t05s01.htm#edu

United Nations. (1992). *The Rio declaration on environment and development*. Rio de Janeiro: UN.

Wilkinson, R. G., & Pickett, K. (2009). *The spirit level: Why more equal societies almost always do better* (Vol. 6). London: Allen Lane.

Wood, G. D. A. (2012). What Is Sustainability Studies?. *American Literary History*, *24*(1), 1–15. doi:10.1093/alh/ajr044

Wu, Y. C. J., Shen, J. P., & Kuo, T. (2015). An overview of management education for sustainability in Asia. *International Journal of Sustainability in Higher Education*, *16*(3), 341–353. doi:10.1108/IJSHE-10-2013-0136

Yaduvanshi, N. R., Myana, R., & Krishnamurthy, S. (2016). Circular economy for sustainable development in India. *Indian Journal of Science and Technology*, *9*(46). doi:10.17485/ijst/2016/v9i46/107325

Yiu Yuen, P., & Cheong Cheng, Y. (2000). Leadership for teachers action learning. *International Journal of Educational Management*, *14*(5), 198–209. doi:10.1108/09513540010373144

KEY TERMS AND DEFINITIONS

BSTE (Bounded Socio-Technical Experiment): Attempts to introduce a new technology, service, or a social arrangement on a small scale. They are carried out by coalitions of diverse actors, and are driven by long term and large-scale visions of advancing society's sustainability agenda.

Community Outreach: An activity of providing services to any populations who might not otherwise have access to those services. Outreach is often meant to fill in the gap in the services provided by mainstream (often, governmental) services, and is often carried out by non-profit, nongovernmental organizations.

Curriculum Transformation: A process whereby faculty in universities and schools study their curricula in order to incorporate the insights and information from their research into their courses.

Double Loop Learning: Occurs when error is detected and corrected in ways that deeply challenge the existing norms, policies and objectives. It leads to effective modification of existing organizational norms and offers radically new solutions.

Institutional Social Responsibility (ISR): Institutional Social responsibility pertains to Institutional conscience, Institutional citizenship, social responsibly or responsible business. ISR is one of the most important business ethics quality for an institution to become successful.

Interdisciplinary Research: A mode of research by teams or individuals that integrates information, data, techniques, tools, perspectives, concepts, and/or theories from two or more disciplines or bodies of specialized knowledge to advance fundamental understanding or to solve problems whose solutions are too complex to be answered satisfactorily by any one discipline.

Peer Advising: Generally refers to programs in which students assist other students. A formal definition is "Peer advising is an educational process in which students are intentionally connected with other students to support learning and success" (Koring and Campbell, 2005).

Single-Loop Learning: If the outcomes of a learning routine are not desirable, then reflections and modifications of existing practices are designed to restore outcomes to expectations. Single-loop learning seems to be present when goals, values, frameworks and, to a significant extent, strategies are taken for granted.

Students' Self-Governance: Or autonomy, is an abstract concept that applies to several scales of organization. Mahatma Gandhi's term 'Swaraj' is a branch of this self-rule ideology.

Transformational Leadership: Transformational leadership is a style of leadership where a leader inspires subordinates to transcend their own self-interests and who are capable of having profound effects on them. Subordinates identify needed change, creating a vision to guide the change through inspiration, and executing the change in tandem with committed members of the group. It involves psychological, convictional and behavioral changes.

Chapter 9
Teacher Learning and Effective Professional Development:
How One School Implemented, Modified, Stretched, and Extended the Principles of Action Learning

Bronwyn Mehorter
University of Wollongong, Australia

ABSTRACT

Action Learning has been and is regularly implemented within Australian schools as a platform for teacher professional development and professional learning. The following chapter reports on a decade-long association between one Australian government primary school, an Academic Partner and the process of Action Learning. Initially, Action Learning was implemented in 2005 in the form of a small-scale, more traditional Action Learning project; In 2009, Action Learning was then modified and stretched to involve the whole school's teaching, support and executive staff; In 2013, the principles of Action Learning were extended as the school executive and teaching staff began to take ownership over their own learning. This case study demonstrates that Action Learning can be implemented on a school-wide basis for the effective professional development of the whole school's teaching and executive staff. Recommendations are made for similar schools who are considering extending Action Learning across the whole school.

INTRODUCTION

Action Learning is a process of learning that enables organisations to adapt to change (Revans, 2011). Action Learning has been accepted within educational and schooling contexts as a form of teacher professional development and professional learning. In Australia, Action Learning has been utilised both frequently and effectively as a platform for improving teacher knowledge, skills and understanding. The case study presented within this chapter highlights how the principles of Action Learning can be implemented, modified, stretched and extended as a way of enabling the effective professional development

DOI: 10.4018/978-1-5225-2642-1.ch009

and improved learning of the whole school's staff. The objectives of this chapter are to: emphasise that the aim of Action Learning is teacher learning, highlight Action Learning as a platform for effective professional development and discuss how Action Learning may be implemented and extended within other similar schools for the effective professional development of the whole school's staff.

BACKGROUND

Action Learning

The purpose of Action Learning is learning: a type of learning, which enables organisations to adapt during times of change. In *The ABC of Action Learning,* Revans[1] (1983: 2011) highlights how important it is for organisations to be able to adapt in times of rapid change. An adaptation, he emphasises, which can only occur through learning: a type of learning he describes as:

Being able to do tomorrow that which might have been unnecessary today, or to be able to do today what was unnecessary last week. (Revans, 2011, p.1).

Revans (2011) argues that this type of learning (L) is a combination of programmed knowledge (P) and questioning insight (Q): thus 'The Learning Equation', L = P + Q. Revans (2011) highlights that for this type of learning, and thus adaptation, to occur questioning insight must compliment programmed knowledge. Questioning insight or the quest for insight, is the ability to ask fresh questions, "to pose increasingly insightful questions from an origin of ignorance, risk and confusion" (Revans, 2011, p.11). Furthermore Revans (2011) explains that action learning aims to: make useful progress on real-world problems; enable managers to learn how to best approach challenges that currently exist without solution, and to encourage teachers and those in management to engineer the conditions in which managers are able to learn from and with each other. Therefore Action Learning becomes a process by which real-life problems are studied and resolved through learning and action.

Action Learning programmes are by very nature, flexible and adjustable (Brook, Pedler & Burgoyne, 2012; Revans, 2011). This is because the design of the programme responds to the individual needs of organisations. Revans (2011, p.17) emphasises this when he says, "the study of real life problems or opportunities by real managers in real time admits to a wide range of programme designs". As such, he intentionally omits logistical details. However, he does provide the essential logistics, of which some include:

- The participation of individuals or teams,
- A focus on posing questions to problems which currently defy solution,
- A set involving four to six fellows or group members who meet together regularly to discuss their progress and setbacks,
- Induction exercises, and
- Four programme phases (diagnosis, six sequential phases {analysis, development, procurement, construction or assembly, application and review}, intermediate invigoration and therapeutic).

To add to this Revans (2011) presents twenty assumptions of Action Learning, those of particular importance to this chapter are that:

- Learning is cradled in the task,
- Problems require Insightful Questions,
- Learning involves doing,
- Learning is voluntary,
- Urgent problems or enticing opportunities provide the spur for learning,
- The risk is imperative,
- The central importance of the set,
- The place of expertise,
- Learning with and from each other, and
- The facilitator role.

These form the basis on which Action Learning programmes are implemented within organisations for the purpose of adapting to change through learning.

Experiential Learning is a theory of learning that underpins Action Learning. Beard & Wilson (2006) describe Experiential Learning as a fundamental and natural process of learning that can occur on an individual basis or as a team. They state that Experiential Learning is "the sense-making process of active engagement between the inner world of the person and the outer world of the environment" (p.19). Adding to this, Dick (1997) draws links between Experiential Learning and the learning component of Action Learning. He describes Experiential Learning as a process in which learning is drawn from experience, or more commonly, as learning from experience. A major component of this learning theory are the ongoing cycles of action and reflection, which can be described as the Experiential Learning Cycle, during which a learner learns by reflecting on their action which causes a change in behaviour and future action (Dick, 1997). Both Experiential Learning and Action Learning find their basis in Constructivism[2], a theory of learning that views the learner as being an active participant in their learning: active, as they create new knowledge by reflecting on what they already know and the ideas or knowledge with which they come into contact with (Agarwal & Gautam, 2011; Fosnot, 1996; Pelech & Pieper, 2010; Richardson, 1997; Yoders, 2014).

Since its origins, Revans' Action Learning (1983) has flourished[3] and as a result has been implemented within a wide range of fields including: healthcare, business, education and government. So much so, that Brook et al., (2012) proclaim, "Action Learning has now wide acceptance (across the world) as a learning and development approach" (p.270). They also discuss the impact these variations have had on Revans (1983) original form and they highlight that it is due to its flexible and adaptable nature, that Action Learning has been able to diverge and develop (Brook et al., 2012). However, they also strongly state that, "attention must be paid to its core essences whilst allowing for growth and experimentation" (Brook et al., 2012, p.272). Such growth and experimentation has enabled Action Learning to be utilised within the field of education both around the world.

Action Learning Within Australian Schools

Action Learning is widely accepted and utilised within Australian educational settings as a form of teacher professional development and professional learning[4]. When implementing Action Learning in schools, the Academic Partner advises following a series of five steps, which are:

1. Identifying a broad professional issue or domain of professional concern
2. Planning the project
3. Implementing an action plan
4. Observing, reflecting, recording, discussing and reflecting upon the action
5. Evaluate and celebrate (Academic Partner, 2017).

On the use of Action Learning in schools for the purpose of teacher professional development, Ewing et al., (2010, p.5) states that:

An Action Learning (Revans, 1983) approach to professional development is based on the belief that professionals working together have the ability to ask meaningful questions and searching questions about their own practice and formulate answers to those questions.

Beveridge & Hinde McLeod (2009) report on an Action Learning project, which aimed to deepen the school communities understanding of Australia's Indigenous culture and history. Six volunteer teachers participated in the project, which resulted in closer ties being developed between the school and Indigenous community groups. Findings also showed that the teachers and students perceived the learning which occurred as part of the project as being significant, and also that the awareness and knowledge of teachers about Australia's Indigenousness people and their culture increased. Similarly, Kilpatrick & Loughland (2012) report on an Action Learning project that focused on the use of Interactive Whiteboards (IWBs) to support literacy learning. This project occurred over three quarters of a school year and involved the participation of two classroom teachers and an academic partner. As a result of this program teachers developed new understanding and pedagogical practices around the use of IWBs within literacy learning, and students expressed an expectation that these quality teacher pedagogies would be used in other content areas. Furthermore Warhurst et al., (2010) reported on an Action Learning project, which was implemented within a suburban primary school. A small group of teachers met with an academic partner in order to seek answers to several questions about what it meant to be a critically literate reader. Findings from the project showed that the participating teachers developed a greater willingness to share with other teachers what was developed from the project, and students developed a more confident approach to their learning, increased their ability to listen to and value contributions made by their peers, gained a deeper understanding of narrative, and developed their creativity and imagination. Findings from these three reports demonstrate the effective and successful way in which Action Learning is being utilised within Australian schools.

The Australian Government Quality Teaching Program (AGQTP), Quality Teaching Action Learning (QTAL) Initiative

An initiative funded by the Australian Government Quality Teaching Program (AGQTP) was the Quality Teaching Action Learning (QTAL) initiative. The QTAL initiative sought to "connect teacher professional learning needs to a comprehensive change process focused on improving student learning" (Ewing et al., 2010, p.4). This saw Action Learning, as this 'comprehensive change process', implemented across both government and independent primary and high school within the state of NSW. The QTAL initiative provided the support and funding for Action Learning to be utilised as a school-based process of teacher professional learning for the purpose of promoting school-based change in teacher pedagogy and school culture (Ewing et al., 2010).

During the four phases of the QTAL, 135 Action Learning projects were implemented into schools or school-clusters, between 2003 & 2007 with a further 57 Action Learning projects being implemented between 2008-2009 (Ewing et al., 2010). The aim of these Action Learning projects was "to strengthen the currency and depth of teachers' pedagogical knowledge and skills and to encourage productive engagement in collegial networks that extend and support knowledge and skills" (Ewing et al., 2010, p.). These projects were characterised by:

- Teams of teachers (usually 6-8),
- Project-based learning,
- Teacher directed learning,
- Cycles of inquiry, discussion and reflection, and
- Learning partnerships (Ewing et al., 2010).

Along with funding from the overarching AGQTP, each Action Learning project was provided with the support of a project officer from the Department of Education and was allocated a university academic partner, who acted as critical friend to the Action Learning team during the project (Aubusson et al., 2005; Dinham et al., 2006; Ewing et al., 2010).

An evaluation of the QTAL initiative took place during its initial stages in July 2005. A team of researchers investigated the "the conditions which influence teachers' implementation of an inquiry-based approach to changing teacher practice" (Aubusson et al., 2005, p.2). Their research spanned 50 Action Learning projects, which had received funding from the AGQTP. Findings from this evaluation demonstrated that the QTAL initiative:

- Promoted collaboration within schools and between schools,
- Created movement around achieving stated outcomes and creating quality teaching resources,
- Fostered reflection,
- Facilitated ongoing change through the action-reflection cycle,
- Cultivated quality teaching,
- Encouraged changes in teaching practice,
- Improved attainment of student learning outcomes, and
- Advanced teacher understanding.

The main recommendation stemming from this evaluation was for the continued support and expansion of Action Learning within schools (Aubusson et al., 2005).

Reports on the individual Action Learning projects demonstrate the success of the QTAL model, however they also highlight the need to modify and stretch the Action Learning process to a whole school level. Beveridge (2010) reports on an Action Learning project called 'The Thinking School Project'. This project focused on increasing staff knowledge about the Quality Teaching Framework (QTF) and creating a shared understanding of the QTF within the school. This project went for two years and consisted of six volunteer teachers and an academic partner. Findings from this study reveal that teachers' knowledge of the QTF increased significantly as a result of participation in the Action Learning project and that the teachers were able to use this knowledge in their programming and lesson delivery. This project however did not continue in its existing form, as a there was concern by teachers about the dichotomy between the participating teachers and the rest of the school staff. This project was therefore expanded the following year to the whole school, with teachers being organised into three Action Learning teams.

Ewing et al. (2010) conducted a meta-analysis of the QTAL initiative after its completion. Their meta-analysis covered all four phases of the initiative from 2003-2009 and 192 individual Action Learning projects. Overall the QTAL programme was found to be a great success. However, success was stronger for the Action Learning projects implemented within the later stages of the QTAL initiative as it was seen that they benefited from the opportunity to learn from the stories and strategies of the earlier projects: a finding which the authors strongly argue should be replicated in similar, future programs. Additionally, the meta-analysis identified certain elements of the QTAL model as being enablers of school-based change. Funded opportunities for release from teaching gave teachers significant time for collaborative professional discussion, during which teachers reflected, investigated and evaluated current teaching practice. The appointment of academic partners to each QTAL project was seen as an important success of the overall initiative as it was found that they brought a voice to the Action Learning projects, which was external to the school. On the longevity of these changes, the authors report:

Sustainability of those changes initiated during the different phases of the QTAL program and how long they will be sustained is a matter for which there is less evidence and more aspiration. (Ewing et al., 2010, p.71)

The following report describes how one NSW government primary school, Western Public School[5] (WPS) implemented, modified, stretched and extended the principles of Action Learning, to develop a more empowering professional learning culture within their school.

ACTION LEARNING FOR TEACHER LEARNING AND EFFECTIVE PROFESSIONAL DEVELOPMENT

Western Public School

Western Public School (WPS) is a P-6 (Preschool-Year 6), NSW government school located within a metropolitan area in NSW. The school's student body is comprised of 371 students, made up of up of 21% of Aboriginal, 32% of students who speak English as an Additional Language/Dialect (EAL/D), a two-unit preschool and 4 support classes for students with intellectual disabilities. The school has an Index of Community Socio-Educational Advantage (ICSEA) of 850, which, when compared to the national average of 1000, demonstrates the level of educational disadvantage that these students come to school with (ACARA, 2015). As a result of these complex issues facing the school on a daily basis, the school has made the professional learning of its teachers a high priority since 2004.

A Small-Scale, More Traditional Action Learning Project

In 2005, a small-scale, more traditional Action Learning project was implemented within the school. This Action Learning project was funded by the AGQTP QTAL initiative and involved the ongoing support and mentoring of an academic partner. During an Australian literacy conference a leading literacy expert was approached and asked to come on-board as academic partner to the school. This small-scale, more traditional Action Learning projected aimed to increase student engagement during the Literacy Block: A time which had been identified as an unhappy time for teachers and students, a time in which teachers continually fought students for control (Academic Partner, personal communication, Sep 7, 2016).

This AGQTP QTAL project became known as "The Engagement Project". Four teachers volunteered to be involved in this small-scale Action Learning project. During a period of ten weeks the Academic Partner introduced the teachers and their students to project-based literacy learning. Using this form of Constructivist learning the students were able to select a pop-culture song/video clip, which they redesigned and presented within a CD cover that they also designed themselves: their spelling, comprehension and grammar were integrated into this pop-culture inquiry-based project. At the end of the ten weeks the students, parents and teachers of these four classes celebrated their work by watching each group of students, 3-4 minute video clips (Academic Partner, personal communication, Sep 7, 2016).

This more-traditional Action Learning project resulted in: an increase in teacher knowledge, increased student engagement during the Literacy Block, and the substantial promotion of teacher professional learning within the team members who participated (Flick, 2011). However soon after the completion of this project the four participating teachers left the school for positions within other schools, and thereby taking their learning and the impetus created by this Action Learning project with them. As such there was little-to-no spill over of learning to the rest of the school's staff (Academic Partner, personal communication, Sep 7, 2016). This result is similar to findings of Beveridge (2010) who also indicated that little spill over occurred to the rest of the staff as a result of a more traditional Action Learning project.

At this time there was a change of leadership within the school and the new and incoming executive were motivated by this lack of spill-over of learning to the rest of the staff, to ask the Academic Partner whether the principles of Action Learning could be modified and stretched to a whole school level (Flick, 2011). This is an example of what Revans describes as Questioning Insight and as a result of this Action Learning was modified and stretched to include the whole school's teaching, support and executive staff. The Academic Partner volunteered to stay on in this role: offering his expertise, time and knowledge and understanding of the Action Learning process to support the school as they stretched the principles of Action Learning across the whole school.

A Whole School Action Learning Project

In 2009, a Whole School Action Learning Project was implemented and the principles of Action Learning were then stretched to involve the mandatory participation of the whole school's teaching, support and executive staff. The aim of this Action Learning project was to bring about a change in the school ethos through long-term engagement in whole school professional development (Flick, 2011). In order to achieve this the Whole School Action Learning Project sought to: create an infrastructure within the school that would enable the entire teaching staff to meet together, to receive input, to engage in professional development activities focused on literacy, and to enable teachers to translate the knowledge and skills acquired during the Action Learning process into effective practice during their Literacy Block's. This Action Learning project sought to answer the unresolved problem that the staff were asking, "How can we restructure our Literacy Blocks to create a school wide culture of effective and avid reading behaviour?" (Flick, 2011): this is an example of Questioning Insight.

Topics addressed during the initial two-year period of the Whole School Action Learning Project included:

- Cambourne's Conditions of Learning,
- Developing a common language for talking about classroom practice,
- Teaching 'episodes',
- S.T.A.R.T comprehension strategy,
- Miscue analysis,
- Developing a pool of Teaching Learning Activities (TLA's),
- Rich Task as a strategy for mean-making, and
- A book study of 'The Book Whisperer' by Donald Miller (Flick, 2011).

The structure of the WSALP included:

- **Input Days:** these half-day opportunities for whole school professional development occurred twice a term (once every five weeks) and involved the whole teaching staff and support staff meeting together with the Academic Partner. The aims of the Input Days were: to enable teachers to receive pedagogical and theoretical information, to encourage teachers to trial, use and assess pedagogical practices in their classrooms, and to promote personal, professional reflection by the whole staff. Each Input Day was structured as such: the Academic Partner established the agenda for the day, then the content and learning of the previous session was revisited and reviewed, the Academic Partner then presented the teachers with a range of resources, which included: fact sheets, summaries, activities and response sheets, which they read and discussed collaboratively. Between each of these Input Days teachers implemented their new learning and teaching strategies into their classroom, reflected on this process, and recorded these reflections into a diary.
- **Stage Team Meetings:** These teams consisted of a Stage Team Leader and 3-4 teachers (or support staff) from the same stage/grade level, each member of the teaching and support staff were a part of a team. A Stage Team Leader facilitated these meetings every Tuesday afternoon, after the students had left school for the day. The Stage Team Leaders used a 'Stage Leader Guide' to run the meetings and deliver the content to their team. These guides provided the Stage Team Leaders with: content areas to focus discussion on, further questions to explore in-depth, and further ideas to implement within each teachers' classroom.
- **Academic Partner:** The Academic Partner, initially self-appointed to the school during the AGQTP QTAL initiative, continued to act as academic partner during the stretching of this Action Learning project to the whole school level. During this Action Learning project, he was engaged in: planning and facilitating the Input Days, deciding on the content of these days, designing or locating relevant resources, designing the Stage Team Leader Guides and identifying possible problems within the school be addressed by the Whole School Action Learning Project. As well as this the Academic Partner at WPS provided opportunities for teachers to: discover links between their own teaching philosophies, theory and practice, to share, reflect on and clarify their own teaching and learning philosophy, and to learn about other teaching and learning philosophies (Flick, 2011).

For the purpose of her Honour's Thesis, Flick (2011) conducted an evaluation into the Whole School Action Learning Project in 2011. Flick (2011) found that the Action Learning process used during this Action Learning project resulted in: quality input from the academic partner being received by the teachers, time to engage in professional dialogue with colleagues, and that this discussion supported and promoted teachers to implement change in pedagogy within their classrooms. She also noted how the modification and stretching of Action Learning to the whole school enabled the development of: positive collegial relationships among the whole teaching and support staff, the development of a common language and understanding amongst the whole school's staff, and the construction of shared vision between the staff. Flick (2011) also discovered that the strengths of the Whole School Action Learning Project were: the expertise and quality of the academic partner, the trust established between the academic partner and the school executive, opportunities for collaboration between staff, the quality leadership of the school, and opportunities to celebrate success, which took the form of presenting their learning at national conferences.

Overall, the Whole School Action Learning Project resulted in the professional development of the whole staff at WPS (Flick, 2011). Evidence of this can be seen through:

- The development of a common language for discussing classroom practice among staff,
- An increase in teachers literacy knowledge and practice,
- The development of a new and engaging literacy culture among the school,
- The development of a pool of quality literacy teaching and learning resources,
- The implementation of a range of new literacy and learning strategies within the school and reflection on their success,
- Teachers' feelings of empowerment to improve their pedagogy and practice,
- Learning and understanding by the teachers, and
- The growth of the language and vocabulary of teachers and students.

Successful teacher professional development can be clearly identified as being the main outcome, which occurred as a result of stretching the Action Learning process to a whole school level.

Several features of the Whole School Action Learning Project were found to inhibit these positive outcomes. Flick (2011) identified the most prominent hindrance was the demand the Action Learning project placed on teachers time. Teachers found that the length of time allocated to this Whole School Action Learning Project was very demanding. Related to the theme of the demands on teachers' time were the consequences that occurred as teachers were released from face-to-face teaching to attend the Input Days, and their class being allocated a substitute teacher. This release from face-to-face teaching time was described by some as being disrupting and distracting for the students who, due to the inherent nature of the school, required consistent routine and structure (Flick, 2011). In addition to this the teachers reported that during their absence from class, student behaviour would become challenging and disruptive and that their professional development opportunities were occasionally disrupted as they were called back to their classrooms to deal with student behaviour (Flick, 2011). Furthermore, teachers reported that even if they weren't called back to class, they would be constantly worrying about how their students were behaving for their substitute teacher and about having to deal with student behaviour once they returned to their classrooms (Flick, 2011). The hindrance of releasing teachers from face-to-face teaching time demonstrates the inherent difficulties in implementing professional development experiences within a school. Other inhibitors included: repetition of content, delivery methods used during the Input Days and staff turnover which resulted in the new, incoming staff needing to be updated with the professional learning that had occurred as a result of the Whole School Action Learning Project.

In her evaluation of this Action Learning project, Flick (2011) recommended that for schools wishing to stretch and modify Action Learning across the whole school, they address the issues of: the quality and expertise of the academic partner, the strength of leadership within the school, the ability of this leadership to provide access to and distribute resources in such a way that would enable participation in these professional development opportunities. Specific to WSP, Flick (2011) recommended that the teachers be given: more time for free-flowing professional collaborative discussions, more reading time, grouping of teachers in such a way that enabled new ideas to be fostered and teachers to be pushed further in their thinking. These recommendations display the complexity of teacher professional development (Opfer & Pedder, 2011).

A 12-Week Staff-Directed Action Learning Project

In 2013, the staff of WPS and the Academic Partner attended a national literacy conference in Sydney, in which, well-known Australian author and literacy advocate, Mem Fox (2013), encouraged the notion of teachers' writing themselves as a way of supporting student writing in the classroom (Academic Partner, personal communication, Sep 8, 2016). Primarily it was this call to write, combined with a number of other indicators that motivated the teachers to ask the Academic Partner if the focus of their next Action Learning project could be developing themselves as writers and as teachers of writers. These other indicators included:

- **Assessment Results:** Standardised test results demonstrated that the gap between students writing achievement widened as they went through the primary school years.
- **The Need to Develop an Avid Writing Culture:** The executive staff began to notice that while the focus had been on developing an avid reading culture, the surface had only been skimmed in terms of developing an avid writing culture.
- **Teacher Anxiety:** In the beginning of the project some staff members displayed an initial nervousness about the process of pursuing their own writing. This nervousness decreased when the teachers were put into groups to write as they were supported by their group and got swept up in the excitement with their colleagues.
- **Timing of the Project:** In September, 2013 the staff at WPS expressed to the Academic Partner that they would prefer that the project be completed prior to the end of the school year and so they made a strong commitment to the project.
- **Literature:** The Academic Partner explained to the executive a link in literature between the development of an avid writing culture and teachers own ability to write (Academic Partner, personal communication, July 26, 2016). He commented that:

A common underlying cause of the failure to develop an avid writing culture, namely, because of the way they (teachers) were taught to write, very few teachers consider themselves effective or 'avid' readers, nor do they enjoy the act of writing. (Academic Partner, personal communication, July 26, 2016)

The staff decided that this Action Learning project was organised to run for the final 12 weeks of the school year. During this Action Learning project the teachers would work together in small groups in order to construct a text, which would be presented, to the rest of the staff at the end of the project. The Action Learning project became known as 'The Text Factor' and was structured with:

- **An Initial Input Day:** During this full day session that occurred at the beginning of Term 4, the Academic Partner and selected facilitating staff would receive an overview of the Action Learning project.
- **A Whole-School Meeting:** During which the rest of the staff were introduced to the project, given a timeline and asked to establish their own groups, in which they were to brainstorm possible ideas for writing and set up parameters for their group to work within.
- **Between Session Tasks:** The Academic Partner set these tasks, which focused on writing improvement. Resources used in these tasks included: summarised readings, handouts, news articles, writing guides and relevant video clips. The staff were expected to complete these tasks between their scheduled group meetings.
- **Regular Group Meetings:** During a period of five weeks the groups would meet regularly to work on their piece during which they would craft, argue, debate and reflect on the Between Session Tasks. During these meetings, one member of each team would take notes on the process they were undertaking in order to write their text.
- **A Debrief Day:** This whole day event occurred at the end of the school year and involved the facilitating teachers, the executive staff and the Academic Partner, debriefing their experience of the project.
- **A Staff Development Day:** This day occurred at the end of the school year, after the students had gone on holidays. On this day groups presented and shared their work and discussed what they learnt from this Action Learning project and reflected on what the process meant to the students in their classroom (Academic Partner, personal communication, July 26, 2016).

On completion of the project all the published texts were compiled and presented into an Anthology of Teacher Writing. The texts covered a range of text types and topics, including:

- A parody on a traditional fairy-tale,
- Two short stories,
- Two play scripts, and
- An information text about what the team wished they had known before they began to teach.

According to the Academic Partner the main outcome of this Action Learning project was the teacher learning which occurred around the process of writing (Academic Partner, personal communication, July 26, 2016).

A Significant Turning Point

Taking Ownership of Their Professional Learning

Both the Academic Partner and the executive staff regard this 12-Week Whole School Action Learning Project as being a significant turning point. Historically the focus and form of each Action Learning project was decided by Academic Partner, however during this 12- Week Action Learning project on teacher writing, the staff at WPS began taking ownership over their own professional learning by taking responsibility for deciding on the focus and form that this Action Learning project would take. The school executive were responsible for identifying the need to focus on teacher writing in this Action

Learning project and the staff committed to the timing and duration of this Action Learning project. As well as this the made the decisions to work in groups and to have the teachers select their own groups and the genre and topic of their writing (Academic Partner, personal communication, July 26, 2016).

Changing Roles

Not only did this 12-Week Whole School Action Learning Project on Writing mark a subtle change in ownership and responsibility of the professional development, this time also marked a considerable change in the role of the Academic Partner to the school. Initially the Academic Partner's role was to identify the professional development needs of the school by identifying the problems and issues that they were facing and then plan how this problem could be addressed. He comments:

As academic partner it was me ('the expert') who was controlling and making suggestions about the actual content of workshops and/or inquiries the team leaders would implement. I would invite the executive and team leaders to make explicit their worries, issues and problems, (which they did) and then they would wait for me to identify and give name to their problems and issues, and then design some workshop, inquiry or task for them to apply to their groups. (Academic Partner, personal communication, May 4, 2016)

He reports how his role subtly changed during this 12-Week Action Learning Project on Writing as the staff began to take responsibility for the form and focus of their own professional development and he would then assist them in these decisions. Of this change in role he comments:

I help them stay true to the principles of Action Learning, and how to record, evaluate and interpret, what I call, the professional meaning-making that results. (Academic Partner, personal communication, May 4, 2016)

The Academic Partner is very supportive of this change in role as it links to his own Constructivist philosophy. On this he states:

My own learning theory, ideology, and values kept telling me they would be much more empowered if they had more control over making decisions about the content and nature of the professional learning they wanted to undertake. (Academic Partner, personal communication, May 4, 2016)

An Equal Partnership

After the completion of this project the school executive and staff continued to take responsibility for their own professional development and the Academic Partner continued to support and assist them in this, in what could be described as an equal partnership. Early the next year, the executive staff contacted the Academic Partner and suggested:

We want to learn about Constructivism and we want to do it in a way in which we are in control, where we make the decisions. (Academic Partner, personal communication, July 26, 2016)

Since this 12-Week Action Learning Project on Writing, the staff at WPS have continued to take ownership of and responsibility for their professional learning and professional development. To this the Academic Partner responds:

That little 12-week Action Learning project was one in which a lot of the values and beliefs that you have now about professional learning started to emerge as a consequence of the project. (Academic Partner, personal communication, Sep 8, 2016)

ACTION LEARNING FOR EFFECTIVE PROFESSIONAL DEVELOPEMENT

Current literature has much to say about what makes professional development effective. Professional development refers to:

The vast range of activities and interactions that may increase (teachers) knowledge and skills, and improve their teaching practice, as well as contribute to their personal, social, and emotional growth as teachers. (Desimone, 2009, p.182)

Desimone (2009) and Darling-Hammond et al. (2009) identify the components of effective professional development as being: content-focused, active learning, coherence, duration and collective participation. Other researchers, such as Guskey (1991), Mansour et al., (2014), indicated a number of other key features of effective professional development, these being: the role of change, personal feedback and continued support, occurring within the teachers' context, and considering the needs of participating teachers. Empirical studies also highlight features that contribute to the effectiveness professional development (see - Miller et al., 2015; Owens, 2015; Qablan et al., 2015; Smith, 2015).

The staff at WPS are being offered professional development which demonstrates these components. This has occurred as a result of: their long and ongoing association with Action Learning, the way in which Action Learning was modified, stretched and extended during this time, and in they way they were enabled to take responsibility of the form and focus of their own professional development. Some of the components or features of effective professional development which are demonstrated within this school are: collaboration, active participation, sufficient duration, differentiated and situated, content focus and mentoring.

Collaboration

'Collaboration' refers to the "disciplined and purposeful approach to collaborating to solve the challenges that are most important to improving student outcomes" (Australian Institute for Teaching and School Leadership, 2012, p.5). In a study conducted by Mansour et al., (2014), teachers explain how collaboration was effective in enabling them to implement the content of the professional development into their classroom, and also in enabling them to develop new teaching ideas and resources. Smith (2015) also found that the teachers in his study viewed collaboration as an important way of combating teacher isolation, and as a way of providing the support they needed to develop as learners. Furthermore, in a study conducted in America teachers agreed that collaboration enabled them plan, review and reflect on their learning, as well as enabling them to benefit from the expertise of other teachers. The teachers in this study all agreed that collaboration was one of the most important components of 'effective' professional development (Quick et al., 2009). Collaboration occurs both formally and informally within the Whole School Action Learning Project and the Whole School Action Learning Project and enables the teachers to: develop a common language for discussing classroom practice among staff, establish positive collegial relationships, and have opportunities for in-depth discussion and reflection on both the content and their implementation of new content and teaching strategies in their classrooms.

Active Participation

'Active Participation' refers to teachers being active participators in their professional development: this can occur through the teachers being involved in such activities as: observing, engaging in discussion, planning for implementation or reviewing student work. Wilkie & Clarke (2015) found that when teachers actively participated in analysis of student work and collaborative debriefs and discussions, their content and pedagogical knowledge increased. Similarly Smith (2015) found that when teachers were actively engaged in reflective practice, including regular feedback, their practice changed, which resulted in an increase in students' attitudes towards the subject area. Furthermore, Garet et al., (2001) found that professional development which features active learning relates to the enhanced knowledge and skills of teachers. Within the professional development at WPS teachers were actively involved in: implementing new teaching strategies and learning into the classroom and reflecting upon this implementation, engaging in collegial discussions, applying new knowledge, reflecting on their learning from Input Days, constructing texts and presenting at conferences and Staff Development Days.

'Active Participation' can also be seen at WPS in the form of control over their own professional learning. Since the 12-week Whole School Action Learning Project on Writing, the teachers are actively involved in making decisions about the structure and content of their professional development. According to Stanley et al., (2013) active learning may also take on the form of teacher choice, in which teachers are able to choose the content and style of delivery of the professional development. And Hammel (2007) suggests that teachers can be given control over their own learning by choosing the goals and objectives they believe are appropriate. Teacher control of their own learning is one of the major findings stemming from the history with and ongoing use of Action Learning at WPS and demonstrates one of the features of effective professional development.

Sufficient Duration

'Sufficient Duration' refers to both the time spent in a professional development program and the overall length of the program (Darling-Hammond et al., 2009). Research shows that when teachers are able to experience professional development that is of a sufficient duration, the following results are likely to occur: teacher practice is more likely to change or improve, as improvement is seen as a continuous and ongoing process (Darling-Hammond, et al., 2009; Desimone, 2009; Guskey, 1991, p. 244; Qablan et al., 2015; Whitworth & Chiu, 2015), they will be more likely to successfully implement the strategies, knowledge or skills taught within the professional development (Guskey, 2010; Smith, 2015) and there will also be an increase in their learning, skills (Bayar, 2014) and in their motivation and commitment, as they view the 'bigger picture' of their professional development (Hunzicker, 2010). The ongoing whole-school application of Action Learning enables teachers to be involved in professional development that is of a sufficient duration, both in terms of time within the professional development and the length of the program.

Differentiated and Situated

'Differentiated and situated' refers to professional development which is differentiated to meet the individual needs of teachers and situated within their teaching context. Hunzicker (2010) reports that by being embed in context of teachers, professional development is relevant and authentic, and can be seamlessly integrated into teachers practice. Similarly, in their study, Mansour et al., (2014) report that teachers preferred professional development programs which were situated at school and which directly meet the needs of the teachers (Mansour et al., 2014). Both the Whole School Action Learning Project and the 12-Week Whole School Action Learning Project on Writing were differentiated according to the professional development needs of the staff at WSP and situated within the school itself and directly met the needs of the teaching staff.

Content Focus

'Content Focus' refers to focusing on both knowledge of subject matter and knowledge of how students learn content (Garet et al., 2001). Past research has linked professional development with a content focus with increases in teacher knowledge and skills and improved teacher practice (Desimone, 2009; Whitworth & Chiu, 2015). Miller et al., (2015) support this, as they report on the positive effect the content-focused, physical activity professional development program had on developing students' fundamental movement skills and increasing the Year 6 students' in-class physical activity. Likewise, Smith (2015) reported that the content-focused professional development program implemented in the West of Ireland was effective in changing teacher practice and an increase in students' attitudes. The content of the professional development was relevant to the problems experienced at WPS and focused on both the subject matter and knowledge of how students learn that content.

Mentoring

'Mentoring' refers to a type of continued support structure, for example: classroom visits, engagement with a critical friend or in-school study groups. Empirical research also supports the use of mentoring within professional development. Quick et al., (2009) found that "participation in coaching/mentoring sessions was positively associated with increased amounts of instruction around higher-level meaning of text" (p.62). In a quantitative study on the effects of a physical activity professional development program, Miller et al., (2015) discovered that mentoring enabled teachers to understand the structure and purpose of the lessons, and to be active in implementing, observing and troubleshooting the content of the program within their teaching context. Evidence for the benefits of mentoring can also be seen in Owens' (2015) research. She links mentoring to the ability of teachers to share Indigenous intellectual knowledge and culture and explains how teachers were empowered to initiate small step changes in their teaching practice through mentoring (Owens, 2015). Mentoring within the professional development at WPS occurred through the partnership between the Academic Partner and the school itself and through the relationship, which the Stage Team leaders purposively established with the teams they lead.

Coherent

'Coherent' refers to the degree that the professional development teachers participate in aligns with school, state and national educational goals, policies and initiatives (Quick et al., 2009; Surrette & Johnson, 2015; Whitworth & Chiu, 2015). Empirical research demonstrates that professional development, which is coherent with larger educational goals, increases teacher quality and student outcomes. The professional development program studied by Smith (2015) demonstrated coherence and connection through an alignment of the program content with syllabus content. Coherence was furthermore achieved through the linking of specific pedagogy with state goals, as well as with learning theories. As a result of this program, teacher practice was developed and student attitudes towards the subject content improved (Smith, 2015). The professional development reported on in this chapter demonstrates coherence to: the Quality Teaching model, the English K-6 Syllabus, the goals of the school, Department of Education requirements and the state goals of teacher professional development.

The Complexity of Professional Development

Opfer & Pedder (2011) argue for a complex conceptualisation of professional development, which recognises that the professional development cannot be viewed in terms of cause and effect, nor the form of professional development as a predictor of teacher learning. They reason that there are various dynamics, mechanisms and systems at work, and that the conditions in which teachers learn must be acknowledge, so as to be able to "understand under what conditions, why, and how teachers learn" (Opfer & Pedder, 2011, p.378).

It is not this author's intent to compare the form and feature of the professional development at Western Public School to its success. Only to present that through their long and ongoing association with Action Learning, this particular school has been able to embed within its structures and processes features identified by literature as those which enable effective professional development.

ISSUES, CONTROVERSIES, AND PROBLEMS

Teacher Learning

Action Learning for the purpose of teacher learning is one of the main issues emerging out of the use of Action Learning as a form of teacher professional development at WPS. The focus on learning is supported by Revans (2011) who emphasises that learning (L = P + Q) as that which enables organisations to adapt to change. In addition to this, the Academic Partner supports this view of learning as being the focus of Action Learning projects, when he states:

One of the problems I faced in getting Action Learning to work at WPS was to keep reminding them that the core of Action Learning was teacher learning, and NOT student learning. (Academic Partner, personal communication, May 4, 2016)

Voluntary Participation vs. Mandatory Participation

As Action Learning was modified and stretched at Western Public School, the participation of teachers changed from being voluntary to being mandatory in nature. While it is widely recognised that a feature of Revans' original Action Learning program was voluntary participation, in order to 'work' on a whole school basis, teachers weren't given this allowance. While this may have been an issue early on, once a culture of Action Learning and teacher professional learning was established within the school, teachers were more than willing to be involved. This suggests that, while there may be some initial teething issues when implementing Action Learning on a whole school basis, the benefits of this in the long-term far outweigh these initial challenges.

SOLUTIONS AND RECOMMENDATIONS

Applying Action Learning to the Whole School

In 2009, Action Learning at Western Public School was extended to include the whole school's teaching, support and executive staff. While research demonstrates that the use of small-scale, more traditional Action Learning projects is both widespread and frequent, there are fewer case studies like this one, which describe such a long, ongoing association with and adaptation of Action Learning for the purpose of teacher learning. This author would like to recommend Action Learning be utilised on a school-wide basis for the professional development of a whole school's staff. Furthermore, the author would like to suggest that the challenges and issues faced during this process are to be welcomed as an opportunity for further reflection, action and learning – such is the very nature of Action Learning.

The Support of an Academic Partner

To endeavour to implement Action Learning on a school-wide basis requires support. The partnership between this particular school and the Academic Partner clearly demonstrates how important it is to have this kind of support, facilitation and guidance in initiating and continuing this process. With this in mind, it must be stated, that the success or otherwise of an Action Learning should not be dependent on either the academic partner or the school's relationship with that person. Clearly, time can be seen as an important factor in the development of a working relationship between the school and the academic partner.

Teacher-Directed Action Learning

The executive and school staff began taking on the responsibility for their own professional learning during the 12-Week Whole School Action Learning Project on Writing. To this day, the staff continue to take the responsibility for and direct their own professional learning. This author suggests that not only do other Action Learning projects enable teachers and executive staff to direct their own learning, but any attempts at professional development do also. This should be a consideration for any other school or organisation who currently utilise, or who are planning on implementing or extending the Action Learning process.

FUTURE RESEARCH DIRECTIONS

Schools, like many organisations and businesses, are constantly faced by change. For any school, this may include: changes in staff, changes in school leadership, changes in curriculum, changes in policies and programs implemented within the school, changes in practices, or changes within the demographic of the school itself. As mentioned in the Background section of this chapter, Action Learning is a process that enables, through learning, organisations to be able to adapt to this type of change. As such, this author recommends that, the ongoing research into Action Learning as a form of school-based professional development should continue. It is also imperative that future research examines the value and merit of implementing Action Learning on a school-wide basis and that case studies, such as this one, continue to be presented, examined and discussed in literature.

CONCLUSION

In presenting the case study of how Western Public School has implemented, modified, stretched and extended the principles of Action Learning over the course of a decade, it is hoped that more schools, similar to this one, will take up on the challenge of engaging teachers in professional development on a school-wide basis.

ACKNOWLEDGMENT

The author of this chapter would like to acknowledge the contribution and assistance of Dr Brian Cambourne for both his expertise and support in constructing this chapter.

REFERENCES

Academic Partner. (2017). *A teacher friendly summary of Action Learning principles*. Unpublished report.

ACARA. (2015, Aug 5). *MySchool*. Retrieved from: https://www.myschool.edu.au/

Agarwal, R., & Gautam, A. (2011). Effect of Constructivist methods of teaching on academic achievement of primary school students. *GYANODAYA*: *The Journal of Progressive Education*, *4*(1).

AITSL. (2012). *Australian charter for the professional learning of teachers and school leaders*. Australian Institute for Teaching and School Leadership. Retrieved from: http://www.aitsl.edu.au/docs/default-source/default-document library/australian_charter_for_the_professional_learning_of_teachers_and_school_leaders

Aubusson, P., Brady, L., & Dinham, S. (2005). *Action learning: what works*. NSW Department of Education and Training. Retrieved from: http://sydney.edu.au/education_social_work/professional_learning/resources/papers/AGQTP_reports/2005_Action_Learning-What_Works.pdf

Bayar, A. (2014). The Components of Effective Professional Development Activities in Terms of Teachers' Perspective. *Online Submission*, *6*(2), 319–327.

Beard, C. M., & Wilson, J. P. (2006). *Experiential learning: A best practice handbook for educators and trainers*. London, UK: Kogan Page Publishers.

Beveridge, L. (2010). Implementing pedagogical reform through action learning: Emerging issues from the local experience. *ALAR: Action Learning and Action Research Journal*, *16*(1), 39.

Beveridge, L., & Hinde McLeod, J. (2009). Action learning through Indigenous literature. *Intercultural Education*, *20*(2), 187–197. doi:10.1080/14675980902922218

Boshyk, Y., & Dilworth, R. (2010). *Action learning: History and evolution*. Springer. doi:10.1057/9780230250734

Brook, C., Pedler, M., & Burgoyne, J. (2012). Some debates and challenges in the literature on action learning: The state of the art since Revans. *Human Resource Development International*, *15*(3), 269–282. doi:10.1080/13678868.2012.687926

Darling-Hammond, L., Wei, R. C., Andree, A., Richardson, N., & Orphanos, S. (2009). State of the Profession: Study Measures Status of Professional Development. *Journal of Staff Development*, *30*(2), 42.

Desimone, L. M. (2009). Improving impact studies of teachers professional development: Toward better conceptualizations and measures. *Educational Researcher*, *38*(3), 181–199. doi:10.3102/0013189X08331140

Dick, B. (1997). *Action Learning and Action Research*. Retrieved from: http://www.aral.com.au/resources/actlearn.html

Dinham, S., Aubusson, P., & Brady, L. (2006). *Distributed leadership through action learning, keynote address*. Paper presented at the Fifth International Conference on Educational Leadership, Australian Centre for Educational Leadership, University of Wollongong, NSW.

Ewing, R., Groundwater-Smith, S., Mockler, N., Loughland, T., Simpson, A., Smith, D., & Books, D. et al. (2010). *Meta analysis of quality teaching action learning project*. Sydney, NSW: University of Sydney.

Flick, E. (2011). *Creating a Culture of Avid Readers: The Western School Literacy Project* (Unpublished honors dissertation). University of Wollongong, NSW, Australia.

Fosnot, C. T. (1996). *Constructivism: Theory, Perspectives, and Practice*. New York, NY: Teachers College Press.

Garet, M. S., Porter, A. C., Desimone, L., Birman, B. F., & Yoon, K. S. (2001). What makes professional development effective? Results from a national sample of teachers. *American Educational Research Journal*, *38*(4), 915–945. doi:10.3102/00028312038004915

Guskey, T. R. (1991). Enhancing the effectiveness of professional development programs. *Journal of Educational & Psychological Consultation*, *2*(3), 239–247. doi:10.1207/s1532768xjepc0203_3

Hammel, A. M. (2007). Professional development research in general education. *Journal of Music Teacher Education*, *17*(1), 22–32. doi:10.1177/10570837070170010106

Hunzicker, J. (2010). Effective professional development for teachers: A checklist. *Professional Development in Education, 37*(2), 177-179.

Kilpatrick, L., & Loughland, T. (2012). Critical literacy and interactivity year 1 in a classroom. *Scan: The Journal for Educators, 31*(3).

Mansour, N., Heba, E.-D., Alshamrani, S., & Aldahmash, A. (2014). Rethinking the theory and practice of continuing professional development: Science teachers perspectives. *Research in Science Education, 44*(6), 949–973. doi:10.1007/s11165-014-9409-y

McGill, I., & Beaty, L. (2001). *Action Learning: a guide for professional, management & educational development.* Psychology Press.

Miller, A., Christensen, E. M., Eather, N., Sproule, J., Annis-Brown, L., & Lubans, D. R. (2015). The PLUNGE randomized controlled trial: Evaluation of a games-based physical activity professional learning program in primary school physical education. *Preventive Medicine, 74,* 1–8. doi:10.1016/j.ypmed.2015.02.002 PMID:25668220

Opfer, V. D., & Pedder, D. (2011). Conceptualizing teacher professional learning. *Review of Educational Research, 81*(3), 376–407. doi:10.3102/0034654311413609

Owens, K. (2015). Changing the teaching of mathematics for improved Indigenous education in a rural Australian city. *Journal of Mathematics Teacher Education, 18*(1), 53–78. doi:10.1007/s10857-014-9271-x

Pelech, J., & Pieper, G. W. (2010). *The comprehensive handbook of constructivist teaching: From theory to practice.* Charlotte, NC: Information Age Pub.

Qablan, A., Mansour, N., Alshamrani, S., Aldahmash, A., & Sabbah, S. (2015). Ensuring effective impact of continuing professional development: Saudi science teachers' perspective. *Eurasia Journal of Mathematics, Science & Technology Education, 11*(3), 619–631.

Quick, H. E., Holtzman, D. J., & Chaney, K. R. (2009). Professional development and instructional practice: Conceptions and evidence of effectiveness. *Journal of Education for Students Placed at Risk, 14*(1), 45–71. doi:10.1080/10824660802715429

Revans, R. W. (2011). ABC of action learning. Farnham, UK: Gower.

Richardson, V. (1997). *Constructivist teacher education: Building new understandings.* Bristol, PA: Psychology Press.

Smith, G. (2015). The Impact of a Professional Development Programme on Primary Teachers Classroom Practice and Pupils Attitudes to Science. *Research in Science Education, 45*(2), 215–239. doi:10.1007/s11165-014-9420-3

Stanley, A. M., Snell, A., & Edgar, S. (2013). Collaboration as effective music professional development: Success stories from the field. *Journal of Music Teacher Education.*

Surrette, T. N., & Johnson, C. C. (2015). Assessing the Ability of an Online Environment to Facilitate the Critical Features of Teacher Professional Development. *School Science and Mathematics, 115*(6), 260–270. doi:10.1111/ssm.12132

Warhurst, J., Crawford, K., Ireland, J., Neale, D., Pickering, J., Rathmell, C., Watson, G., & Ewing, R. (2010). Improving literacy pedagogy and outcomes through teaching imaginatively. *Practically Primary, 15*(3).

Whitworth, B. A., & Chiu, J. L. (2015). Professional development and teacher change: The missing leadership link. *Journal of Science Teacher Education*, *26*(2), 121–137. doi:10.1007/s10972-014-9411-2

Wilkie, K. J., & Clarke, D. (2015). Pathways to professional growth: Investigating upper primary school teachers' perspectives on learning to teach algebra. *Australian Journal of Teacher Education*, *40*(4), 6.

Yoders, S. (2014). Constructivism Theory and Use from 21st Century Perspective. *Journal of Applied Learning Technology*, *4*(3).

KEY TERMS AND DEFINITIONS

Literacy Block: A period of time (about 2 hours) set aside each day for the teaching of: reading, writing, speaking and listening.

ENDNOTES

1 The author wishes to acknowledge the origins of Action Learning as being in the work of Professor Reg Revans in the early 1980's. The description of Action Learning above has been intentional as pertaining to Revans' original description. Further referencing has been used later on in describing how Action Learning has flourished since its origins.

2 As do Action Research, Inquiry-based learning, Problem-based learning, Professional Learning Communities and Communities of Practice.

3 For further treatments of Action Learning see McGill & Beaty (1995) and Boshyk & Dilworth (2010).

4 The terms, professional development and professional learning, while often differentiated, will be used in this chapter interchangeably.

5 Pseudonyms are used through this chapter to de-identify individual participants and the school.

Chapter 10
Volunteer Involvement in an Academic Camp in the Dominican Republic:
An Examination of Participation in a CBPR Project

Karie Jo Peralta
University of Toledo, USA

ABSTRACT

This chapter examines the expansion of volunteer participation in a participatory research project, which explored possibilities to enhance volunteer involvement in an academic camp operated by a non-governmental organization in the Dominican Republic. Although there are numerous studies that identify how organizations should involve volunteers, these studies typically represent an outsider perspective. Therefore, the concerns, ideas, and agency of volunteers—all of which may facilitate improvements in their participation—are often overlooked. For the purpose of this chapter, volunteers are elevated as key actors in the development of the organization. Specifically, community-based philosophy's concept of participation is applied to identify key moments in the project when the broadening of volunteer involvement was important for achieving a greater understanding of the issue of interest. The objective is to advance an appreciation of how projects with volunteer participation, albeit in fluctuating levels, can result in the co-creation and elevation of local knowledge.

INTRODUCTION

Community-based projects are important for addressing various educational issues in order to increase the life-chance opportunities of marginalized populations (see, e.g., Irby, Mawhinney, & Thomas, 2013; Johnson-Burel, Drame, & Frattura, 2014). These endeavors value community members as central figures in planning and implementing strategies for change (Israel et al., 1998). What underpins this appreciation is the belief that these initiatives should be grounded in the views of the individuals affected by them

DOI: 10.4018/978-1-5225-2642-1.ch010

(Minkler & Wallerstein, 2005). Participation is thus a key element needed to promote projects that stem from a community's knowledge and interests (Murphy, 2014).

Community-based Participatory Research (CBPR) is a form of action research that emphasizes the participation of community members (Israel et al, 2003). According to McNiff and Whitehead (2002), "action research has to be participatory because the practice we are investigating is always in relation with other people" (p.36). Action-oriented studies, however, often have varying degrees of participation (Greenwood, Whyte, & Karkavy, 1993). Along these lines, projects may encompass greater participation in later stages (McTaggart, 1991). Moreover, there may be different types of participation over the course of a project (Pyles, 2015), which range from controlled to self-directed involvement (Pretty, 1995).

Although numerous studies describe participatory processes (see, e.g., Balakrishnan & Conforth, 2013; Dahl, 2014), there is a lack of research that gives explicit attention to the shifts in participation throughout an action research project. To fill this gap, this chapter focuses on how participation was expanded in a CBPR project (Hacker, 2013) that involved a non-governmental organization and its community of volunteers in the Dominican Republic and explored possibilities for enhancing volunteer participation in an academic camp. Informed by community-based philosophy's (Murphy, 2014) concept of participation and motivated by the need to attend to how action research is participatory (McNiff & Whitehead, 2002), the author identified key moments in the project when the broadening of participant involvement was important for achieving a greater understanding of the issue of interest. What is illustrated is how mutual understandings informed decisions to include new participants throughout the project and how the participatory process led to the recognition of new communal knowledge and possibilities. The objective is to advance an appreciation of how projects with volunteer participation, albeit in fluctuating levels, may result in the co-creation and elevation of "local knowledge" (Fals Borda, 1988).

CBPR AND COMMUNITY-BASED PHILOSOPHY

A CBPR approach is "designed to ensure and establish structures for participation by communities affected by the issue being studied, representatives of organizations, and researchers in all aspects of the research process to improve health and well-being through taking action, including social change" (Viswanathan et al., 2004, p. 3). Influenced by the action research tradition, CBPR researchers take seriously a community's perspective in order to understand social issues (Fals Borda, 1988). Thus, CBPR is guided by the contributions that community participants make to the research process (Hacker, 2013).

Underpinning this approach is community-based philosophy (Murphy, 2014), which is founded on the rejection of Cartesian dualism (Bordo, 1987). Specifically, this philosophy recognizes the interpretive element of local knowledge that can be used by communities to direct their own progress and make appeals for support. The path opened by contemporary philosophers via the "linguistic turn" entails the development of a new understanding for how language does not simply reflect but rather forms reality (Wittgenstein, 1958). In terms of a community, language mediates how participation results in solidarity. Furthermore, "communicative competence" (Habermas, 1970) makes possible cohesion and mutual understanding because persons can "enter the realities of others" (Murphy, 2014, p. 42) through dialogue.

Achieving dialogue is important for fostering the social relationships among action research participants. When seeking knowledge together, action research unifies participants in joint learning. This process involves collaboration that calls for participants to engage the views of others in order to reach a shared understanding about social life (van Beinum, 1999).

From a community-based perspective, participation creates the social world (Murphy, 2014). Community-based initiatives are thus shaped by "popular knowledge" (Fals Borda, 1988) that is understood by participating (Murphy, 2014) and acquiring insight into a community's biography (Berger & Luckmann, 1966). Such knowledge is the result of the fusion of meanings that group members give to the history and current conditions of their community, as well as to the product of their blended interests in possibilities for social change.

Community order, similar to this knowledge base, is constructed by "praxis," the interplay of the reflection and action of community members, stemming from their creativity (Fals Borda, 1979). Opposed to the realism that dominates the Western Tradition, and maintaining that individuals must be restrained by an absolute foundation, persons are capable of using praxis as a basis for the organization of a community. This effort requires a type of "joint action" (Blumer, 1969) that requires persons to take into account their own, and each other's, interpretations and meanings of actions and circumstances. The assumption is that persons are able to invent and coordinate themselves, which involves participation.

In action research, "people can generate their own knowledge from their experience of living and learning" (McNiff & Whitehead, 2002, p 18). Given this epistemological stance and the methodological emphasis on reflection (McTaggart, 1994), community-based philosophy is consistent with action research and its various forms, particularly CBPR. Recognizing these theoretical and practical connections between CBPR and community-based philosophy is important for analyzing participant involvement in a project.

In the remainder of this chapter, the development of a volunteer community's understanding of its members' collaboration and involvement in relation to the expansion of participation in a CBPR project in the Dominican Republic is traced. While this chapter could bring attention to other concerns, this specific issue emerged during the initial stage of the project and remained a concern until the end. Thus, the role of participation with respect to this issue can be analyzed at various points throughout the project.

BACKGROUND OF THE CBPR PROJECT

Local Educational Context

Low-level access to quality education is a major concern in the Dominican Republic, particularly in poor, rural regions. Studies show that grade repetition rates have increased from 3% to 7% since 1999 (UNESCO Institute for Statistics, 2012), while primary school students have the lowest levels of proficiency in math and reading among all students in Latin America (Regional Bureau for Education in Latin America and the Caribbean, 2011). UNESCO reports several areas in need of improvement in order for the Dominican Republic to reach the goal of free, high-quality, and compulsory education. Specifically, teaching methods, learning materials, infrastructures, use of instructional time, professional teachers, and financial investments should be priorities (Gajardo, 2007).

Organizations that strive to improve education in the Dominican Republic recognize that barriers to accessing and attaining education include issues resulting from poverty, teenage pregnancy, and violence (UNICEF, 2012). An enduring problem in Dominican society is racism against Haitians (United Nations Human Rights Council, 2008; United Nations Human Rights Council, 2010). Children born of undocumented Haitian migrants in the Dominican Republic are prevented from claiming Dominican citizenship. As a result, they lack basic rights, such as access to education and healthcare (Kosinski, 2009), as well as "protection against trafficking, child labor, or early marriage" (Wooding, 2008, p.

368). Furthermore, child prostitution is prevalent, particularly in coastal towns and urban centers (U.S. Department of State, 2013).

Description of the Organization

Educate Everyone[1] became a nonprofit organization in 2002 after a period of seven years, during which numerous students from American universities made short-term trips to the Dominican Republic to bring educational supplies and build classrooms. These efforts resulted in increased support from tourists, businesses such as hotels and restaurants, and local communities; eventually, entire schools were being built. These previous smaller projects were expanded upon and transformed into fully developed long-term programs.

Currently, Educate Everyone operates 14 programs in 15 communities and targets children and adolescents who are at risk of dropping out of school. The vision of the organization is that equal learning opportunities should be provided to all children and youth who live in the Dominican Republic. Aligned with this outlook, Educate Everyone carries out initiatives that focus on early childhood and primary education, as well as on youth and community development. These programs are promoted to be evidence-based and replicable in other underprivileged communities (Davies, 1999).

The guiding philosophy of Educate Everyone is that work is pursued with local communities until these communities can successfully run the programs on their own. Compatible with this view is the organization's emphasis on dignity, acceptance of cultural differences, and holistic approaches to program development. These elements are consistent with a community-based framework (Murphy, 2014).

The organization is advised by 12 executive board members and led on the ground by a multicultural staff of 13 administrators. The various educational programs are run by 34 staff members who are mostly local Dominicans. Presently, there are 11 long-term volunteers who support the organization's work in various communities. Numerous short-term volunteers work primarily during the summer academic camp.

Summer Academic Camp

Each year, in three or more communities, Educate Everyone operates a five-week educational camp, which also includes one week of volunteer orientation. This endeavor has been academically focused for the past eight years: the aims are to advance the math, literacy, and life skills of participants between the ages of 8 and 12, and to train local teachers in instructional strategies that they can apply during the regular academic year under the guidance of "expert" educators.

According to census information gathered in 2012 by Educate Everyone and its collaborating partners, the neighborhood that serves as Educate Everyone's epicenter has approximately 800 households. Eighth grade is the average level of education attained by adults. The main sectors of employment are tourism, services, and agriculture, but unemployment is a concern.

The largest camp took place at this central site, with more than 195 children out of a total of 420 attending. Two smaller camps operated out of local public schools in neighborhoods within ten miles from the main site. Although there is no data available on these neighborhoods, they have likely fewer households but similar characteristics.

In 2013, when the project was carried out, more than 60 volunteers were recruited to become teacher trainers, teachers, or counselors. A total of 45 volunteers resided locally, while 16 volunteers were from the United States. The local volunteers included 29 teachers and 16 counselors. International volunteers consisted of five teacher trainers and 11 counselors.

International volunteers applied for one of two unpaid positions – teacher trainer or counselor. The teacher trainers had the tasks of demonstrating instructional techniques to local teachers and mentoring them as they planned lessons and implemented strategies during the camp. However, the teacher trainers also guided activities. The counselors were responsible for supporting the learning process and for managing discipline. In order to be considered, an applicant had to have at least an intermediate understanding of Spanish, since all camp activities are carried out in the local language. Beyond this language requirement, teacher trainers had to have at least three years of teaching experience in the United States, and counselors needed to have experience working with children.

Local volunteers applied to teach or to be a counselor. Although the counselors did not receive monetary compensation, the teachers earned a modest stipend.[2] Teachers who applied to facilitate the core curriculum, such as math and reading, were responsible for leading instruction by utilizing the techniques that they were learning from the teacher trainers. Those teachers who led the life skills component were trained facilitators of their particular curriculum, and, therefore, did not receive training from the teacher trainers. Local counselors had to express interest in the position and have some leadership skills to be considered. The task of assisting teachers during lessons was shared between the international and local counselors.

There was one academic director who supervised all three camps, and there were two directors at each site. Two of these directors were Dominican. For camp activities, volunteers were divided into teams that included two local teachers, one teacher trainer, and two or three camp counselors. Most teams included at least one local and one international counselor. These teams carried out activities with the same group of participants throughout the camp. While teachers and teacher trainers shared time leading classroom instruction, counselors assisted by supervising activities, preparing materials, and handling discipline.

The Orienting Concern of the Project

Previous research has illustrated the wide-ranging academic, social, emotional, and psychological impacts of community-based and after-school programs (Hirsch, Deutsch, & DuBois, 2011; Mahoney, Larson, & Eccles, 2005). As Quinn (2005) points out, there is recognition of the importance of volunteer participation in these programs. Ensuring that volunteers feel engaged in organizational activities (Vecina et al., 2012) and well supported by other volunteers (Skoglund, 2006) is crucial to maintain their commitment and satisfaction. More research on these issues relating to the volunteer experience, however, is needed (Wilson, 2012). Although numerous studies identify how organizations can involve volunteers, these studies typically represent an outsider's perspective (Hamovitch, 1997; Hirsch et al., 2011). Therefore, the concerns, ideas, and agency of volunteers, all of which may facilitate improvements in their participation, are often overlooked.

Community-based Participatory Research (Hacker, 2013; Wallerstein & Duran, 2003) allows an organization's volunteer community to discover and utilize knowledge that may strengthen member participation in organizational programs (MacLeod, Skinner, & Low, 2012). This effort is contrary to the typical means-end rational action of traditional organizations, which involves upper level staff members mostly as the key decision-makers in terms of how organizational activities should operate (Weber, 1968). From a community-based perspective, this strategy overlooks the praxis of volunteers and their creativity and reflective capacities to figure out how to contribute to the organization's success. This project thus took a participatory approach that is crucial in action research (Hacker, 2013; McIntyre, 2008); Educate Everyone provided all volunteers the opportunity to draw from their own experiences to make organizational advancements.

The ideal participatory project develops from the concerns of the participants (Rogers Stanton, 2014). However, the researcher enters the field often with a broad issue as a starting point, which, in the present case, resulted from conversations with an organizational staff member (Somekh, 2006). Consequently, this CBPR project stemmed from an interest in enhancing the participation and collaboration of volunteers during Educate Everyone's summer academic camp.

METHODS

The project followed a cyclical process that is typically used in action research (for examples, see Hyland, 2009; Kemmis & McTaggart, 2005; Rai, 2012). In this study, this approach engaged participants in the following activities: reflection on the research questions, collaborative discovery and analysis of the data through reflection and discussion, planning for action, identifying possibilities for further inquiry, and repeating this process until action was carried out and refined (McIntrye, 2008). Important to note is that the Institutional Review Board[3] at the author's university approved the project at the time.

The author led the participants in the gathering and reflective analysis of data through participant observation (DeWalt & DeWalt, 2002) of camp activities, five discussion groups, three focus groups (McIntrye, 2008), and 30 interviews (Skinner, 2012). Discussion groups had less structure than the focus groups and included 14-27 participants. Focus groups had 5-6 participants. Discussion and focus groups were conducted in a bilingual fashion, but interviews were carried out in the native language of the participant. Discussion and focus groups were approximately one hour and 30 minutes in duration. Interviews lasted from 40 minutes to one hour. Discussion groups were at the hotel where most of the international volunteers and several local volunteers were staying. The hotel was located in the community where many local volunteers live. The last discussion group was in a classroom at one of the camp locations. Focus groups were facilitated in a back area of a local restaurant, while interviews took place at the camp locations and at a local café.

Other ethnographic methods, such as informal conversations and the examination of organizational archival documents (Berg & Lune, 2011), provided the opportunity to gain additional insight. The types of documents that were reviewed were the camp's orientation pamphlets, counselor guidebook, teacher evaluations, and past camp director evaluations. Field notes were taken for two weeks during camp preparation, orientation week, and throughout the four weeks of camp; these field notes served as the primary point of reference during the project.

Although individual and group interviews were digitally recorded, there was not sufficient time during the camp for participants to transcribe and analyze all conversations in a collective (McIntrye, 2008) yet traditional manner (Saldaña, 2012). Therefore, the main form of analysis during the project occurred through ongoing collective reflection with participants during these events (McIntrye, 2008), which is consistent with action research strategies (Tavares, 2016). After the activities, the author summarized the discussion, jotted down interesting comments that participants made, and recorded themes that participants deemed to be meaningful. Themes were used to guide the group in further exploration and inform decisions about whose viewpoints needed to be integrated into the project. One discussion group, however, involved an activity in which participants worked in small groups to identify themes in a focus group interview, which the author transcribed.

In order to examine the expansion of participation over the course of the project, the author revisited data from field notes and group and individual interviews transcribed after the project. First, the author identified points in the project when participation expanded. Next, she pinpointed data that showed why she thought there was a need to include new viewpoints, so that a more holistic understanding of the issue could be obtained. Specifically, she took note of how co-constructed knowledge continuously informed her decisions to include new participants.

Participant Recruitment

Participants were initially recruited during orientation week. After providing an explanation of the study to the group of counselors, volunteers were asked to attend the first discussion meeting if they were interested in participating. The invitation was open to participants who attended the first discussion group to also participate in future research activities. As later described, the need to include the perspectives of local teachers, teacher trainers, and staff members was perceived quickly. Therefore, the author eventually approached these individuals to see if they would be interested in participating in different activities in the study.

Project Participants

This project involved 55 participants in the 2013 summer academic camp: ten staff members and 45 volunteers. 22 of these participants were international and 33 were local. Volunteer participants consisted of 16 local teachers, 5 teacher trainers, 11 international counselors, and 14 local counselors. Most foreign counselors were in high school or college, while the local counselors were in high school. The teacher trainers and local teachers were of adult age. The local participants were Dominican, but some were of Haitian, French, or English decent. All international participants were from North America. Overall, participants ranged in ages from 16 to 30. There was a wide gap in socioeconomic status between the international and local volunteers, since most of the Dominicans were residents of the poor communities where the organization worked and many of them participated in the camp as youths.

The Role of the Researcher

The author's role as the principal researcher was to explain the investigative process to participants, coordinate research activities, encourage participation, and facilitate dialogue (Kindon et al., 2007). Furthermore, she supported collective learning through reflection (Mhina, 2009). Having had significant experience living, working, and volunteering in the Dominican Republic for more than four years, she offered insights into cultural differences, the Dominican educational system, and the volunteer experience in order to provoke critical inquiry and awareness.

ANALYSIS OF THE EXPANSION OF PARTICIPATION

The following sections illustrate how the community of volunteers elaborated on their understanding of the central issue, which, in turn, informed the author's perspective on the need to expand participation throughout the process. To highlight the shifts in participation over the course of the project, the author will focus on three stages of the initiative. The first stage is the *inception of the project*, which is when the author, along with a staff member, pinpointed the study's main concern of the organization prior to entering the field. *Understanding the concern*, the second stage, is when the author worked at the site with the volunteer community to investigate the issue and ways to address it. The third stage, *confirming a solution*, is when she was about to leave the field and, therefore, made a last effort to authenticate her understanding of the group's viewpoint and to convey participants' ideas to the organization.

Inception of the Project

The present project began after the author contacted one of Educate Everyone's staff members and expressed interest in collaborating in action-oriented research that would meet a need of the organization. During the conversation, the author revealed her previous experience living, working, and volunteering in the Dominican Republic. Additionally, she shared her understanding of the challenges of working in local communities as an international volunteer, such as overcoming cultural barriers and the perception of her as an outsider.

This initial communication was followed by two phone conversations, during which the strengths and weaknesses of various programs were considered from this staff member's perspective. According to this individual, over the last several years, most of the effort to improve the camp had been devoted to refining logistics and developing the curriculum, while volunteer involvement had received less focus. In past camps, volunteer participation and collaboration had been challenging to maintain mainly due to cultural differences and the low level of motivation of the volunteers. Because each camp involves numerous volunteers from distinct countries and with different abilities, experiences, and language skills, the socialization of volunteers (Haski-Leventhal & Bargal, 2008) into the camp has not always been easy (Studer & von Schnurbein, 2013). Additionally, volunteers often hope to use their skills and feel a sense of fulfillment, like many persons engaged in volunteer work (Musick & Wilson, 2008). Since the organization was seeking volunteers with profiles similar to previous years, these issues were anticipated.

Based on initial discussions, the author and staff member identified volunteer participation for the annual summer camp to be an area in need of greater attention. The author had expected, however, that the focus might be narrowed, changed, or possibly even abandoned once participants became involved (Kapoor & Jordan, 2009). Given this anticipation, she was open to the volunteer community's views and desires, which is in keeping with community-based philosophy and action research (Murphy, 2014). Thus, she was mindful of her position as an international collaborator and a "professional" researcher, and was careful not to impose her perspective on the initiative at its outset.

At this early stage, the participation of a single staff member may be viewed to be minimal. However, this person acted as a critical guide (Berg & Lune, 2011) by giving key insight into the organization and by facilitating a dialogue about local issues that could be addressed through the collaboration. Without this staff member's participation and willingness to take a risk on an initiative with a flexible plan, the project may have never gotten off the ground. Although entering into a project without firm objectives and recruited participants evoked a feeling of uncertainty, the author had confidence that the process would lead to success (Cook, 1998).

UNDERSTANDING THE CONCERN

Pre-Arrival of Volunteers

At the field site, the author spent two weeks prior to the arrival of the volunteers engaging in participant observation. During this time, she accompanied camp directors in their meetings and reviewed documents in the organization's digital archive. She was able to have several informal conversations with directors, who affirmed the value of focusing on volunteer participation.

Field notes on these interactions captured the directors' concerns of how counselors would be helpful during activities, remain motivated during the camp, and handle classroom discipline. The author recorded that one of the directors supervised a camp location the year before and two other directors were counselors in the past. Given this information, she noted how their insight was likely based on their past experiences and, perhaps, informed by directors' evaluations of previous camps. These evaluations included directors' perceptions of how, at times, some volunteers lacked efficacy and motivation. In terms of community-based philosophy (Murphy, 2014) and action research epistemology (Cornwall & Jewkes, 1995), the local knowledge base is grounded in these experiences and perceptions.

Recognizing the logistical constraints of the project, such as time and resources, the author knew that the project needed a narrower, yet adaptable, focus. In order to be consistent with action research principles (Greenwood & Levin, 1998), she made a conscious effort to be attentive to how she facilitated the development of the project. Therefore, she sought the guidance of the directors to decide how to make the project as feasible as possible. In field notes, the author described how their insight was crucial to determining a plan for when and where to hold project activities as well as what themes may be viewed as relevant by volunteers. Given their shared ideas, they decided to do a close examination of the counselor role and the challenges of fulfilling its responsibilities. However, they knew that the project would take this direction only if there were volunteers who both found this aim to be important and agreed to participate.

Post-Arrival of Volunteers

In order to prepare the volunteers for the camp, Educate Everyone provided five days of orientation. The sessions in this training focused on an array of issues, including team problem solving, managing discipline, and Dominican culture. During this time, the author participated in activities and workshops alongside the volunteers. Toward the end of orientation, she invited the directors and counselors to participate in the study and asked them to attend the first discussion group, which was held the following week. At this discussion group, there were 19 participants (one local director, seven local counselors, and 11 international counselors). The aim of this discussion was for volunteers to reflect on their initial camp experiences and how they felt about their roles.

Among international volunteers, there was a shared sentiment of fulfillment in being able to help the children, despite the presence of language barriers. One international volunteer said:

Sometimes, it's frustrating because I feel like I'm handicapped because I don't speak good Spanish, and I don't know if I'm communicating accurately with the kids...and it's frustrating and it makes me nervous and I feel bad, but... um... at the end of the day, they still, they don't care and I think that's what I like the most is that this girl, who I only met yesterday, she brought me a lollipop that said, "I love you," and then during the activity in class she wrote me down [sic] and said I was her best friend. So, even though I don't think that maybe I'm doing the best job possible, I'm clearly affecting someone's life, and I guess that's what I take home at the end of the day.

This storytelling is an important form of participation in action research projects, because sharing stories allows for volunteers to reflect critically on their experiences (Santos, 2016). What quickly became apparent through observations of camp activities, and which participants corroborated during collective reflections (McTaggart, 1991), was the significant role that counselors play in keeping the camp operating smoothly. Along these lines, a local counselor noted:

I have learned that you have to be really responsible and always be respectful and firm, and always keep the kids within sight, because if you don't, they aren't respectful...There will always be disorder.[4]

However, some participants in the discussion group were confused as to exactly what the counselors' responsibilities entailed during the camp (Merrell, 2000; Taylor, Mallinson, & Bloch, 2008). This confusion was attributed to the inconsistency between the expectations they formed during orientation and their actual experiences. Mainly, counselors did not expect to provide instruction, but the program format required that counselors take on active teaching responsibilities.

This camp was different from previous camps, due to its heavy reliance on a stations model, which was implemented as an instructional and classroom management strategy. The children worked in small groups on specific activities at different stations, and they rotated stations every 10 to 15 minutes (Teacher Quality Enhancement Center, 2009). Although this stations model was utilized in prior camps, greater emphasis was placed on designing more lessons with stations in the present year. This modification often required that counselors thoroughly engage in teaching tasks that were outside of their prescribed role of supporting learning processes. Therefore, there were moments when counselors were actually teachers.

At this point, it is important to note that the perspectives of the participants shed light on how the structure of the camp should be taken into account when considering their behaviors (Taylor et al., 2008). In other words, a counselor who appears disengaged may not be taking initiative simply because he or she thinks doing so would be stepping outside of the counselor role. This lack of proactivity should thus not be assumed to be an individual deficiency such as low intrinsic motivation. This example reflects how the participatory nature of a CBPR approach allows for nuanced understandings and clarification of volunteer issues (MacLeod, Skinner, & Low, 2012).

In the end, most counselors agreed that the camp's structure required that they go beyond their defined roles (Netting et al., 2005). Nevertheless, some participants enjoyed being thoroughly involved in learning activities and recognized the importance of their support. An international counselor asserted:

It's nice to be, like, a part of it, and I also feel like it's like, the perfect example of what a classroom should be like because well, at least in my group, because like, I know like, a lot of things we've talked about in the classrooms, like the struggles, and there is only one teacher. It's like, what can you do when you have all these students? But like, we have three counselors and two teachers. So, like everybody's getting enough attention and there's lots of like, singing and getting the kids excited and standing up and not just sitting in a desk.

When concluding the discussion group, the author asked participants to suggest topics that should receive further attention in the first focus group. In this way, the volunteers participated as co-designers of the research (Hacker, 2013) and provided direction to the project, which is aligned with community-based philosophy (Murphy, 2014) and CBPR (Raymaker, 2016). An international counselor proposed:

How to better communication between like, from, like, not only child to counselor, but also counselor to teacher…just like the whole flow of communication when something goes wrong, if you're not sure about something, especially for like, the international people, who don't necessarily speak Spanish, because there is a lot of miscommunication that happens.

The conclusion that was drawn from this discussion, including the aforementioned suggestion, was that the research participants needed to get an idea of how the teachers and teacher trainers viewed and related to the counselor role (Wilson, 2012). Therefore, the author decided to invite teacher trainers to a discussion group,[5] and interview them and the local teachers. Although some teachers and teacher trainers discussed their frustrations with collaborating with counselors during interviews, they developed an understanding and appreciation of the counselor role. At first, according to many teachers and teacher trainers, counselors were visualized to be part of the background; they were there to support, but not necessarily to lead.

One local teacher, however, expressed how pleased she was with the counselors with whom she worked, but recognized that these particular counselors may be unique. She stated:

For example, almost always when our counselors arrive, they go to the classroom and do an activity with the kids, and the others don't do that. They are always around to…I don't know how it is in other groups, but others tell me, "Oh, they don't do that." The activities that they do, for example, are always related to the class. They use math. They use numbers. It's incredible. They work really well.[6]

Similarly, data from other interviews and discussion groups show that teachers and teacher trainers began to view the counselors as "teaching assistants." This acknowledgement meant that counselors were viewed to have a more active role in the learning experience than had been originally defined. One local teacher, in fact, expressed that the Spanish term used for the counselors, "consejeros," was not appropriate, given their responsibilities. For this teacher, a "consejero" is someone who is older and gives advice, not someone who works with students in a classroom. This teacher proposed the term, "orientador," which could be translated into English as "guide" or "coach." Although many of the local counselors were not critical of the term "consejero," the international counselors did not feel the work that they were doing corresponded to their image of a camp counselor.

Integrating teachers and teacher trainers as participants in the project was thus important for obtaining a more complete understanding of the issue. Aspects of the camp's design, such as the one in the example mentioned previously, could have easily gone unexamined without the involvement of the teachers and teacher trainers. Therefore, in this case, the co-constructed knowledge that resulted from the widening of participation helped to ensure that the project would be relevant, not only to counselors, but also to the broader volunteer community.

CONFIRMING A SOLUTION

As the camp came to a close, the author sought confirmation of her understanding of what project participants believed would improve future camps with respect to the counselor role. This corroboration was important to obtain before preparing a preliminary report of the project for the organization, which she committed to providing before leaving the field site. Because participants left within days of the end of the camp, they were not available to be involved in the writing and dissemination of the report (Sommer, 2009). While the full spiral of phases of action research, including planning, acting, observing, and reflecting (Kemmis & McTaggart, 2005), was not completed with regard to this particular issue, other project concerns were explored by the author and participants following the cyclical action research process. Presenting how those issues were examined and how research efforts resulted in change, however, is beyond the scope of this chapter.

During a break at one of the camps, the camp directors allowed the author to hold a discussion group that involved counselors, local teachers, and teacher trainers. As a result, she facilitated a conversation that included multiple perspectives on the counselor role as it resulted from the counselors' expanded participation. One local teacher emphasized the improvement he saw in how counselors supported classroom activities over the course of the camp. He shared that he believed the lack of initiative he perceived on the part of counselors was due to their shyness. However, an international counselor made it known that their hesitancy was not necessarily attributable to timidity. Rather, she asserted that they were afraid of "stepping on the teacher's toes." What was revealed was that teachers and teacher trainers did, in fact, want the counselors to take a more active role.

Through the exchange of ideas, and as a result of expanded participation, the author observed that there was a shared understanding of the need to reconceptualize the counselor role. Specifically, what the participants deemed important to take into account was how counselors work with local teachers and teacher trainers. The general view was that counselors take on responsibilities that are comparable to those of teaching assistants and, consequently, prospective volunteers should be made aware before applying to volunteer that they will be expected to support student learning. Counselors, nevertheless, viewed this new understanding of the role as an opportunity for Educate Everyone to attract a different type of volunteer, particularly one who is interested in getting classroom experience. Along these lines, project participants suggested that the organization focus recruitment efforts on students who are pursuing education degrees.

This "popular knowledge" (Fals Borda, 1988) was included in the preliminary report of the project, which the author discussed with the same staff member who supported the initiative from the beginning. What became clear to this staff member was how this project, given the participation of a traditionally-overlooked group of people, exposed some of the ways in which Educate Everyone had not based practices on local knowledge and culture (Kassam, 2002). For example, the staff member noted that the Americans, who originally established the camp, likely translated the English term, "counselor" directly into Spanish without questioning whether or not the translation was contextually appropriate, and the organization had continued to use this term without examining its applicability.

The willingness of the staff member to listen, and take seriously, the perspective of the volunteer community is important to recognize. According to community-based philosophy (Murphy, 2014), the appreciation of different views is crucial in furthering the goals of a group. Moreover, this person's participation in the dissemination phase of the project is an additional way in which participation continued to be inclusive, even late in the endeavor (McTaggart, 1991).

CONCLUSION

By examining participation as a dynamic factor in a project, this chapter shows how the initiative under study honored local knowledge through a process of co-construction (Gergen & Gergen, 1991). Moreover, this process highlighted how participation facilitated co-learning, which is critical in modes of action research (Israel et al., 1998). The author recognizes that the participation of the volunteers, to an extent, depended on her ability and commitment to provide opportunities for their involvement, as well as on her reflexivity (Steier, 1991). Because she was sensitive to her position and power as a researcher, she was careful not to impose her views or to undermine the views of others. Also, she avoided pushing individuals to participate. Encouraging participation without forcing it is a recognized concern of action researchers (James, 2016).

In the case examined in this chapter, a CBPR approach helped to ensure that participants were the central actors in generating ideas for ways that volunteer participation could be improved (MacLeod, Skinner, & Low, 2012; White & Maddox, 2012). Through increased volunteer involvement, and based on the experiences and reflections of these volunteers, participants redefined the role of the counselor. Additionally, the integrated insights of volunteers revealed new possibilities for the future of the camp and of the organization.

After analyzing the fluctuation of participation in a particular initiative, what became clear is that this aspect of the process may reveal opportunities for co-learning. Furthermore, a lesson learned was the need for the research process to be flexible enough to attract new participants. Therefore, in future action research, investigators should view changing levels of participation in a project as something potentially beneficial instead of negative.

ACKNOWLEDGMENT

I would like to thank the organization in the Dominican Republic that provided me the opportunity to carry out this project and the volunteers who collaborated with me. Additionally, I would like to thank Dr. John W. Murphy for his comments on earlier drafts of this chapter. Finally, I would like to thank the Center of Latin American Studies and the Department of Sociology at the University of Miami in Coral Gables, Florida for providing funding to support this endeavor.

REFERENCES

Balakrishnan, V., & Conforth, S. (2013). Using working agreements in Participatory Action Research: Working through moral problems with Malaysian students. *Educational Action Research*, *21*(4), 582–602. doi:10.1080/09650792.2013.832347

Berg, B. L., & Lune, H. (2011). *Qualitative research methods for the social sciences* (8th ed.). Upper Saddle River, NJ: Pearson.

Berger, P. L., & Luckmann, T. (1966). *The social construction of reality*. New York, NY: Doubleday.

Blume Dahl, K. K. (2014). From worse to better: How Kenyan student-teachers can use Participatory Action Research in health education. *Educational Action Research*, *22*(2), 159–177. doi:10.1080/0965 0792.2013.859089

Blumer, H. (1969). *Symbolic interaction: Perspective and method*. Englewood Cliffs, NJ: Prentice-Hall, Inc.

Bordo, S. R. (1987). *The flight to objectivity: Essays on Cartesianism and culture*. Albany, NY: State University of New York Press.

Brown, P., Morello-Frosch, R., Brody, J. G., Gasior Altman, R., Rudel, R. A., Senier, L., & Simpson, R. et al. (2010). Institutional Review Board challenges related to Community-based Participatory Research on human exposure to environmental toxins: A case study. *Environmental Health*, *9*(1), 39–50. doi:10.1186/1476-069X-9-39 PMID:20637068

Cook, T. (1998). The importance of mess in Action Research. *Educational Action Research*, *6*(1), 93–109. doi:10.1080/09650799800200047

Cornwall, A., & Jewkes, R. (1995). What is participatory research?. *Social Science & Medicine*, *41*(12), 1667–1676. doi:10.1016/0277-9536(95)00127-S PMID:8746866

Davies, P. (1999). What is evidence-based education? *British Journal of Educational Studies*, *4*(2), 108–121. doi:10.1111/1467-8527.00106

DeWalt, K. M., & DeWalt, B. R. (2002). *Participant observation: A guide for fieldworkers*. Walnut Creek, CA: Altamira Press.

Fals Borda, O. (1979). *El problema de como investigar la realidad para transformarla: Por la praxis*. Bogotá, Colombia: Ediciones Tercer Mundo.

Fals Borda, O. (1988). *Knowledge and people's power: Lessons with peasants in Nicaragua, Mexico and Colombia*. New York, NY: New Horizons Press.

Gajardo, M. (2007). *Country profile prepared for the Education for All Global Monitoring Report 2008 Education for All by 2015: Will we make it? Dominican Republic case study*. Retrieved May 3, 2015, from http://unesdoc.unesco. org/images/0015 /001555/155539e.pdf

Gergen, K. J., & Gergen, M. M. (1991). Toward reflexive methodologies. In F. Steier (Ed.), *Research and Reflexivity* (pp. 76–95). Newbury Park, CA: SAGE Publications, Inc.

Greenwood, D. J., Foote Whyte, W., & Karkavy, I. (1993). Participatory Action Research as a process and as a goal. *Human Relations*, *46*(2), 175–192. doi:10.1177/001872679304600203

Greenwood, D. J., & Levin, M. (1998). *Introduction to Action Research: Social research for social change*. Thousand Oaks, CA: SAGE Publications.

Habermas, J. (1970). Towards a theory of communicative competence. *Inquiry: An Interdisciplinary Journal of Philosophy*, *13*(1-4), 360–375. doi:10.1080/00201747008601597

Hacker, K. (2013). *Community-based Participatory Research*. New York, NY: SAGE Publications, Inc.

Hamovitch, B. A. (1997). *Staying after school: At-risk students in a compensatory education program*. Westport, CT: Praeger Publishers.

Haski-Leventhal, D., & Bargal, D. (2008). The volunteer stages and transitions model: Organizational socialization of volunteers. *Human Relations*, *61*(1), 67–102. doi:10.1177/0018726707085946

Hirsch, B. J., Deutsch, N. L., & DuBois, D. L. (2011). *After-school centers and youth development: Case studies of success and failure*. New York, NY: Cambridge University Press. doi:10.1017/CBO9780511994913

Hyland, N. E. (2009). Opening and closing communicative space with teachers investigating race and racism in their own practice. *Action Research*, *7*(3), 335–354. doi:10.1177/1476750309336717

Irby, D., Mawhinney, L., & Thomas, K. (2013). Re-examining Participatory Research in dropout prevention planning in urban communities. *Educational Action Research*, *21*(2), 267–283. doi:10.1080/09650792.2013.789732

Israel, B. A., Schulz, A. J., Parker, E. A., & Becker, A. B. (1998). Review of community-based research: Assessing partnership approaches to improve public health. *Annual Review of Public Health*, *19*(1), 173–202. doi:10.1146/annurev.publhealth.19.1.173 PMID:9611617

Janes, J. E. (2016). Democratic encounters? Epistemic privilege, power, and community-based participatory action research. *Action Research*, *14*(1), 72–87. doi:10.1177/1476750315579129

Johnson-Burel, D., Drame, E., & Frattura, E. (2014). Participatory research in support of quality public education in New Orleans. *Educational Action Research*, *22*(3), 288–305. doi:10.1080/09650792.2013.872576

Kapoor, D., & Jordan, S. (Eds.). (2009). *Education, Participatory Action Research, and social change: International perspectives*. New York, NY: Palgrave Macmillan. doi:10.1057/9780230100640

Kassam, A. (2002). Ethnotheory, ethnopraxis: Ethnodevelopment in the Oromia regional state of Ethiopia. In P. Sillitoe, A. Bicker, & J. Pottier (Eds.), *Participating in development: Approaches to indigenous knowledge* (pp. 64–81). New York, NY: Routledge. doi:10.4324/9780203428603_chapter_4

Kemmis, S., & McTaggart, R. (2005). Participatory Action Research: Communicative action and the public sphere. In N. K. Denzin & Y. S. Lincoln (Eds.), *Strategies of qualitative inquire* (3rd ed.; pp. 271–330). Thousand Oaks, CA: Sage Publications.

Kindon, S., Pain, R., & Kesby, M. (Eds.). (2007). *Participatory Action Research approaches and methods: Connecting people, participation, and place*. New York, NY: Routlege.

Kosinksi, S. (2009). State of uncertainty: Citizenship, statelessness, and discrimination in the Dominican Republic. *Boston College International and Comparative Law Review*, *32*(2), 377–398.

MacLeod, A., Skinner, M. W., & Low, E. (2012). Supporting hospice volunteers and caregivers through Community-based Participatory Research. *Health & Social Care in the Community*, *20*(2), 190–198. doi:10.1111/j.1365-2524.2011.01030.x PMID:21978371

Mahoney, J. L., Larson, R., & Eccles, J. S. (Eds.). (2005). *Organized activities as developmental contexts: Extracurricular activities, after-school and community programs*. Mahwah, NJ: Lawrence Erlbaum Associates.

McIntyre, A. (2008). *Participatory Action Research*. Thousand Oaks, CA: Sage Publications, Inc.

McNiff, J., & Whitehead, J. (2002). *Action research: Principles and practice* (2nd ed.). New York, NY: Routledge Falmer. doi:10.4324/9780203199961

McTaggart, R. (1991). Principles for Participatory Action Research. *Adult Education Quarterly*, *41*(3), 168–187. doi:10.1177/0001848191041003003

McTaggart, R. (1994). Participatory action research: Issues in theory and practice. *Educational Action Research*, *2*(3), 313–337. doi:10.1080/0965079940020302

Merrell, J. (2000). Ambiguity: Exploring the complexity of roles and boundaries when working with volunteers in well woman clinics. *Social Science & Medicine*, *51*(1), 93–102. doi:10.1016/S0277-9536(99)00442-6 PMID:10817472

Mhina, C. H. (2009). Research and agency: The case of rural women and land tenure in Tanzania. In D. Kapoor & S. Jordan (Eds.), *Education, Participatory Action Research, and social change: International perspectives* (pp. 155–168). New York, NY: Palgrave MacMillan. doi:10.1057/9780230100640_11

Minkler, M., & Wallerstein, N. (2005). Improving health through community organization and community building: A health education perspective. In M. Minkler (Ed.), *Community organizing and community building for health* (2nd ed.; pp. 26–50). New Brunswick, NJ: Rutgers University Press.

Murphy, J. W. (2014). *Community-based interventions: Philosophy and action*. New York, NY: Springer. doi:10.1007/978-1-4899-8020-5

Musick, M. A., & Wilson, J. (2008). *Volunteers: A social profile*. Bloomington, IN: Indiana University Press.

Netting, F. E., OConnor, M. K., Thomas, M. L., & Yancey, G. (2005). Mixing and phasing of roles among volunteers, staff, and participants in faith-based programs. *Nonprofit and Voluntary Sector Quarterly*, *34*(2), 179–205. doi:10.1177/0899764005275204

Pretty, J. N. (1995). Participatory learning for sustainable agriculture. *World Development*, *23*(8), 1247–1263. doi:10.1016/0305-750X(95)00046-F

Pyles, L. (2015). Participation and other ethical considerations in Participatory Action Research in post-earthquake rural Haiti. *International Social Work*, *58*(5), 628–645. doi:10.1177/0020872815581912

Quinn, J. (2005). Building effective practices and policies for out-of school time. In J. L. Mahoney, R. Larson, & J. S. Eccles (Eds.), *Organized activities as developmental contexts: Extracurricular activities, after-school and community programs* (pp. 479–495). Mahwah, NJ: Lawrence Erlbaum Associates.

Rai, R. K. (2012). A Participatory Action Research training initiative to improve police effectiveness. *Action Research, 10*(3), 225–243. doi:10.1177/1476750312439901

Raymaker, D. M. (2016). Intersections of Critical Systems Thinking and Community-Based Participatory Research: A learning organization example with the autistic communities. *Systemic Practice and Action Research, 29*(5), 405–423. doi:10.1007/s11213-016-9376-5 PMID:27833398

Regional Bureau for Education in Latin America and the Caribbean. United Nations Educational, Scientific and Cultural Organization. (2011). *Regional report on Education for All in Latin America and the Caribbean.* Retrieved May 1, 2015, from http://unesdoc.unesco.org/ images/0019/001916/191666e.pdf

Rogers Stanton, C. (2014). Crossing methodological borders: Decolonizing Community-Based Participatory Research. *Qualitative Inquiry, 20*(5), 573–583. doi:10.1177/1077800413505541

Saldaña, J. (2012). *The coding manual for qualitative researchers* (2nd ed.). Thousand Oaks, CA: SAGE Publications.

Santos, D. (2016). Re-signifying participatory action research (PAR) in higher education: What does P stand for in PAR? *Educational Action Research, 24*(4), 635–646. doi:10.1080/09650792.2015.1103658

Skinner, J. (Ed.). (2012). *The interview: An ethnographic approach.* New York, NY: Bloomsbury Academic.

Skoglund, A. (2006). Do not forget about your volunteers: A qualitative analysis of factors influencing volunteer turnover. *Health & Social Work, 31*(3), 217–220. doi:10.1093/hsw/31.3.217 PMID:16955660

Somekh, B. (2006). *Action Research: A methodology for change and development.* New York, NY: Open University Press.

Sommer, R. (2009). Dissemination in Action Research. *Action Research, 7*(2), 227–236. doi:10.1177/1476750308097028

Steier, F. (1991). Introduction: Research as self-reflexivity, self-reflexivity as social process. In F. Steier (Ed.), *Research and reflexivity* (pp. 1–11). Newbury Park, CA: SAGE Publications, Inc. doi:10.1017/CBO9780511553547.001

Studer, S., & von Schnurbein, G. (2013). Organizational factors affecting volunteers: A literature review on volunteer coordination. *Voluntas, 24*(2), 403–440. doi:10.1007/s11266-012-9268-y

Tavares, C. (2016). Human rights education and the research process; action research as a tool for reflection and change. *Educational Action Research, 24*(4), 617–634. doi:10.1080/09650792.2015.1124045

Taylor, T., Mallinson, C., & Bloch, K. (2008). Looking for a few good women: Volunteerism as an interaction between two organizations. *Nonprofit and Voluntary Sector Quarterly, 37*(3), 389–410. doi:10.1177/0899764007310420

Teacher Quality Enhancement Center. (2009). *Co-teaching strategies & examples.* Retrieved May 1, 2015, from http://www.tacoma.uw.edu/sites/default /files/global/ documents/education/co-teaching_definitions_x_examples.pdf

UNESCO Institute for Statistics. (2012). *Opportunities lost: The impact of grade repetition and early school leaving*. Retrieved September 1, 2015, from Library/Documents/global-education-digest-opportunities-lost-impact-grade-repetition- early-school-leaving-2012-en.pdf

UNICEF. (2012). *Dominican Republic: Country programme document 2012-2016*. Retrieved May 1, 2015, from http://www.unicef.org/about/execboard/files/ Dominican_Republic_final_aproved_2012-2016_English_10Feb2012.pdf

United Nations Human Rights Council. (2008). *Report of the Special Rapporteur on contemporary forms of racism, racial discrimination, xenophobia and related intolerance, Doudou Diène, and the independent expert on minority issues, Gay McDougall: addendum: Mission to Dominican Republic*. Retrieved May 1, 2015, from http://www.unhcr.org/refworld/docid/47e23b252.html

United Nations Human Rights Council. (2010). Report *submitted by the Special Rapporteur on the human rights of migrants, addendum: Communications sent to governments and replies*. Retrieved May 1, 2015 from http://www.unhcr.org/ refworld/docid/ 4c29bb2b2.html

U.S. Department of State. (2013). *Dominican Republic: Country reports on human rights practices-2012*. Retrieved May 1, 2015, from http://www.state.gov/j/drl/ rls/hrrpt/humanrightsreport/index.htm#wrapper

Van Beinum, H. (1999). On the design of the ACRES program. In D. J. Greenwood (Ed.), *Action Research: From practice to writing in an international action research development program* (pp. 3–24). Philadelphia, PA: John Benjamins Publishing Co. doi:10.1075/dowi.8.03bei

Vecina, M. L., Chacón, F., Sueiro, M., & Barrón, A. (2012). Volunteer engagement: Does engagement predict the degree of satisfaction among new volunteers and the commitment of those who have been active longer? *Applied Psychology*, *61*(1), 130–148. doi:10.1111/j.1464-0597.2011.00460.x

Viswanathan, M., Ammerman, A., & Eng, E. (2004). *Community-based participatory research: Assessing the evidence*. Rockville, MD: Agency for Healthcare Research and Quality.

Wallerstein, N., & Duran, B. (2003). The conceptual, historical, and practice roots of Community Based Participatory Research and related participatory traditions. In M. Minkler & N. Wallerstein (Eds.), *Community-Based Participatory Research for health* (pp. 27–52). San Francisco, CA: Jossey-Bass.

Weber, M. (1968). *Economy and society: An outline of interpretive sociology* (G. Roth & C. Wittich, Eds., E. Fischoff, Trans.). New York, NY: Bedminister Press. (Original work published 1921)

White, D., & Maddox, J. (2012). Organization development in Guatemala: Rapid engagement with long term impact. *Organization Development Journal*, *30*(4), 32–43.

Wilson, J. (2012). Volunteerism research: A review essay. *Nonprofit and Voluntary Sector Quarterly*, *41*(2), 176–212. doi:10.1177/0899764011434558

Wittgenstein, L. (1958). *The blue and brown books*. New York, NY: Harper and Row.

Wooding, B. (2008). Contesting Dominican discrimination and statelessness. *Peace Review*, *20*(3), 366–375. doi:10.1080/10402650802330279

KEY TERMS AND DEFINITIONS

Collaboration: A cooperative practice, whereby individuals unify their efforts and work together to achieve a shared goal.

Community-Based Participatory Research: A form of action research that centers community participants in a cyclical process that involves collaborative phases of reflection and action.

Community-Based Philosophy: A set of principles that guide projects to become community-based.

Dominican Republic: A country that shares the island of Hispaniola in the Caribbean.

Local Knowledge: A kind of knowledge that stems from community and personal experiences, everyday routines, and popular understandings.

Participation: The act of becoming involved, being fully present, and recognizing others in activities.

Volunteer: A person who participates in activities of his or her own accord.

ENDNOTES

1 "Educate Everyone" is a pseudonym that is used to maintain the confidentiality of the organization. The description about the organization includes information that was retrieved from Educate Everyone's website. For confidentiality purposes, the website is not cited.

2 Despite being paid, teachers were considered to be volunteers by the organization.

3 Brown et al. (2010) present the challenges of upholding CBPR principles when confronted with Institutional Review Board requirements.

4 The author translated this statement from Spanish.

5 Due to logistical constraints, local teachers were unable to attend discussion groups, except for one that was held during camp hours at the camp location.

6 The author translated this statement from Spanish.

Chapter 11

Action Research Driving a Scaffolded Soil Science Curriculum:

Within the Context of Post-First Year Academic Underpreparedness

Milton Milaras
University of South Africa, South Africa

Tracey McKay
University of South Africa, South Africa

ABSTRACT

A case study of a Soil Science teaching intervention for second year university students is presented here. The purpose was to address post-first year student underpreparedness for higher education. The curriculum was redesigned to iteratively develop students' academic writing skills: with particular regard to environmental report writing, whilst imparting Soil Science content. To this end, students had to integrate practical and theoretical learnings into a soil erosion modelling report. 'Micro action research dialogues' were used to drive the development and delivery of this scaffolded curriculum. Both the process of curriculum development, and the curriculum itself, are detailed here. Recommendations were arrived at by the means of action research reflection: primarily, academics should engage formally with education as a discipline if curricula are to be successfully redesigned. Secondary recommendations are: a deep approach to learning requires a less dense curriculum; tutors need to be more formally capacitated; and administrative hurdles should not be underestimated.

DOI: 10.4018/978-1-5225-2642-1.ch011

INTRODUCTION

Soil Science is an important and topical discipline with 2015 being declared the United Nation's 'Year of Soil' (Keesstra et al, 2016). Soil is a non-renewable resource, with 1 cm taking 1 000 to 10 000 years to form, and yet 98% of humanity's food stuffs are dependent thereon (Wakatsuki & Rasyidin, 1992; Vitousek, Mooney, Lubchenco, Melillo, 1997). Despite this, some 66% of arable soil worldwide is classified as degraded mostly due to human activities (Vitousek et al, 1997). Knowledge of Soil Science is therefore a pressing need (Keesstra et al, 2016). Soil Science in fact consists of two sub-disciplines: Pedology and Edaphology. Pedology deals with the chemical and physical properties of soils, *in situ* soil formation and classification; while Edaphology is primarily concerned with how soils affect plants (Bockheim, Gennadiyev, Hammer & Tandarich, 2005). Nevertheless the terms Soil Science and Pedology are often used interchangeably (Shaw, 1930). Within this context a second year Soil Science module was offered at a South African university.

The module was developed using two main approaches: that of community of practice and that of action research. To meet the needs of the post-first year student, the module described here was designed using best-practice teaching principles with a strong focus on curriculum structure. In particular, the aims and objectives went beyond standard 'soil science' knowledge outcomes, to embrace explicit academic development outcomes, such as academic writing and team work. As the literature is surprisingly silent on the need to assist post-first year students achieve academic success, the intervention contributes to the literature on epistemological access and underprepared students for second year university (Topping, 1996; Dochy, Segers & Sluijsmans, 1999; Struyven, Dochy, & Janssens, 2003). The class of ~145 students were diverse in terms of: race, socio-economic status, gender and the degrees for which they were registered (science, humanities and education). That said, the majority were black, male, roughly 20 ½ years of age, South African, came from Gauteng and matriculated at rural or township schools that had relatively good matriculation pass rates. In the literature such students are regarded as 'underprepared' for tertiary study due to the generally poor quality of secondary schooling (Jacobs, 2007; Chokwe, 2013; Pineteh, 2014; McKay, 2016). Epistemological support to post-first year students is essential to address the systemic throughput and dropout challenges which universities face.

The curriculum was developed using action research, in particular micro action research dialogues. Micro action research dialogues are a series of critical conversations that drive the larger action research project forward. In this instance these dialogues took place between mentor and mentee. Thus, the authors were participants in the intervention: the lead author as the lecturer-researcher and mentee, and the secondary author as the mentor. This chapter was written as part of the reflective facet of the action research and also comprised an iterative process between the mentor and mentee. The chapter firstly presents the theoretical framework which underpinned the intervention. The personal teaching philosophy is then briefly described. The methodology is then outlined and the intervention described in depth. The chapter concludes with reflections and recommendations for the next action research cycle.

LITERATURE REVIEW

The Theoretical Framework: Community of Practice

The study was informed by the community of practice theoretical framework (Grant, 2005; Davids, 2014). Within this framework, insiders (the lecturers, tutors) of the discipline work with newcomers (the students) in a way that enables newcomers to become full practitioners of the discipline. The onus falls on the insiders to assist newcomers to gain mastery over the knowledge and skills of the discipline and eventually become insiders of the community of practice. The approach is also draws on the work of Vygotsky (1978), as mediation is done by more knowledgeable insiders (Lave & Wenger 1991; Wegner, 2000). Within this approach, newcomers are regarded as legitimate peripheral participants so as to acknowledge their rights and legitimize their expectations with regards to membership of the community.

Soil Science, like all academic disciplines has its own 'way of doing', such as specific terminologies, ways of writing academically (appropriate phrases and nomenclatures), citing, and mapping. These ways of doing, or what Wenger (1998) calls 'reification', must be overtly taught to newcomers. In this regard, the practical component of the module for example, was designed to guide students, using a feed-forward approach, to write an academically appropriate Soil Science report. The practical sought to share the routines, tools and symbols that comprise the practice of Soil Science within a safe, supportive learning environment (Wenger, 1998; Sutherland, 2009; Aluko & Hendrikz, 2012). Lave and Wenger (1991) argue that this is essential as students entering university seldom possess such knowledge, and may unfairly be deemed 'unsuccessful' because of this knowledge gap. So, the cultural system, in this case how to write a Soil Science report and the like, must be brokered by the academics (Wenger, 1998). The ability to broker the artefacts of the discipline in the way that empowers students and builds their disciplinary repertoire, is however not innate. Ways to acquire these educational skills are for lecturers to attend workshops or enrol for qualifications in educational practice, or undertake self-guided reading.

Action Research in Higher Education

In addition, an action research approach was adopted to enable the lecturer-researcher to formally and actively interrogate or reflect on their teaching practice. Action research involves iterative cycles of action and reflection, supported by planning and observation. It is an engaged and simpatico approach to research, with an underlying purpose of social emancipation (Brydon-Miller, Greenwood & Maguire, 2003). By its cyclical nature, action research lends itself not only to *iteration*, which develops deepened understanding and more effective action, but also to the simultaneous *layering* of the methodology upon itself – from the micro, through the meso, and to the macro. Harvey, Ambler & Cahir (2017) also used action research in a higher education setting in Australia. In particular, similar to our study, action research was not only used to improve teaching practice, but also to measure the impacts of a mentor-mentee relationship. They argue that such a stance is an innovative way to explore university communities and new ways of 'doing' for academics. Furthermore, Amir, Mandler, Hauptman & Gorev (2017) argue that action research in an education setting is an excellent way to improve teaching practice. They indicate however, that one must take cognisance of the difficulty of conducting research on one's own practice and indicate that little academic work has been done on this field.

The Leitch & Day (2000:180) definition of reflection was used, namely "*a metaphor for representing a process of learning from experience*". The process followed that of Russell & Munby (1992:3) and involved the "*systematic and deliberate thinking back over one's actions*". In undertaking this process, the lecturer reflectively interrogated their own (documented) educational practice, through the lens of the literature, and consequently deepened their understanding of their own teaching practice. Over time, the teaching practice was adapted and improved so as to work towards best practice. In this regard, the study also adopted what is known as design-based research. This type of research is common in educational settings (Simpson & McKay, 2013). It involved the lecturer-researcher designing a curriculum and various teaching and learning interventions, implemented them, evaluating them and subsequently adapting them 'on the go' so as to derive "*research findings from formative evaluations*" so as to directly and positive impact on the learning process (Collins, Joseph, & Bielazcyc, 2004:16). Thus, what concerned us was "*how, when and why*" student learning was taking place (Design-Based Research Collective, 2003:5). Thus, the entire intervention fell under a broad but flexible action research plan designed to foster student epistemic access to the discipline (Lave & Wenger, 1991; Becher & Trowler, 2001).

One of the core principles of enabling epistemic success is the creation of a scaffolded curriculum (as defined by Wood, Bruner & Ross, 1976; Morrow, 2007). In this case, scaffolding is breaking down tasks into their basic components, so as to make the skills and knowledge required to complete the task more explicit and provide students with direct learning support where necessary (McKay, 2013). Scaffolding helps students move from the known to the unknown, the simple to the complex. For example, for the Soil Science report, the various components of the report were disaggregated and students were assisted to complete each component. The notion of constructive alignment also influenced the curriculum design (Biggs, 1996). In this case this was achieved, by making the assessment criteria explicit, and offering direct support for assessment tasks in lectures, practicals and tutorials. This ethos and methodology of the constructivist philosophy aligns itself to action research. In light of this theoretical framework, the personal teaching philosophy of the lecturer-researcher will now be discussed.

PERSONAL TEACHING PHILOSOPHY

Constructivist Approach

A constructivist teaching philosophy and approach was adopted with emphasis on student-centred-learning (Jones & Brader-Araje, 2002). That is, the lecturer-researcher holds the view that knowledge is constructed internally, not acquired by rote (Ertmer & Newby, 1993). Thus, the focus was on facilitation, encouraging students to "*use appropriate learning strategies*", and enabling deep-learning (Ertmer & Newby, 1993:52). The student is viewed as an active agent where the lecturer facilitates self-directed learning (Banning, 2005). Consequently, students engage in problem solving activities and build on their prior knowledge to make conceptual leaps and connections. Facilitation tools included project-based learning and active class participation (Makgato, 2012). But all constructivist classrooms need 'ground rules' to ensure students respect each other, the process, and take responsibility for their own learning. The classroom must be a 'safe space' where students can ask questions and make suggestions without fear (Barbara, Gayle, Cortez & Preiss, 2013). Such 'ground rules', focused on respect and engagement, and were established from the start of the module and aimed to reduce the kinds of disruptions to which large classes can be prone.

Rationale

The module formed part of an Environmental Management and Geography degree. As such, the module needed to teach students job specific skills such as report writing: as environmental managers are called upon to write audit reports, specialist environmental reports, and project motivation reports (de Villiers, 1998; Dewar & Shippey, 2002). As 'environmental considerations' are seldom prioritized in the mining and construction industries for example, their reports must be well structured and solidly argued to have socio-environmental efficacy (Lotz-Sisitka, 2004; Amechi, 2009). To write well, students need academic self-confidence. They also need to be active participants in the writing process and be self-directed learners. However, most had underdeveloped critical thinking and academic writing skills; consequently, they were academically underprepared for the demands of the module and the professional demands of industry. Thus, the module had the following *pedagogical aims*:

1. To create contextualised meaning for students within the discipline's setting;
2. To impart and develop a positive attitude towards scientific rigour;
3. To model 'real world, professional world' skills;
4. To teach academic writing skills within the discipline's setting.

The intervention was also an opportunity for the personal professional development of the mentor and mentee. For the mentor, it sought to enhance their understanding of action research, the community of practice approach and mentoring an academic colleague. For the mentee, it was an opportunity to practice and hone higher education teaching skills and develop research. There was already an element of reflection and informal action research for the mentee and mentor to reflect on the students' academic needs, because both had taught the now second year students in their first year. Thus, this intervention could be considered a 'second cycle of teaching and action research', with the first cycle being undocumented.

The 'micro action research dialogue' cycles, between the lecturer-researcher (mentee) and the mentor which underpinned developing this pedagogical intervention, are described next under methodology.

Methodology

In the case of this intervention, within the iterative macro cycles of action and reflection there were meso or 'sub-cycles' of action research comprising shorter cycles of action-reflection. The meso cycles drive the macro cycle forward and generate thick, rich reflective data. Nevertheless each meso cycle is also a self-contained whole. Together, however, within the macro cycle, they collectively and interactively contribute to the outcome. The 'energy' of these cycles was generated by the mentor-mentee micro action research dialogues which nested in the meso and macro cycles. In this regard we were informed by Feldman (1999) who highlights the importance of dialogue in driving action research. Micro action research 'projects' however, have only been informally described in non-academic literature, as a methodology congruent with the philosophy of 'marginal learning gains' (see Elder, 2012). But the term is used quite differently here and so it is termed 'micro action research dialogues'. These are successive conversations that critically discuss a current issue/task within the meso and/or macro action research cycle, with the aim of advancing it. Each micro dialogue itself involves action (speaking, drafting) and reflection (listening, reviewing, editing). These dialogues were ongoing and not 'once off'. Rather they need to take place between the same parties each time, and must underpin and drive meso action-reflection cycles forward, and consequently develop the larger macro action research project.

Figure 1 illustrates this conceptual model, outlining how the action research process starts at the meso level (with for example, the first drafting of a curriculum outline) then reflecting thereupon, followed by consequent revisions (further action). At this point, a micro action research dialogue would occur, where we critically discussed the intervention so as to inform the next meso reflection cycle, before the meso action cycle of revising the curriculum again was taken up. Thus action follows reflection follows action etc. Micro action research dialogues however, spurred on both the action and reflection stages

Figure 1. Micro action research dialogues driving meso level action research methodology within a macro action research project

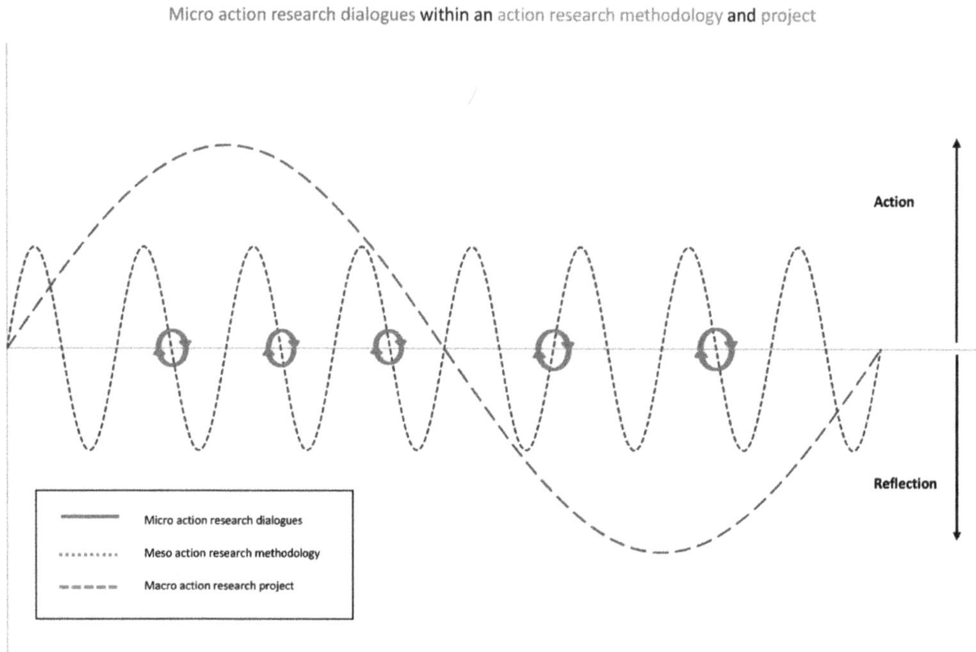

and sometimes occurred more than once within each macro cycle. We found that multiple micro action research dialogues are vital to both drive and refine the intervention, as well as being essential in eliciting significant and useful data. Thus, the development and delivery of the curriculum was a 'lived process'. The micro action research dialogues also occurred in the macro reflection cycle of the project; namely, the writing of this chapter (whereas the meso action cycle was developing and delivering the intervention itself). In summation, micro action research dialogues occur *within*, and in support of, a layered and overarching action research methodology, driving meso and macro cycles of action and reflection. This intervention is unpacked in detail under the next section.

THE INTERVENTION

Introduction: The Context

Particular constraints had to be dealt with. Firstly, the contact time with students was limited. The 7 week module was only allocated four lecture periods of 45 to 50 minutes (3h05m/week) and four practical sessions per week (3h05m/week). This was a total of 56 periods. In order to increase time on task under the guidance of a more able peer (as per the Vygotskian approach (1978)), two periods a week (one from the theory, one from the practical) were designated as tutorial sessions. Moreover, due to high student numbers and limited number of tutors (who were final year or Honours students), students were only required to attend one tutorial per week. Thus, each student had 7 contact sessions a week. Thus, each week's lecture content was reinforced with activities in the tutorials.

Secondly, a textbook, namely: Ashman, M.R. and Puri, G. (2002). *Essential Soil Science.* Oxford: Blackwell Publishing, had already been prescribed. Thus, the lecturer was obliged to use it. Fortunately it did provide a step-wise framework to the fundamentals of Soil Science, starting with characteristics of soil structure, advancing to soil chemistry, soil classification and ultimately land use and degradation issues. However, the textbook assumed certain prior scientific knowledge. This was despite the authors stating that "This book has been written for students who need to understand quickly the basic principles of soil science. [...] Every attempt has been made to assume no previous scientific knowledge, to ensure new concepts are explained simply, often using examples of everyday items" (Ashman & Puri, 2002: vii). Even a cursory glance, however, of the textbook demonstrated this to be far from the truth as "mineral materials and organic matter [...] percolating water [...][and] soluble compounds" are referred to (Ashman & Puri, 2002: 1). As scientific facts, these require more than a 'basic' scientific knowledge to understand.

Thirdly, the textbook however, was written from a British perspective. As a result, the significant differences in climate, soil types, slope, agricultural practices and economic capacity of Europe compared to Africa, meant that certain sections were not contextually appropriate. Hence some sections were omitted or adapted to take South African soil conditions into account. The textbook also only used United Kingdom (UK) and United States of America (USA) examples, with the authors regularly referring to regionally specific soil issues (e.g. water logging) and climatic patterns. Such examples and contexts are foreign, remote and unhelpful to South African students. Furthermore, classification systems such as: defining soil particles, textures or types, were often provided in USA or UK schema – ignoring the localised South African body of knowledge and set of schema. Important Southern Africa issues such as irrigation, salinity and erosion were neglected, and no attention paid to subsistence agriculture practices whatsoever. Even the 'further reading' section of the textbook only referred students to Western universities and Soil Research Institutes. As both authors hail from UK based research institutes, it is unsurprising that they overlooked the nuances and importance of developing country contexts; as well as how soils are affected by climates, topographies and locale specific environmental pressures found outside of Europe and the USA.

Fortunately the lecturer had the academic freedom to partially adapt the module outcomes. Thus, the lecturer could contextualise issues and deviate from the textbook where necessary. This may be a tentative step towards Africanisation. As Letsekha (2013:1) so aptly puts it, Africanisation in the context of higher education is "a call to adapt curricula and syllabuses to ensure that teaching and learning are adapted to African realities and conditions." Accordingly, South African examples were provided to replace the textbook's Euro-centric cases. Such examples included erosion issues, subsistence agriculture issues, the effects of mining and agricultural land use conflicts. Fey's (2010) South African soil classification system (which correlates with the World Reference Base (WRB) soil classification system) was also emphasised. Nevertheless, the fundamentals of soil structure, chemistry and nutrient cycling were not adapted as these soil processes operate globally, simply at faster or slower rates depending on the locale's geology, topography and climate. In overview, then, the outcomes of the module were for students to:

- Have a thorough knowledge of basic Soil Science.
- Be able to explain the complex interrelationships between different facets of soil and the agents (including humans) which affect it.
- Be able to source valid, peer-reviewed scientific articles online, paraphrase and integrate the information into a scientific report and cite the sources correctly.

- Produce a well-structured and coherent scientific report on soil erosion and erosion modelling.

Accordingly, these outcomes were shared with the students in the first lecture (see Cornbleth, 1984).

Core Knowledge Concepts

In terms of knowledge and skills sequencing, the module commenced with basic Soil Science concepts and the physical properties of soils, then, progressed to soil chemistry and nutrient exchanges. This was because such foundational knowledge is necessary to understand how soils behave under various conditions and uses. Focus then shifted to soil erosion and its severity in South Africa. This explicitly dovetailed into the practical component. Attendant land use issues, with a specific focus on South African agriculture, were then explored alongside appropriate soil conservation measures. Briefly, soil taxonomic systems, and finally soil contamination and remediation measures were considered. There was an overt attempt to explicitly link each section to the previous one, in a step-wise manner, to build meaning and create a holistic view. Accordingly, the theory and practical portions schedule of work were structured as follows (Table 1):

How this curriculum was delivered over the seven weeks is now reflected upon.

Teaching Approach

To foster discussion and engaged learning a *subverted* lecture/PowerPoint format was used in class. That is, instead of dispensing information to passive students, each lecture slide posed questions to which students were asked to suggest answers. Thus, the material was covered by posing questions rather than merely imparting content. This approach was informed by Harland (2012:47) who reasoned that that such a 'discussion approach' enables a lecturer to: *"(1) enhance knowledge, understanding and critical thinking; (2) help students develop [discussion] values; (3) induct and educate students into a disciplinary community; (4) educate students in the art of discussion."*. Accordingly, students had to prepare for theory lectures by reading an assigned chapter from the prescribed textbook. Students had to either volunteer answers, or individual students were selected arbitrarily by the lecturer and asked to venture an answer. Thus, students were challenged to have done the reading, and make an effort to think of an answer. This was designed to take students out of their comfort zone, into a learning zone (Vygotsky, 1978). Unfortunately for many this was uncomfortable and many resisted this format. In part this may be due to the unnerving experience of no longer being able to passively 'accumulate' information dispensed in lectures, but rather involved being forced to critically engage with the material – to both understand it and answer the questions posed. Some also disliked or were unused to voicing their views in public. For some, a demand for them to be an active participant was a totally new experience.

Consequently, a safe learning environment had to be created to alleviate fears of humiliation in front of one's peers and/or the lecturer. For example, if a student answered incorrectly, the lecturer-researcher gently said the answer was "not quite" or rephrased the question ("what if…"), or provided leading context ("think about…"). This was to encourage students to examine the topic and find the answer in an exploratory way by internally reformulating their answer and exercising their own learning agency. Additionally, students were allowed to hold 'one minute discussions' with their immediate neighbour.

Table 1. Schedule of work for lectures and practicals

	Schedule of Work	
	Theory Sessions	**Practical Sessions**
Week 1	• Importance of Soil Science • Genesis of soil • Basic concepts • Basic properties of soil: o Structural, Chemical and soil air, soil water, soil temperature	• Sciences Librarian demonstrates online academic databases in computer lab. • During session, students look up "Soil Erosion in South Africa" articles. • Students find and save 4 – 5 citable resources.
Week 2	• Soil composition o Structural • Soil taxonomy (International and South African) • Overview of South African soils o South African soil issues	• Academic report writing session with 'Academic Development Centre'. o Students develop a 2 to 3 page "Overview of Soil Erosion in South Africa"; the report must include: (1) Rates, regions and reasons for soil erosion in South Africa; (2) Socio-economic impacts of soil erosion in South Africa; (3) Environmental impacts of soil erosion in South Africa; and (4) Proper references.
Week 3	• Soil Chemistry o Nutrient cycling o Ion exchange o Plants and organisms	Plagiarism lecture and exercises to raise awareness, learn how to avoid plagiarism.
Week 4	• Erosion o Mechanics of erosion o Types of erosion o Factors affecting soil erosion o South Africa and erosion o Basic erosion modelling	Demonstration of how to calculate a soil erosion modelling example.
Week 5	• Soils and land use o Fitness of land use based on soils (construction, grazing agriculture, etc.) o Soil pollution o Soil management o Fertilisers o Salinisation of soils o Human dependence on soils o Soils and land use in ZA	In groups students calculate erosion modelling scenarios with data provided.
Week 6	• Soils and wetlands o Gleyed soils o Wetlands in South Africa o Ecological importance	• Write up findings and conclusions. • Students write an academic report integrating: an overview of soil erosion in SA, erosion modelling calculations, findings and conclusions. • References to be included
Week 7	• Soil pollution • Soil quality and measuring • Revision	Weathering and Erosion revision.

That is, the lecturer posted a question and students could caucus with a peer sitting next to them, before venturing an answer. This was designed to reduce the anxiety by making students feel less vulnerable and more willing to share views as part of a pair. Once the right answer was arrived at through such class discussion, the lecturer then explained concepts in more detail, and encouraged students to ask for clarification if necessary. Overall, the aim was to make students active participants, by speaking about and engaging with the content. They were also forced to take notes in class as the lecturer only made abbreviated lecture slides available.

The Tutorial Sessions

All students had to attend one tutorial a week and write an informal 'test' on subjects covered in the lectures that week, so as to consolidate content understanding and foster writing skills. Tutors marked these tutorial tests (moderated by the lecturer), recorded the results (which were low stake, and did not contribute much to the final semester mark, see Table 2), and returned the tests to the students. Students received the memo the following week to promote assessment *for* learning. The tests were open book, students could discuss answers amongst themselves and consult the tutors. Thus, tutorial tests followed Carroll's (1989) 'time on task' model, where a prescriptive 'time on task' assignment, in a controlled environment, sought to complement and reinforce the theory lectures. The marks enabled the lecturer to track student performance, and making the task for marks encouraged tutorial attendance (as many students would adapt a surface approach to learning and skip the tutorial if the task was not for marks) (McKay, 2016). The structuring and weighting of the assessments is discussed under another section.

Practical Component

The practical component focussed on soil erosion due to its severity, and the consequent social and environmental repercussions, in South Africa (Le Roux, Newby & Sumner, 2007). Unfortunately, the university in question no longer had a soil laboratory[1], so the practical activities took the form of a mini-scientific report. This practical component was laid out over seven weeks as shown in Table 1.

In Week 1, the practical session was conducted in a computer lab with one student per computer. In the session, the Sciences Faculty Librarian demonstrated how to access online academic journals and e-book databases in a step-by-step manner. Students had to find and save at least five relevant journal articles on soil erosion in South Africa during this session. Thus, this skills-based activity was highly contextualised to make for meaningful learning. The same Librarian also set up an academic resources page for the students to reference[2]. Week 2 was a coaching session on scientific academic writing by the 'Academic Development Centre' (ADC), during which students drafted a 2 - 3 page 'Overview of Soil Erosion in South Africa' essay using their sources from Week 1. The students were further encouraged to use the ADC throughout the module so as to receive continued professional feedback on their academic writing and development. The third practical entailed class exercises on plagiarism and how to cite and reference sources.

Thus, Weeks 1 to 3 laid the foundation of good scientific report writing. In Week 4, Wischmeier and Smith's (1978) Universal Soil Loss Equation (USLE) was presented. This very widely used equation (Kamuju, 2016) takes numerous soil, climate, slope and vegetation factors into account, and calculates a value of 'soil lost to erosion' in *tons per hectare per year*. The lecturer-researcher had prepared a scenario, based in the Eastern Cape of South Africa of a specific hillside. This detailed 'real world' example of applying the equation was then worked through with the students. For example, the gradient was calculated from an orthophoto map, the land use and vegetarian cover similarly, while the soil type was ascertained from a soil map, which in turn allowed soil texture and erodibility indices to be estimated. Students were shown (step by step) how to calculate certain USLE variables from maps, and how to determine other required USLE indices using nomographs.

In Week 5 the students were split into 30 groups of 4-5, *not* of their own choosing. The purpose was to build teamwork skills and the ability to work with a range of people. The groups were provided with 'abstracted' forms of data such as maps with transects and soil types, from which they had to calculate the factors for input into the USLE, and arrive at the correct answer for their group's data set. Each group was given a different data set. Each group had a USLE task to complete during the session. The task contributed to their semester mark (see Table 2). The following week the USLE task had to be incorporated into their methodology and results sections of their academic report. Week 6's session entailed the group writing up their findings and conclusions during class, with the lecturer-researcher and student assistants facilitating were necessary. Week 7 was a revision session with a small reflective class test.

The practical was designed with various pedagogical aims in mind. Firstly, it sought to teach intermediate scientific research report writing. That is, whilst the task was more complex than what would be expected at first year, it was also not at the level demanded of a final year student. As can be discerned from the scaffolded approach, Figure 2 is a schema which was provided to the students so as to make explicit the 'hidden curriculum' of a scientific report. The hidden curriculum are the assumptions and expectations of a curriculum which are undeclared (Cornbleth, 1984). Making them explicit aids students in navigating the curriculum more successfully (Portelli, 1993). The schema also highlighted the logical step-wise argument construction and the accompanying use of sub-headings. The purpose was to ensure that students were aware of how the report would be assessed so as to assist their academic success.

Assessment Plan

The assessment plan for the module was structured and weighted as per Table 2.

Assessment was driven by the 'Assessment *for* Learning' philosophy (Clark, 2008; Tan, 2013). Accordingly formative assessment was used to support learning and not simply to test content knowledge retention. Thus assessment opportunities were placed at key points in the module, with a particular focus on giving timely feedback, to promote deep learning. Starting from Week 1, there were 7 tutorial

Figure 2. A deconstructed report writing tool provided to students

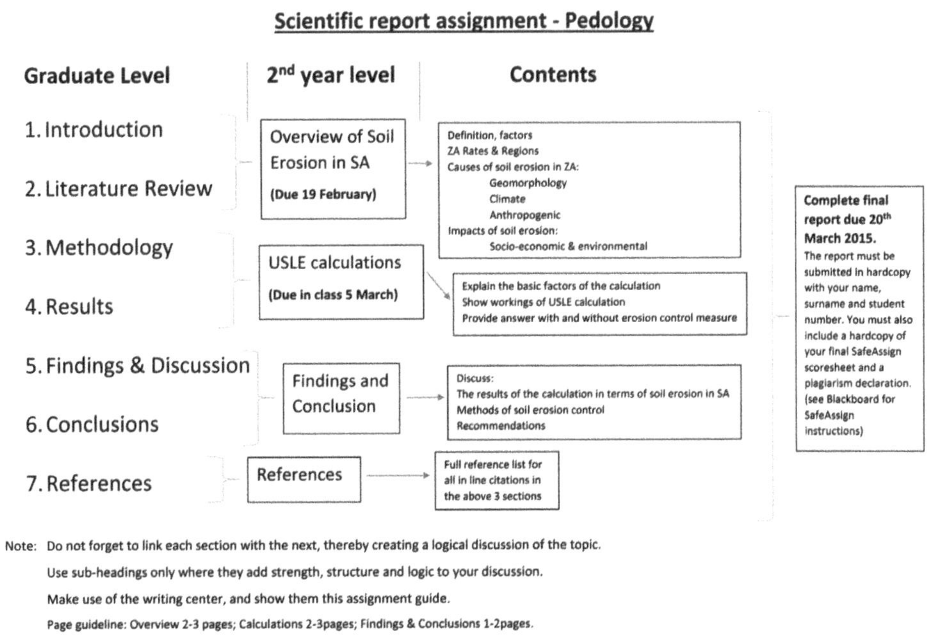

Table 2. The seven week assessment plan for the module

Week	Practical		Theory	
	Description	Weight of 100%	Description	Weight of 100%
Week 1	-	-	Tutorial 1	4%♦
Week 2	-	-	Tutorial 2	4%♦
Week 3	Submit report 'An overview of Soil Erosion in South Africa'	10%	Class test	20%
			Tutorial 3	4%♦
Week 4	-	-	Tutorial 4	4%♦
Week 5	Submit group USLE calculation done during session	20%	Tutorial 5	4%♦
Week 6	-	-	Semester Test	60%
			Tutorial 6	4%♦
Week 7	Application & reflection class test	10%	Tutorial 7	4%♦
	Submit final soil erosion report	60%		
Semester weighting	**40%**		**60%**	

♦The best five of seven tutorials counted 20% of the theory mark

assessment tasks, although only the best five contributed to the semester mark (to enable students to recover from a poor performance or absence). In total, the tutorial assessments counted 20% toward the final term mark.

In Week 3 was a 30 minute class test on the theory covered in Weeks 1 and 2. This class test counted 20% towards the theory mark. The marked class test was returned with the memorandum in Week 4. The test itself was also unpacked in class. By this stage 'at risk' students were identified and were personally emailed warning them that they were in danger of failing the module, and were encouraged to make use of student support services, or consult the lecturer. All students were encouraged to use the upcoming assessment opportunities to improve their marks. The remaining 60% of the term's theory mark comprised a 1 ½ hour semester test in Week 6 which covered the content of Weeks 1 to 4. With Weeks 1 and 2 assessed twice, they were excluded from the final examination.

The final term mark comprised 60% theory and 40% practical. The practical component comprised the production of a scientific report on soil erosion in South Africa. The two to three page '*Overview of Soil Erosion in South Africa*' counted 10% of the term's practical component. This introduction was marked according to the following rubric (Table 3). As can be seen significant weighting (53%) was given to writing skills in line with the module rationale.

By the end of Week 4, students had been given feedback and a mark on the introductory section of their soil erosion report using the rubric in Figure 3. This is in line with a the assessment philosophy of 'feeding-forward' in which: "in order to emphasise continuous learning—feeding back to feed forward [...] students need to have opportunities to make mistakes and to learn from them prior to summative assessment (through formative assessment and feedback)" (Wheatley, McInch, Fleming & Lord, 2015:1). Using formative assessment in Weeks 3 and 5 was to encourage and enable re-drafting before the final submission at the end of Week 7. However, the activities were not entirely formative in nature, as the marks counted towards the final mark. In Week 5, groups worked in class on a given USLE equation which counted 20% of the final practical mark. Again feedback was provided such that students could

Table 3. Marking rubric for Week 3's formative assessment of 'An Introduction to Soil Erosion in South Africa'

Criteria	Not achieved	Partly achieved	Almost achieved	Achieved	Achieved with distinction	Weighting	Score
Introduction to soil erosion • **Definition, factors** • **ZA Regions** • **Rates**	Very poor – language needs work, ineffective, unsuitable = 0	Weak introduction, needs a lot of revision = 2	Partly effective, but language and logic needs revision = 4	Clear, effective and logical = 6 to 8	Excellent, sets the scene well, logical and effective = 10 to 12	15%	12
Causes of soil erosion in ZA: • **Geomorphology** • **Climate** • **Anthropogenic**	Not addressed at all = 0	Addressed superficially = 2 to 6	Addressed part of the time = 7 to 8	Addressed much to most of the time = 9 to 10	Fully addressed = 11 to 14	18%	14
Logical discussion	No logical discussion, essay is almost impossible to read = 0 to 5	Essay is readable but needs a lot of revision and editing = 6 to 8	Discussion can be followed, needs some editing = 9 to 10	Easy to read, logical and coherent = 11 to 12	A pleasure to read, beautiful, logical, coherent discussion = 13 to 16	20%	16
Referencing: correct information that is supported with accurate referencing (incl. reference list)	No or very little evidence provided = 0 to 1	Some evidence provided 2 to 4	Part of the necessary evidence required was provided = 5 to 7	Lots of evidence provided = 8 to 9	Sound argument was backed by strong evidence = 10 to 11	14%	11
Conclusion • **Impacts of soil erosion: socio-economic & environmental**	Very poor – language needs work, ineffective, unsuitable = 0 to 2	Weak conclusion, needs a lot of revision = 3 to 4	Partly effective, but language and logic needs revision = 5 to 6	Clear, effective and logical = 7 to 9	Excellent, wraps up the essay very well, logical and effective = 10 to 12	15%	12
Written language skills (academic English)	Many spelling mistakes = 0 to 1 Ineffective, incorrect use of punctuation = 0 to 1 Weak sentence construction = 0 to 1 Poor, weak vocabulary = 0 to 1 Lack of cohesion = 0 to 1			Few spelling mistakes = 2 Good use of punctuation = 2 Okay sentence construction = 2 okay vocabulary, turn of phrase, register = 2 Partly coherent = 2	No spelling mistakes = 3 Excellent use of punctuation = 3 Good sentence construction = 3 Good vocabulary, turn of phrase, register = 3 Coherent = 3	19%	15

Student name: **Total mark:** **80**

Student number: **Percentage:**

Comments:

learn from and correct their errors. This methodology and results were to be included in their report following on the introduction (See Figure 3). The remaining two weeks of practical were allocated to giving groups an opportunity to refine their report and write up the discussion and conclusion. During this time students were given access to the online plagiarism-checking tool SafeAssign ™[3]. Students were encouraged to upload their reports (any number of times) to the tool to check if they may have committed plagiarism. However, students faced mark penalties should their final report score over 30% plagiarism on SafeAssign™. Students were thus incentivised to lower their SafeAssign™ score. Although automated, this too is a feed-forward approach which aided students in their learning to avoid plagiarism. The final academic report comprised the 60% of the practical component of the module. This too was marked to a rubric (Table 4).

There was a small final practical class test in Week 7 which focused on two aspects. Firstly students had to demonstrate they had synthesized knowledge gained in the practical with that of the theory lectures. In this regard, certain factors such as climate, topography and soil type were provided and students had to assess the area's erosion potential and provide reasons therefore. Secondly was a reflection activity

Table 4. Rubric for final research report

Criteria	Not yet achieved	Achieved	Achieved with distinction	Weighting	Score
Introduction: An overview of soil erosion in SA	•Introduction not give a useful overview of: erosion definition, factors (soil, climate, geomorph, veg, human), ZA regions, rates, socioeconomic & enviro impacts (0 - 6)	•Introduction considers & partly discusses: erosion definition, factors (soil, climate, geomorph, veg, human), ZA regions, rates, socioeconomic & enviro impacts (7 - 10)	•Introduction discusses & explains in logical meaningful detail: erosion definition, factors (soil, climate, geomorph, veg, human), ZA regions, rates, socioeconomic & enviro impacts (11 - 14)	14%	14
Methodology: USLE shown & explained	•Methodology does not adequately explain the research methods (0 - 6)	•Methodology explains the research method, but may lack detail or clarity (7 - 10)	•Methodology explains the research method excellently & in detail (11 - 14)	14%	14
Logical discussion: Throughout essay	•The report does not follow a logical academic discusison •Points and sections do not build on each other (0 - 10)	•The report generally follows a logical academic discusison •Points and sections attempt to build on each other (11 - 16)	•The report sussinctly follows a logical academic discusison •Points and sections build and link well to each other (17 - 22)	22%	22
Referencing: correct information supported with accurate referencing & list	•Little evidence provided •Poor & incorrect referencing standards used (0 - 8)	•Significant evidence provided •Few mistakes in referencing standards (9 - 13)	•Sound argument was backed by strong evidence •Impecable referencing standards (14 - 16)	16%	16
Findings & conclusions: Impacts of soil erosion control methods, recommendations	•Findings & conclusions are weak and poorly interrogated (0 - 8)	•Findings & conclusions show some insight & are adequatly discussed (8 - 12)	•Findings & conclusions show strong insight & investigation with adequate backing (14 - 18)	18%	18
Written language skills: Academic English	•Spelling mistakes; •Poor punctuation; •Weak sentence construction; •Weak vocabulary; •Lack of cohesion; •Inconsistent layout & formatting (0 - 6)	•Few spelling mistakes; •Good use of punctuation; •Adequate sentence construction & vocabulary; •Fairly coherent; •Few inconsistencies in layout & formatting (7 - 11)	•No spelling mistakes; •Excellent use of punctuation; •Good sentence construction & vocabulary; •Coherent; •No inconsistencies in layout & formatting (12 - 16)	16%	16

Student name:		**Total marks:**	**100**
Student number:		**Percentage:**	
Comments:			

where students had to detail what they had learnt in the module, as well as detail the challenges they had encountered and overcome. This assessment encouraged deep learning by encouraging students to synthesis and reflect on their seven weeks' worth of efforts at producing a single scientific report. This made up the remaining 10% of the practical component's mark.

An imperative of action research is reflection, and the authors now turn to a reflection on both the intervention and their personal professional academic learnings.

RESULTS, LIMITATIONS, AND RECOMMENDATIONS

Reflection afforded the mentee certain insights on the development of this Pedology curriculum. These range through: insights into the writing and the delivery of the module, to professional academic learnings on educational theory and action research practice.

Results: Intervention Limitations and Recommendations

In terms of results, it was clear, even during the delivery process, that the initial schedule of work was too ambitious (Table 1). That is, taking a deep interactive approach requires a thinner curriculum to make more time for discovery learning and class interaction. This supports the view of Pritchett & Beatty (2012) who remark that the pace of the curricula should be aligned with the pace of learning. Consequently, the themes 'gleyed wetland soils', as well as 'soil surveys', were culled from the curriculum during the delivery phase. It was also found that too little time was allocated to covering basic soil chemistry concepts. In particular, students enrolled for Bachelor of Arts and Bachelor of Education degrees had no prior knowledge to build on. Hence more time had to be devoted to basic soil chemistry than initially planned. In terms of module delivery, informal student feedback revealed that students were uncomfortable with the subverted lecturing style. They preferred the more familiar, traditional formal lecture style, which rests on the behaviourist 'instruction paradigm' which seeks to passively 'deliver' information to students (Barr & Tagg, 1995). Thus, the lecturer-researcher's expectation of responses and comments from students was met with recalcitrance. Perhaps 'clickers'[4], as a means to depersonalise the process would be appropriate in future (Caldwell, 2007).

Results: Tutors, Tutorials, and Their Limitations

The tutors and tutorials are another aspect which bear reflection. Logistical and administrative hurdles in arranging tutorials meant that they had to be slotted into times take from both the practical and theory component. Thus, the tutorials did not increase overall time allocated to Soil Science in terms of the timetable. Another serious challenge was finding venues on a campus that is catering for up to three times as many students as it was originally designed for. Furthermore, due to financial and contractual constraints, two of the four tutors were only available for three hours per week, instead of nine hours per week. This added to the logistical problems. Other than the tutors having to be third or fourth year Geography students, their appointment was arbitrary. Not all tutors had a Bachelor of Science, although all had completed an undergraduate module in Soil Science. Consequently, all four required significant coaching in the content. Additionally, they were new to teaching and so their marking of assessment tasks had to be scaffolded and moderated by the lecturer. Thus, tutors are also 'legitimate peripheral participants', themselves still mastering certain academic skills, and required support and inclusion (Lave & Wenger, 1991). Nevertheless, regardless of the differing teaching aptitudes of tutors, tutors do make a material positive difference to undergraduate students' marks (McKay, 2016). There was significant buy-in by the students with weekly tutorial attendance consistently high (>85%). Much of this can be attributed to the tutorial tasks being for marks. Additionally, marks improved over time. Initially the average tutorial task marks were below 50% but by the end of the module the average was 65%. Despite the challenges of tutorials, they can be said to be academically worthwhile.

Action Research: Lessons Learned

As explained in the start of this reflection section, the lecturer-researcher has come to learn deeply from engaging in the action research process. What is marked is that meso-reflection cycles were less formalised during the macro action phase of the project (Figure 1). So while record was kept of the action/development of the curriculum, the detail (data) on the reflections during the process are thinner.

Moreover, answer sheets from Week 7's reflective test for students were not kept. Hence it is a reflective lesson for this lecturer-researcher to document their meso-reflections with more intentionality. In addition, although all tutors were told to make a summary of the "common and remarkable points" from this reflective test, only one did. The reflective questions in Week 7's tutorial test were as follows:

1. In your opinion, what value did the tutorial sessions add to your learning experience?
2. What challenges did you need to overcome in completing your final practical assignment, and how did you overcome them?
3. What is the most important thing you have personally learnt during this course and why was it valuable to you?

From the one tutor's summary, students who responded to Question 1 found the tutorials offered a "holistic approach"; required students to "think for myself in my own words"; and the revision of that week's theory assisted students. One commented that "Tutorials helped revise my work, making studying for semester tests easier" and this led to improving students' academic confidence, for "Tuts gave me hope for passing the course". These responses show that epistemological access was increased by the tutorial sessions (Morrow, 2007). Although some students commented that tutors were available to assist, others found tutor engagement limited. In response to Question 2, students mentioned that time management, structuring their reports, and lowering their plagiarism score were challenges, but that they garnered assistance from the writing centre, the sciences librarian and by iteratively submitting to SafeAssign™. Question 3's answers also bear reflection. Students highlighted the learnings which they most valued as: "critical thinking"; "To pay attention in class, lecturer's words stick in my head"; "Preparing for lectures help[ed]"; and that "Soil is important for our well-being". These reflections seem to indicate that deep learning took place for these students (Biggs & Tang, 2007).

Reflection on the assessment processes showed that students should have been provided with the research report rubrics (see Figures 3 & 4) *before*, and not just *after* marking. Providing the rubrics to students before submission would have helped to better reveal the 'hidden curriculum' (Cornbleth, 1984; Wheatley et al, 2015). That is to say, it may have helped to reveal the lecturers' expectations and the weightings of various aspects of the report. In particular, the rubrics should have been unpacked and discussed in class ahead of the final submission (Portelli, 1993; Simpson & McKay, 2013).

The lecturer-researcher found that the theoretical underpinnings of the curriculum which was developed, were often more fully understood retroactively. During the action cycle – the development of the curriculum using the micro action research dialogue collaboration approach – the educational theory was only shallowly understood by the mentee. It was only in reflecting on the intervention as a whole that a deeper understanding and, thus, learning was achieved (Biggs & Tang, 2007). In particular, a formal reflection on a module as a whole can promote the scholarship of teaching and learning. The lecturer-researcher found that great personal satisfaction was derived from developing the curriculum, delivering it, researching and reflecting upon it. However, this process would not have been as easily accessed, actioned and engaged with, without the supporting knowledge and encouragement of mentor. Leidenfrost, Strassnig, Schabmann, Spiel & Carbon (2011), identify three mentoring styles, viz. motivating master mentoring, informatory standard mentoring and negative minimalist mentoring. From the perspective of the mentee, who found the mentor-mentee engagement motivating and informative, with little negative mentoring, the relationship could be classified as offering motivating master mentoring.

Reflections on the Mentor-Mentee Relationship

From the perspective of the mentor/co-author, the reflective aspect of action research offered an opportunity to grow as an academic, a researcher and a mentor. Firstly, a lesson learnt for the mentor was that without any official or formal training as a mentor, the mentoring activity could be best described as unconventional and colloquial. Thus, formal training in mentoring would have added value. Secondly, a challenge with respect to mentoring is knowing 'when to guide' and when to allow the mentee to 'find out for themselves' or explore. There is a delicate tension between the two. For a mentor, there are times when their own way of doing things seems 'right' compared to that of the mentee. It is those moments when the mentor must chart a way forward, where the mentor and mentee work together in a partnership in finding a new or 'third way', rather than imposing one's ideas on the mentee or allowing them to 'sink or swim' for example. A third way can offer new insights and create value. In this regard the intervention as a teaching opportunity enabled both the mentor and the mentee to grow. Thirdly, the reflection and research writing exercise can be seen as the outcome of both the mentor and mentee boundary hopping between the two disciplines of higher education teaching and Soil Science (McKay & Simpson, 2013). In this cross-disciplinary approach, we drew on our Soil Science discipline knowledge, and that of teaching in higher education as a discipline, to co-create both the intervention and this research paper. Lastly, the mentoring process, whilst time consuming and requiring high levels of academic engagement, was a creative, exciting and a personally enriching process, whereby the whole project became greater than the sum of its parts. Thus, in the development of this Soil Science curriculum, and the writing of this chapter, the benefits of *both iteratively and simultaneously* implementing an action research approach were reaped.

CONCLUSION

In conclusion, using action research to develop this second year teaching intervention created additional value for the second year students involved, as well as for both the mentor and mentee in a manner beyond what was initially envisaged. The academic engagement associated with action research resulted in the professional growth of both the mentor and the mentee. This study contributes to the literature on access for success by providing details on how this can be achieved at the second year level. The finding that academics must formally engage with boundary hopping between their home discipline (in this case Soil Science) and the theoretical and pedagogical imperatives of Education as a discipline, to construct adequate teaching interventions, is important. That is, no module should be offered through the lens of a scientific discipline alone if epistemological success is to be achieved. In particular, scientific knowledge and skills are best taught if a lecturer can come from an informed pedagogical position informed by cycles of action and reflection.

REFERENCES

Aluko, R., & Hendrikz, J. (2012). Supporting distance education students: The pilot study of a tutorial model and its impact on student's performance. *Progressio*, *34*(2), 68–83.

Amechi, E. P. (2009). Poverty, Socio-Political Factors and Degradation of the Environment in Sub-Saharan Africa: The Need for a Holistic Approach to the Protection of the Environment and Realisation of the Right to Environment. *Law, Environment & Development Journal*, *5*(2), 107–129.

Amir, A., Mandler, D., Hauptman, S., & Gorev, D. (2017). Discomfort as a means of pre-service teachers' professional development–an action research as part of the 'Research Literacy' course. *European Journal of Teacher Education*, 1–15.

Ashman, M. R., & Puri, G. (2002). *Essential Soil Science*. Oxford, UK: Blackwell Publishing.

Banning, M. (2005). Approaches to teaching: Current opinions and related research. *Nurse Education Today*, *25*(7), 502–508. doi:10.1016/j.nedt.2005.03.007 PMID:16140425

Barbara, M. G., Gayle, B. M., Cortez, D., & Preiss, R. W. (2013). Safe Spaces, Difficult Dialogues, and Critical Thinking. *International Journal for the Scholarship of Teaching and Learning*, *7*(2), 5.

Barr, R. B., & Tagg, J. (1995). From teaching to learning – A new paradigm for undergraduate education. *Change*, *27*(6), 12–25. doi:10.1080/00091383.1995.10544672

Becher, T., & Trowler, P. R. (2001). *Academic Tribes and Territories: Intellectual enquiry and the culture of disciplines* (2nd ed.). Buckingham, UK: Society for Research into Higher Education and Open University Press.

Biggs, J. (1996). Enhancing teaching through constructive alignment. *Higher Education*, *32*(3), 347–364. doi:10.1007/BF00138871

Biggs, J., & Tang, C. (2007). *Teaching for quality learning at university* (3rd ed.). Buckingham, UK: Society for Research into Higher Education and Open University Press.

Bockheim, J. G., Gennadiyev, A. N., Hammer, R. D., & Tandarich, J. P. (2005). Historical development of key concepts in Pedology. *Geoderma*, *124*(1-2), 23–36. doi:10.1016/j.geoderma.2004.03.004

Brydon-Miller, M., Greenwood, D., & Maguire, P. (2003). Why Action Research? *Action Research*, *1*(1), 9–28. doi:10.1177/14767503030011002

Caldwell, J. E. (2007). Clickers in the Large Classroom: Current Research and Best-Practice Tips. *CBE Life Sciences Education*, *6*(1), 9–20. doi:10.1187/cbe.06-12-0205 PMID:17339389

Carroll, J. B. (1989). The Carroll Model: A 25-Year Retrospective and Prospective View. *Educational Researcher*, *18*(1), 26–31. doi:10.3102/0013189X018001026

Chokwe, J. M. (2013). Factors Impacting Academic Writing Skills of English Second Language Students. *Mediterranean Journal of Social Sciences*, *4*(14), 377–383.

Clark, I. (2008). Assessment is for Learning: Formative Assessment and Positive Learning Interactions. *Florida Journal of Educational Administration & Policy*, *2*(1), 1–16.

Collins, A., Joseph, D., & Bielazcyc, K. (2004). Design research: Theoretical and methodological issues. *Journal of the Learning Sciences*, *13*(1), 15–42. doi:10.1207/s15327809jls1301_2

Cornbleth, C. (1984). Beyond Hidden Curriculum?. *Journal of Curriculum Studies, 16*(1), 29–36. doi:10.1080/0022027840160105

Davids, M. N. (2014). Traditional tutorial system-fit for purpose or past its sell-by date? University student's pedagogical experiences. *South African Journal of Higher Education, 28*(2), 338–354.

de Villiers, C. (1998). South African Environmental Reporting: What It Is, What It Should Be. *APIRA Conference Proceedings.*

Design-Based Research Collective. (2003). Design-based research: An emerging paradigm for educational inquiry. *Educational Researcher, 32*(1), 5–8. doi:10.3102/0013189X032001005

Dewar, N., & Shippey, K. (2002). Meeting the challenge of change in tertiary level environmental education in South Africa. *International Journal of Sustainability in Higher Education, 3*(4), 324–334. doi:10.1108/14676370210442370

Dochy, F. J. R. C., Segers, M., & Sluijsmans, D. (1999). The use of self-, peer and co-assessment in higher education: A review. *Studies in Higher Education, 24*(3), 331–350. doi:10.1080/03075079912331379935

Elder, Z. (2012). *The MLG Tool: Micro Action Research.* Accessed online on 25 August 2016 at: https://marginallearninggains.com/methodology-micro-action-research/

Ertmer, P. A., & Newby, T. J. (1993). Behaviorism, Cognitivism, Constructivism: Comparing Critical Features from an Instructional Design Perspective. *Performance Improvement Quarterly, 6*(4), 50–72. doi:10.1111/j.1937-8327.1993.tb00605.x

Feldman, A. (1999). The role of conversation in collaborative action research. *Educational Action Research, 7*(1), 125–147. doi:10.1080/09650799900200076

Fey, M. V. (2010). A short guide to the soils of South Africa, their distribution and correlation with World Reference Base soil groups. *Proceedings of: 19th World Congress of Soil Science, Soil Solutions for a Changing World.*

Grant, A. (2005). From a Distance: Tutors as a Community of Practice. *Education As Change, 9*(1), 80–100. doi:10.1080/16823200509487104

Harland, T. (2012). *University Teaching.* London: Routledge.

Harvey, M., Ambler, T., & Cahir, J. (2017). Spectrum Approach to Mentoring: An evidence-based approach to mentoring for academics working in higher education. *Teacher Development, 21*(1), 160–174. doi:10.1080/13664530.2016.1210537

Jacobs, C. (2007). Towards a critical understanding of the teaching of discipline-specific academic literacies: Making the tacit explicit. *Journal of Education, 41*, 59–81.

Jones, M.G. & Brader-Araje, L. (2002). The Impact of Constructivism on Education: Language, Discourse, and Meaning. *American Communication Journal, 5*(3).

Kamuju, N. (2016). Soil Erosion and Sediment Yield Analysis Using Prototype & Enhanced SATEEC GIS System Models. *International Journal of Advanced Remote Sensing and GIS, 5*(1), 1471–1482. doi:10.23953/cloud.ijarsg.39

Keesstra, S. D., Bouma, J., Wallinga, J., Tittonell, P., Smith, P., Cerdà, A., & Fresco, L. O. et al. (2016). The significance of soils and soil science towards realization of the United Nations Sustainable Development Goals. *Soil*, *2*(2), 111–128. doi:10.5194/soil-2-111-2016

Lave, J., & Wenger, E. (1991). *Situated learning: Legitimate peripheral participation*. Cambridge, UK: Cambridge University Press. doi:10.1017/CBO9780511815355

Le Roux, J. J., Newby, T. S., & Sumner, P. D. (2007). Monitoring soil erosion in South Africa at a regional scale: Review and recommendations. *South African Journal of Science*, *103*, 329–335.

Leidenfrost, B., Strassnig, B., Schabmann, A., Spiel, C., & Carbon, C. C. (2011). Peer Mentoring Styles and Their Contribution to Academic Success Among Mentees: A Person-Oriented Study in Higher Education. *Mentoring & Tutoring: Partnership in Learning*, *19*(3), 347–364. doi:10.1080/13611267.2011.597122

Leitch, R., & Day, C. (2000). Action research and reflective practice: Towards a holistic view. *Educational Action Research*, *8*(1), 179–193. doi:10.1080/09650790000200108

Letsekha, T. (2013). Revisiting the debate on the Africanisation of higher. *The Independent Journal of Teaching and Learning*, *8*, 5–18.

Lotz-Sisitka, H. (2004). Environmental education research and social change: Southern African perspectives. *Environmental Education Research*, *10*(3), 291–295. doi:10.1080/1350462042000258143

Makgato, M. (2012). Identifying constructivist methodologies and pedagogic content knowledge in the teaching and learning of technology. *Procedia: Social and Behavioral Sciences*, *47*, 1398–1402. doi:10.1016/j.sbspro.2012.06.832

McKay, T., & Simpson, Z. (2013). The Space Between: Pedagogic Collaboration between a Writing Centre and an Academic Department. *Perspectives in Education*, *31*(4), 27–43.

McKay, T. J. M. (2013). Embedding academic support within an academic discipline: A teaching model. *South African Journal of Higher Education*, *27*(3), 682–695.

McKay, T. J. M. (2016). Do tutors matter? Assessing the impact of tutors on first year academic performance at a South African university. *Journal of Student Affairs in Africa*, *4*(1), 53–64. doi:10.14426/jsaa.v4i1.144

Morrow, W. (2007). *Learning to teach in South Africa*. Cape Town: HSRC Press.

Pineteh, E. A. (2014). The Academic Writing Challenges of Undergraduate Students: A South African Case Study. *International Journal of Higher Education*, *3*(1), 12–22.

Portelli, J. P. (1993). Exposing the Hidden Curriculum. *Journal of Curriculum Studies*, *25*(4), 343–358. doi:10.1080/0022027930250404

Pritchett, L., & Beatty, A. (2012). *The Negative Consequences of Overambitious Curricula in Developing Countries*. CGD Working Paper 293. Washington, DC: Centre for Global Development.

Russell, T., & Munby, H. (1992). *Teachers and Teaching: from classrooms to reflection*. London: Falmer Press.

Shaw, C. F. (1930). Is "Pedology" Soil Science?. Proceedings of: American Soil Survey Association.

Simpson, Z., & McKay, T. J. M. (2013). Assessment Rubrics: Artefacts That Speak In Tongues? *Lingua*, *29*(1), 15–32. doi:10.5785/29-1-539

Struyven, K., Dochy, F., & Janssens, S. (2003). Students' perceptions about new modes of assessment in higher education: A review. In *Optimising new modes of assessment: In search of qualities and standards* (pp. 171–223). Springer. doi:10.1007/0-306-48125-1_8

Sutherland, K. A. (2009). Nurturing undergraduate tutors role in the university teaching community. Mentoring & Tutoring. *Partnership in Learning*, *17*(2), 147–164. doi:10.1080/13611260902860091

Tan, K. (2013). *A Framework for Assessment for Learning: Implications for Feedback Practices within and beyond the Gap*. ISRN Education. doi:10.1155/2013/640609

Topping, K. J. (1996). The effectiveness of peer tutoring in further and higher education: A typology and review of the literature. *Higher Education*, *32*(3), 321–345. doi:10.1007/BF00138870

Vitousek, P. M., Mooney, H. A., Lubchenco, J., & Melillo, J. M. (1997). Human Domination of Earths Ecosystems. *Science*, *277*(5325), 494–499. doi:10.1126/science.277.5325.494

Vygotsky, L. (1978). Interaction between learning and development. In *Mind and Society: The development of higher psychological processes* (pp. 79–91). Cambridge, MA: Harvard University Press.

Wakatsuki, T., & Rasyidin, A. (1992). Rates of weathering and soil formation. *Geoderma*, *52*(3–4), 251–263. doi:10.1016/0016-7061(92)90040-E

Wegner, E. (2000). Communities of Practice and Social Learning Systems. *Organization*, *7*(2), 225–246. doi:10.1177/135050840072002

Wenger, E. (1998). *Communities of practice: Learning, meaning, and identity*. Cambridge, UK: Cambridge University Press. doi:10.1017/CBO9780511803932

Wheatley, L., McInch, A., Fleming, S., & Lord, R. (2015). Feeding back to feed forward: Formative assessment as a platform for effective learning. *Kentucky Journal of Higher Education Policy and Practice, 3*(2).

Wischmeier, W. H., & Smith, D. D. (1978). Predicting rainfall erosion losses—a guide to conservation planning. USDA, ARS, Agricultural Hand-Book #537.

Wood, D., Bruner, J. S., & Ross, G. (1976). The Role of Tutoring in problem solving. *Journal of Child Psychology and Psychiatry, and Allied Disciplines*, *17*(2), 89–100. doi:10.1111/j.1469-7610.1976.tb00381.x PMID:932126

ENDNOTES

1. It was dismantled primarily due to underfunded massification. Firstly, space had to be made for expanded classrooms needed to accommodate the many additional students that the university accepted. Secondly, continuing with hands-on laboratory sessions was deemed financially and logistically impossible.
2. http://ujsciencelibrarian.pbworks.com/w/page/6193647/Geography%202A%20Pedography [accessed 10/02/2017]
3. https://en-us.help.blackboard.com/Learn/Administrator/Hosting/Tools_Management/SafeAssign [accessed 10/02/2017]
4. These Audience Response Systems (ARS) are a technological aid, which allow students in real time to anonymously feedback to a question in class, the results of which can be displayed on the projector screen (Caldwell, 2007).

Chapter 12
Using Transdisciplinary Action Research Toward Sustainable Management of Vineyard Management and Tourism in the Negev Highlands

Noa Avriel-Avni
Dead Sea and Arava Science Center, Israel

Jen M. Holzer
Technion, Israel

Moshe Shachak
Ben Gurion University, Israel

Daniel E. Orenstein
Technion, Israel

Elli E. Groner
Dead Sea and Arava Science Center, Israel

ABSTRACT

Communities often lack a framework to guide research and action by which to mitigate complex socio-ecological challenges in the face of conflicting interests and poorly understood ecological and socio-political mechanisms. In an effort to provide such a framework, this article offers an approach for the systematic analysis of societal interactions with the landscape as well as for the structure and function of the ecosystem. Using an approach informed both by transdisciplinary research (TdR) and participatory action research (PAR), modeling is employed to identify trajectories of human influence on the ecosystem, which is illustrated using a case from the Negev Highlands of Israel. The approach identifies several cascades of effects, allowing diverse stakeholders to better understand the mechanisms by which human activities change the capacity of the ecosystem to support human well-being over time, as well as building capacity for stakeholder cooperation for sustainable management.

DOI: 10.4018/978-1-5225-2642-1.ch012

INTRODUCTION

Coupled Socio-Ecological Systems as a Framework for Sustainable Management

Modeling the human environment as a coupled social-ecological system has become popular as a conceptual tool in recent years (Turner, Georgiou & Fisher, 2008). This view conceptualizes the social system as interlinked with the ecological system, which are both dynamic, due to internal processes and external drivers (Pickett, Kolasa & Jones, 1994; Collins et al., 2011; Landsberg et al., 2011). The benefits the social system derives from the ecological system have come to be known as "ecosystem services". Ongoing pressures and events caused by human behavior change the structure and functioning of the ecological system, and affect the composition and abundance of the ecosystem services it provides (Azapagic, 2010).

Sustainability, according to this theory of socio-ecological systems, is defined as the resilience of the socio-ecological system; namely, its ability to absorb shocks while maintaining its essential functions (Plummer and Armitage, 2007). We use sustainability to refer to the preservation of the system's ability to adapt to change (although not necessarily the preservation of the status quo) (Griggs et al., 2013).

While the meaning of sustainability was always intended to encompass the range of interconnected environmental and social factors, this has long been a challenge, both theoretically and in the context of policy and management (Orenstein and Shach-Pinsly, 2015). One reason for this difficulty has been attributed to the separateness of natural scientists, social scientists, and others and their related disciplines (Liu et al., 2007). Even though it is now accepted that human societies are inextricably linked with the biophysical environment, it is difficult for some to leave behind the dichotomy between "man and nature", long popular as a dominant belief in Western civilization (Pickett, Kolasa & Jones, 1994; Berkes and Folke, 2000; Keiny, 2004). The idea of the coupled socio-ecological system arose concurrently in multiple disciplines as an effort to bridge this disconnect (Singh et al., 2012). This is a conceptual framework in which, ideally, natural scientists and social scientists collaborate in an effort to understand how human behaviors can affect ecosystems and alter their structure and function in a way that damages or enhances its ability to provide ecosystem services; that is, to continue to support the social system (Folke 2006; Azapagic, 2010).

Despite the aspiration to bridge and create a meaningful knowledge exchange among researchers from different disciplines, many socio-ecological models focus primarily on ecosystem processes and relate to social processes as drivers of ecological change without analyzing the underlying causes of individual, social, and political behavior. For example, when a complex socio-ecological issue was modeled using the type of ecological model known as DPSIR (drivers, pressures, states, impacts, and responses), the social drivers that were identified as affecting the greater socio-ecological system (environmental ethics, meat consumption, animal welfare regulations, meat export demand, energy use, and transport, in this study) were identified in a highly simplified fashion (Niemeijer & de Groot, 2008). This kind of analysis, while emphasizing the importance of understanding the entire socio-ecological system, does not provide much insight into the complexity of the social system or the feedback between the two systems. In addition to the links between a given social system and ecosystem, socio-ecological research should explore the underlying motivations, beliefs, and behaviors that drive social phenomena (such as environmental ethics, meat consumption, etc.).

MAIN FOCUS OF THE CHAPTER

In this chapter, we use a case study of a socio-ecological system in Israel's arid Negev Highlands to describe how an environmentally damaging farming practice became the trigger for a new collaboration between scientists and stakeholders. Over the last several decades, as population density increased in central Israel, agriculture expanded to the country's arid zone, which is considered the periphery of the country. When farmers began to have problems of erosion and soil salination, researchers initiated an active learning and participatory action research program in order to tackle socio-ecological problems, leveraging ancient agricultural knowledge whose infrastructure still exists in the form of ancient terraces and dams. One of the motivations behind this collaboration was the desire to create opportunities for greater stakeholder agency in understanding the socio-ecological systems by harnessing knowledge of local residents. We use this case to advocate for a process of transdisciplinary action research that links ecological and social research as well as meaningful stakeholder participation.

Lack of Integrated Knowledge and the Need for a Transdisciplinary Approach

Failure to understand underlying drivers of ecological and social processes may lead to an incomplete understanding of the social-ecological system as a whole, which in turn may lead to ill-conceived approaches to sustainable management (Burns & Worsley, 2015). Unsustainable practices can result for multiple reasons, such as lack of information, a lack of long-term thinking, the prevalence of profit motive over multiple stakeholder interests, and conflicts of interest among stakeholders (Dick, 2011). Transdisciplinary research (TdR) begins with a real-world challenge, and, in an ideal scenario, engages academics, practitioners, and stakeholders to develop new conceptual frameworks, set a research agenda, and conduct research, to produce knowledge needed for policy and management (Mauser et al., 2013). TdR has long been touted as most suitable for complex, high-stakes issues (Funtowicz & Ravetz, 1993). Importantly, TdR encourages scientists not only to work together across disciplines, but to engage other stakeholders as partners in setting the research agenda, and in long-term ongoing research.

Transdisciplinary research can be defined as an approach to problem-solving that coordinates a variety of scientific and non-scientific actors, including stakeholders, to integrate diverse types of knowledge, consider risks and consequences of possible solutions, and generate practical solutions that may be implemented. Ecologists have promoted interdisciplinary (collaborations between academics from different fields) and transdisciplinary (academics working with members of other sectors) research approaches in the hope that it will allow them to better translate scientific knowledge to the realm of policy and management (Orenstein & Groner, 2014). Transdisciplinary action research (TDAR) is a variation of this approach. This term developed out of the planning literature, and describes meta-research -- research that reflects upon a transdisciplinary research process, including the process of developing and conducting research projects as well as translating knowledge into decision making processes and policy (Stokols, 2006).

The Need to Make Stakeholders Real Partners

There has been evidence over recent decades that the most effective solutions to environmental challenges are often those that are researched, developed, and implemented together with stakeholders (Burns and Worsley, 2015; Clark, 2011; Shipley and Utz, 2012). Participatory action research (PAR) developed to better engage local and marginalized people, first to enhance and express their knowledge of an issue affecting them, and then to plan, prioritize, monitor, evaluate and take action on the issue (Chambers 1997, 2008). It is a community development approach in which facilitators work with vulnerable populations to identify problems, and to develop and implement a plan of action to solve those problems (de Negri & Ilinigumugabo, 1998; Swantz, 2008). PAR emphasizes collaboration, equality, and interpretation and criticism, and can be an effective approach for identifying and implementing culturally appropriate solutions to planning in situations of uncertainty (Zuber-Skerritt, 2012). PAR often describes a process in which a non-profit organization works closely with a marginalized group of locals to identify and implement a solution to a problem. By contrast, we use a PAR approach with the aim not only of engaging stakeholders but of allowing them to *drive the process of agenda-setting* for research.

PAR also provides a framework for participants to reflect upon their beliefs and assumptions and how these are linked to social behavior and reality, and by extension, to impacts on the socio-ecological system. This concept is referred to as a theory of action (ToA), in which actors in a social system identify the complex of interrelated (and often unconsciously held) assumptions that govern their behavior in order for the individual or group to achieve their goals (Putnam, 2014). According to Argyris and Schön (1974), creating knowledge that is useful for practice demands a process in which actors reflect upon underlying values, norms, or frames and alter their behavior accordingly. TdR also encourages reflective behavior among researchers -- about the research process -- drawing upon various methods from different disciplines. We demonstrate in the case study below that using ToA to analyze the assumptions held by various stakeholders can be an effective strategy in the context of TdR.

Demand Emerges for a Comprehensive Socio-Ecological Study of the Negev Highlands

In the autumn of 2015, a group of scientists began the process of establishing the Negev Highlands region as Long-Term Socio Ecological Research (LTSER) platform. LTSER platforms are currently being advanced around the world as hubs of interdisciplinary and transdisciplinary research, as part of a global research network focused on producing integrated socio-ecological knowledge about areas representative of global socio-ecological systems (Mirtl et al., 2013). The Negev Highlands LTSER Platform, currently in a start-up phase, is located in the central Negev desert in southern Israel.

Over twenty years ago, a local initiative produced a plan to develop an agro-tourism park in a region adjacent to the town of Mitzpe Ramon (population approx. 5000), in an effort to increase tourism and build up agriculture as a new livelihood in the area. The plan was to set aside 1,200 dunams of rocky desert adjacent to an existing nature reserve, for vacation rentals, vineyards, and olive groves. About a third of the park area was designated for agriculture and tourism entrepreneurs. Much of the land marked for these projects lay within a dry streambed containing ancient relics of Byzantine farms (1400 CE) (Evenari, Shanan & Tadmor, 1982). The project aimed to promote the integration of tourism and agriculture in the desert environment while supporting natural, heritage, and structural conservation. Although this initiative was never formalized, several wineries were established in the designated area. Most of the entrepreneurs are middle class individuals without much experience in agriculture.

Over the years, farmers experienced accelerated soil erosion and salination, and their olive trees and vines were uprooted by floods. Since the Drainage Authority is responsible for regulating the drainage infrastructure in the area, farmers repeatedly requested help from the Drainage Authority to cope with soil degradation processes. The initiative to start this process came from the Drainage Authority, one of the institutional stakeholders of the newly established Negev Highlands LTSER Platform. The Drainage Authority approached platform researchers in the autumn of 2015, which at the time, was composed primarily of ecologists who were eager to expand research collaborations to enable a more interdisciplinary research approach for the platform. The Drainage Authority, with an interest in explicitly socio-ecological research, requested that the scientists conduct an initial survey of farmers to better understand land management practices and the challenges farmers were experiencing in developing agriculture under the hydro-ecological conditions of a rocky desert. Based on observations of water flow in the basin and interviews with stakeholders, the scientists concluded that the issues of erosion and salination were related to interference with the natural water flow pattern in the desert watershed. They determined that farmers contributed to this problem due to their lack of knowledge and experience in managing local runoff systems (Avriel-Avni et al., 2016). The farmers' irrigation methods were effectively degrading the quality (by salinization) and stability (by soil erosion) of the soil. In the language of ecosystem services, farmers had degraded some ecosystem services related to soil function and runoff regulation in pursuit of increasing certain provisioning services, paradoxically harming their livelihood. This presented a potential ethical challenge, because the farmers were privately benefitting from the increased provision services in the short term, while degrading regulating services, which are both a shared resource and are crucial to the farmers' economic wellbeing in the long term.

As a direct result of the research and in recognition of its policy relevance, the Drainage Authority set out to create rules for sustainable management of the environment, which became the trigger for more in-depth social-ecological research. The director of the platform (and the first author of this article) exploited this opportunity to engage stakeholders using a transdisciplinary PAR approach. In other words, at this point, there was recognition by the Drainage Authority that a comprehensive, transdisciplinary study was needed to reconcile economic activities conducted in the watershed with ecological impacts and the social implications resulting from changes in ecosystem services.

Using an integrated approach (which we refer to as TD-PAR) platform scientists queried a range of stakeholders and experts, first in one-on-one interviews, and subsequently in group meetings, in an attempt to define key issues in this arid agro-ecosystem and identify the information and knowledge needed to address these issues. Figure 1 depicts the partnership-building process at the Negev Highlands platform, described in the following section.

The numbers indicate the steps of development of a Transdisciplinary- Participatory-Action Research framework. Numbers in the diagram are elaborated in the description below. (Key: S (1-3) ToA - Stakeholder Theory of Action; LTER- Long Term Ecological Research; LTSER - Long Term Social-Ecological Research platform)

Phase 1: Setting the Foundation for Transdisciplinary Research

Following a request by the Drainage Authority, the LTSER researchers (the authors of this chapter) began to gather physical, ecological and social information about the area (step 1). At the same time, the researchers began to collect information about similar processes around the world, as well as examples for modeling such situations (step 2). They engaged additional scientists for help refining socio-ecological models (step 3). Some of these connections with new scientists would result in their long-term engagement in the process as advisors (step 5). These socio-ecological models described the trajectories of the impact of human activities such as fencing vineyards and planting vines, and how these activities

Figure 1. Building TD-PAR of stakeholders and multi research disciplines researcher

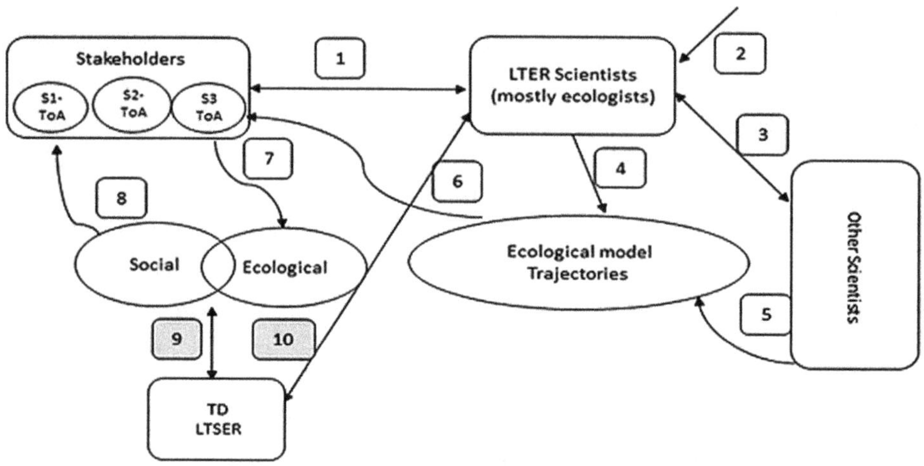

impacted the structure and function of the local ecosystem and affected its continued ability to provide essential ecosystem services.

Phase 2: Establishing Active Learning and Action Research

Data collection relied upon a series of convergent interviews with farmers (step 1). Convergent interviewing is a method that uses a structured process with open-ended questions, and which analyzes interview data as the interviews take place (Dick, 1990). Data analysis from these interviews revealed a larger number of direct and indirect stakeholders than originally assumed. Interviews also showed that each stakeholder (labeled S1 etc. in Figure 1) had a unique understanding of the ideal socio-ecological situation and what it would take to get there; in other words, his own theory of action (ToA). These interviews served to engage stakeholders in dialogue, creating an atmosphere of trust between researchers and local stakeholders (step 6), and the interview process signaled the start of an active learning and action research approach around the issue of vineyard cultivation in the Negev Highlands. As scientists from multiple disciplines checked facts with farmers on the ground, a process of mutual learning began (steps 7 and 8). This learning process enabled the creation of models that visualized the flow of decisions and actions around vineyard cultivation in the region.

An example is illustrated in Figure 2. Planting vineyards involves plowing, removal of native vegetation, and planting rows of grapevines along the flow path of ephemeral river beds (also known wadis, washes, or arroyos) of the desert land. Vineyards were first planted in the area about 20 years ago (notwithstanding agricultural activities of ancient civilizations), reflecting local, national, and global socio-economic influences. Population growth and increased human population density in the center of Israel created an incentive for new farmers to move to desert areas in the periphery of the country. Global economic conditions helped increase demand for boutique wines. Large temperature differentials between summer and winter, low humidity, and lack of pests contributed to a perception of the Ministry of Agriculture, subsequently adopted by farmers, that this area was highly suitable for growing wine grapes. Preparing the land for agriculture was supported in-kind by quasi-governmental organizations like the Jewish National Fund, which helped prepare the land for agricultural uses, and the National

Drainage Authority, which repaired damage that occurred on agricultural land as a result of soil erosion after rain events. Farmers' perceptions seemed to be corroborated when they found the remains of ancient agricultural systems on their land. However, because they had no previous experience using runoff water for agriculture, they ignored the ancient agricultural system, designed to regulate soil erosion and soil salinity (Avriel Avni et al. 2016).[1] Most farmers also chose to remove all natural vegetation, which they understood as competing with their domestic crops for water, but the removal of vegetation had the unintended consequence of increasing soil erosion. As a result of the soil erosion, some farmers stopped planting in the dry riverbed and instead planted their vineyards on the more saline banks. One reason for this was based on the perception that methods from modern agriculture would provide solutions to any problems that might arise. In reality, after a rain event created soil erosion damage, farmers would request aid from the Drainage Authority and would receive it. This social dynamic reinforced the perceptions mentioned earlier, preserving unsustainable agricultural methods. Feedback from various stakeholders was incorporated into the trajectories model, improving the accuracy of the models and creating a forum for dialogue. Systematic mapping was undertaken to identify key actors and their interests, and to begin dialogue among researchers and stakeholders.

The model depicts the interactions (arrows) and components (boxes) of a social-ecological system driven by framers activity (light gray boxes) aiming at ecosystem services of provision of grapes (dark gray box number 1) in a desert based on conventional agricultural methods. Farmers activity affect the dynamics of the ecological system (white boxes) by modulating geodiversity, vegetation pattern and functional connectivity of surface runoff water (solid arrows). This result in modifying a natural ecosystem service of soil erosion regulation (dark gray box number 2). The Drainage Authority's willingness to repair the damage of soil erosion, causing a positive feedback which preserves the existing agricultural method (dashed arrows).

Interviews also revealed that the availability of unlimited water for irrigation (which is expensive, but offered to farmers at a discount), led the new farmers to see themselves as largely independent from the reality of desert conditions, which allowed them to perpetuate their perceptions about the possibilities of farming in this desert ecosystem, including ignoring the ancient agricultural methods for mitigating

Figure 2. The current situation: a socio-ecological model of farming in the Negev Highlands

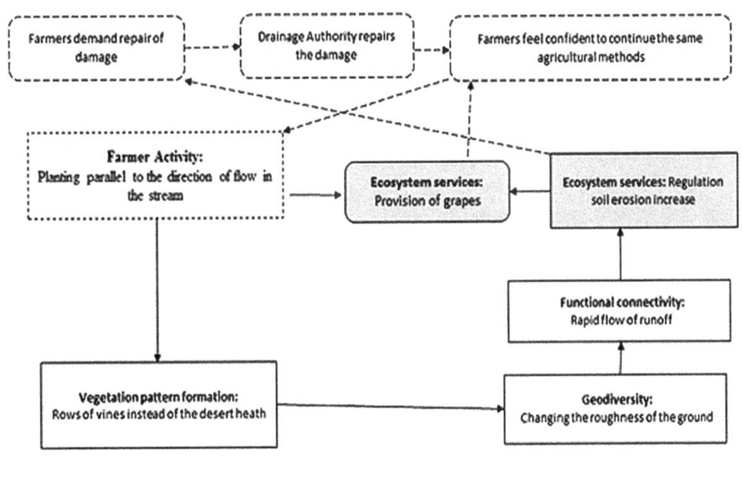

soil erosion and soil salinity. Here, the willingness of the Water Authority to provide farmers with an unlimited supply of water enabled farmers to perpetuate their beliefs and unsustainable practices.

The current understanding of ecologists and social scientists affiliated with the Negev Highlands LTSER platform led them to initiate a series of meetings with farmers in order to figure out together how actors can alter their behaviors and reverse the negative feedbacks that have been perpetuating activities that increase soil erosion.

Phase 3: Using Transdisciplinary-Participatory Action Research (TD-PAR) Toward Sustainable Management

In this phase, collaborative active learning between researchers and farmers led to understanding how to educate actors about the misinformation that perpetuates the ideas and actions that increase soil erosion in farmers' fields. Figure 3 illustrates this new understanding. The light gray boxes and the dashed lines describe the social dynamics that must change in order to achieve the reduction of soil erosion, an increase in the supply of grapes, and a rise in farm productivity.

The model depicts the interactions (arrows) and components (boxes) and feed backs from ecosystem services to the socio-ecological system, aiming at restoration of ecosystem services demerge. Two activities are altered by understanding the social-ecological system: 1) Drainage Authority recovers ancient agricultural terraces to prevent continued soil erosion 2) farmers are planting perpendicular to the direction of runoff flow. As a result the impact cascade of human activity on the ecosystem is altered and the ability of the ecosystem to provide regulation and provision services are restored.

It is important to note that the change in social dynamics requires cooperation with other stakeholders in the area, including Bedouin shepherds, tour operators and the local council. Actors should aim to achieve a shared understanding of the coupled socio-ecological system in order to establish a strong transdisciplinary partnership (steps 9 and 10).

Figure 3. The product of the TD-PAR process: a socio-ecological model of farming in the Negev Highlands

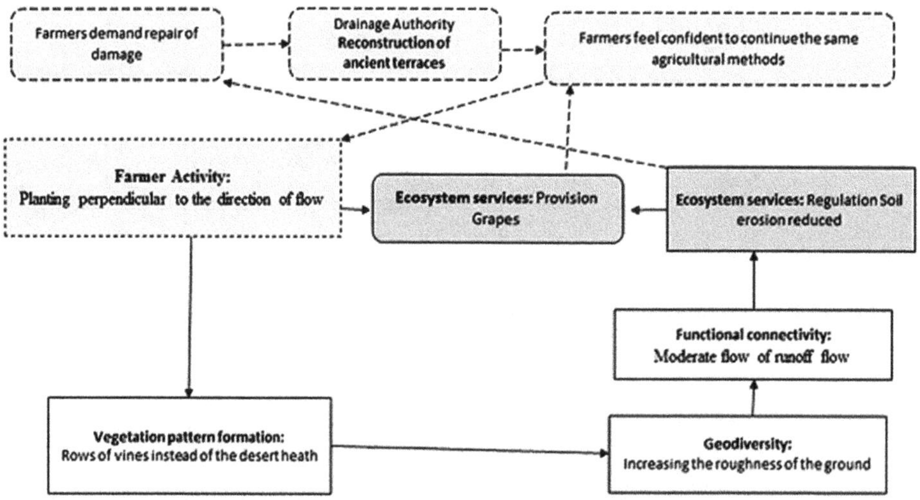

Phase 4: Next Steps

At this point, the Negev Highlands LTSER platform has created structures, processes, relationships, and tools needed for advancing sustainable outcomes in the region. Moving forward, greater attention should be placed on additional regional sustainability issues and their social, political, and economic dynamics that affect ecosystem structure and function. Patterns of thinking, behavioral norms, and economic and political pressures should be analyzed in an effort to identify differences in ToAs between various stakeholders and ultimately, facilitate dialogue that helps stakeholders to create a participatory action theory.

Advanced stages of TD-PAR should result in an integrated social-ecological model, replacing the ecological model. This model should include a detailed analysis of the social dynamics which give rise to particular land use practices, combined with their bio-physical and social trajectories. One of the benefits of an integrated model is that it should aid actors in better understand each other's theories of action, which can instigate a process of developing a common model shared by all collaborators that may better aid progress toward mutual learning, research and management.

CONCLUSION

Using conceptual frameworks drawn from various fields, in particular transdisciplinary research as defined by the field of socio-ecological systems science, and PAR, we have demonstrated how ecologists, social scientists, institutional stakeholders and local stakeholders (such as small-scale farmers) can use an iterative process for building of socio-ecological models to increase understanding of science and local knowledge and practice, becoming familiar with perspectives of diverse stakeholders, building relationships and dialogue, and normalizing active learning.

Ideally, the process of compiling the knowledge needed to build these models, building them, incorporating diverse feedbacks and modifying them to ensure accuracy, and finally using them for educational purposes, sets the groundwork and builds skills to enable diverse actors to collaborate. This itself is a significant precedent, for it can set the stage for collaborative decision making.

However, this type of work is often rife with conflict on several levels (Norris et al., 2016). First, interviews revealed that it can be challenging for ecologists to collaborate with other types of natural scientists because of conflicting scientific paradigms and goals. Further, ecologists may be charged with working directly with stakeholders to revise and refine the socio-ecological model, but ecologists may have a tendency to focus primarily on the ecological model. Ideally, in the case above for example, the research team, now composed of two ecologists, would become a larger, more interdisciplinary group, which might force a more integrated approach to formulating and defining problems and building models. An additional challenge is that institutional and local stakeholders, such as the small-scale farmers in the Negev Highlands, may resist the process in various ways. The central challenge of the process to date has been for LTSER scientists to involve farmers so that the stakeholders would begin to take ownership of the process and take it upon themselves to drive the process forward.

Despite such challenges, the TD-PAR approach has already transformed the collaboration process in the Negev Highlands. It has already facilitated new dialogues across disciplinary divides (among scientists) and across the scientist-stakeholder divide. Reflexive exercises informed by PAR helped actors to identify and understand their respective theories of change, and take steps toward a new, more widely shared theory of change. The Negev Highlands LTSER platform has built social capital, sowing seeds of a shared vision, uncovering facts, and exchanging expertise gleaned from research and experience. In short, the TD-PAR approach has already helped the LTSER scientific team to coordinate a productive process, and moving forward, these authors hope to report transformations not only within the process but also transformations of socio-economic realities, toward sustainable agriculture, tourism, and environmental management in the Negev Highlands.

REFERENCES

Argyris, C., & Schön, D. A. (1974). *Theory in practice: Increasing professional effectiveness.* San Francisco: Jossey-Bass.

Avriel Avni, N., Avni, Y., Pney-Gil, D., & Ginat, H. (2016). *Formulation of management principles of runoff and soil erosion in agricultural projects in upper Zin basin.* Interim Report to Dead Sea Drainage Authority.

Azapagic, A. (2010). Life cycle assessment as a tool for sustainable management of ecosystem services. *Issues in Environmental Science & Technology, 30,* 140–168. doi:10.1039/9781849731058-00140

Berkes, F., Folke, C., & Colding, J. (2000). *Linking social and ecological systems: management practices and social mechanisms for building resilience.* Cambridge University Press.

Bruns, D., & Worsley, S. (2015). *Navigating complexity in international development. Facilitating sustainable change at scale.* Rugby, UK: Practical Action Publishing. doi:10.3362/9781780448510

Chambers, R. (1997). *Whose reality counts? Putting the first last. Intermediate Technology Publications Ltd.* doi:10.3362/9781780440453

Chambers, R. (2008). 'PRA, PLA and pluralism: practice and theory. In P. Reason & H. Bradbury (Eds.), *The Sage handbook of action research: participative inquiry and practice* (2nd ed.; pp. 297–318). London, UK: Sage. doi:10.4135/9781848607934.n28

Clark, S. G. (2011). *The policy process: A practical guide for natural resource professionals.* New Haven, CT: Yale University Press.

Collins, S. L., Carpenter, S. R., Swinton, S. M., Orenstein, D. E., Childers, D. L., Gragson, T. L., & Knapp, A. K. et al. (2011). An integrated conceptual framework for long term social–ecological research. *Frontiers in Ecology and the Environment, 9*(6), 351–357. doi:10.1890/100068

de Negri, B., Thomas, E., & Ilinigumugabo, A. (1998). Community-based program development.: Vol. 1. *Trainer's Manual and Participant's Handbook.* Nairobi: Academic Press.

Dick, B. (2011). Organizational survive and improvement through resilience. In M. J. Celinski & K. Gow (Eds.), *Continuity Versus Creative Response to Challenge: The Primacy of Resilience and Resourcefulness in Life and Therapy*. Nova Science Publishers.

Dick, R. (1990). *Convergent Interviewing*. Brisbane, Australia: Inter-change.

Evenari, M., Shanan, L., & Tadmor, N. (1982). *The Negev: the challenge of a desert*. Harvard University Press. doi:10.4159/harvard.9780674419254

Folke, C. (2006). Resilience: The emergence of a perspective for social–ecological systems analyses. *Global Environmental Change*, *16*(3), 253–267. doi:10.1016/j.gloenvcha.2006.04.002

Funtowicz, S. O., & Ravetz, J. R. (1993). *The emergence of post-normal science. Science, politics and morality* (pp. 85–123). Springer. doi:10.1007/978-94-015-8143-1_6

Griggs, D., Stafford-Smith, M., Gaffney, O., Rockström, J., Öhman, M. C., Shyamsundar, P., & Noble, I. et al. (2013). Policy: Sustainable development goals for people and planet. *Nature*, *495*(7441), 305–307. doi:10.1038/495305a PMID:23518546

Haberl, H., Winiwarter, V., Andersson, K., Ayres, R. U., Boone, C., Castillo, A., & Furman, E. et al. (2006). From LTER to LTSER : Conceptualizing the Socioeconomic Dimension of Long-term Socio-ecological Research. *Ecology and Society*, *11*(2), 13. doi:10.5751/ES-01786-110213

Keiny, S. (2004). *Ecological thinking. A new approach to educational change*. Boston: University Press of America.

Landsberg, S., Ozment, S., Stickler, M., Henninger, N., Treweek, J., Venn, O., & Mock, G. (2011). *Ecosystem services review for impact assessment: Introduction and guide to scoping*. WRI Working Paper. World Resources Institute.

Liu, J., Dietz, T., Carpenter, S. R., Alberti, M., Folke, C., Moran, E., & Taylor, W. W. et al. (2007). Complexity of coupled human and natural systems. *Science*, *317*(5844), 1513–1516. doi:10.1126/science.1144004 PMID:17872436

Mauser, W., Klepper, G., Rice, M., Schmalzbauer, B. S., Hackmann, H., Leemans, R., & Moore, H. (2013). Transdisciplinary global change research: The co-creation of knowledge for sustainability. *Current Opinion in Environmental Sustainability*, *5*(3–4), 420–431. doi:10.1016/j.cosust.2013.07.001

Mirtl, M., Orenstein, D. E., Wildenberg, M., Peterseil, J., & Frenzel, M. (2013). Development of LT-SER platforms in LTER-Europe: challenges and experiences in implementing place-based long-term socio-ecological research in selected regions. In *Long Term Socio-Ecological Research* (pp. 409–442). Springer Netherlands. doi:10.1007/978-94-007-1177-8_17

Mollinga, P. (2013). Boundary concepts for interdisciplinary analysis of water resources management in South Asia. *Crop Science, 50*(S1).

Niemeijer, D., & de Groot, R. S. (2008). A conceptual framework for selecting environmental indicator sets. *Ecological Indicators*, *8*(1), 14–25. doi:10.1016/j.ecolind.2006.11.012

Norris, P. E., O'Rourke, M., Mayer, A. S., & Halvorsen, K. E. (2016). Managing the wicked problem of transdisciplinary team formation in socio-ecological systems. *Landscape and Urban Planning*, 1–8. doi:10.1016/j.landurbplan.2016.01.008

Orenstein, D. E., & Groner, E. (2014). In the eye of the stakeholder: Changes in perceptions of ecosystem services across an international border. *Ecosystem Services*, *8*, 185–196. doi:10.1016/j.ecoser.2014.04.004

Orenstein, D. E., & Shach-Pinsly, D. (2015). *Searching for Spatial Sustainability: One Goal, Multiple Paths?. Annual Conference of the Associations of the European Schools of Planning (AESOP)*, Prague, Czech Republic.

Pickett, T. A., Kolasa, J., & Jones, C. G. (1994). *Ecological understanding. The nature of theory and theory of nature*. San Diego, CA: Academic Press.

Plummer, R., & Armitage, D. (2007). A resilience-based framework for evaluating adaptive co-management: Linking ecology, economics and society in a complex world. *Ecological Economics*, *61*(1), 62–74. doi:10.1016/j.ecolecon.2006.09.025

Putnam, R. W. (2014). Theories of Action. In D. Coghlan & M. Brydon Miller (Eds.), *The Sage Encyclopedia of Action Research* (pp. 778–781). London, UK: Sage Publications Ltd. Retrieved from http://actionsmithnetwork.net/wp-content/uploads/2015/09/Theories-of-Action-Putnam-2014.pdf

Shipley, R., & Utz, S. (2012). Making it count: A review of the value and techniques for public consultation. *Journal of Planning Literature*, *27*, 22–42.

Singh, S. J., Haberl, H., Chertow, M., Mirtl, M., & Schmid, M. (Eds.). (2012). *Long term socio-ecological research: studies in society-nature interactions across spatial and temporal scales* (Vol. 2). Springer Science & Business Media.

Stokols, D. (2006). Toward a science of transdisciplinary action research. *American Journal of Community Psychology*, *38*(1-2), 63–77. doi:10.1007/s10464-006-9060-5 PMID:16791514

Swantz, M. L. (2008). Participatory action research as practice. In P. Reason, & H. Bradbury (Eds.), The SAGE Handbook of Action Research: Participative Inquiry and Practice. Sage Publications. doi:10.4135/9781848607934.n8

Turner, R. K., Georgiou, S., & Fisher, B. (2008). *Valuing ecosystem services: The case of multi-functional wetlands*. London, UK: Earthscan. Retrieved from http://books.google.co.il/books?id=ZacCAWBKHW8C

Zuber-Skerritt, O. (Ed.). (2012). *Action research for sustainable development in a turbulent world*. Bingley, UK: Emerald Group Publishing.

ENDNOTE

1 These 1600-year-old Byzantine agricultural engineering works are comprised of rows of stones laid across the wash, designed to distribute the water across the area and thereby reduce the power of water flow during flood events.

Chapter 13
Participatory Action Research in Supporting HIV/Aids Infected and Affected Learners

Maphetla Magdeline Machaba
University of South Africa, South Africa

ABSTRACT

Contemporary teachers face demands, amongst others fulfilling the diverse and changing roles in their everyday practice. The current roles of teachers are to evaluate teaching programmes. However, teachers often feel ill-equipped to deal with the demands placed on them. Although teachers generally have the will and inclination to help children, they often feel they lack the necessary skills to effectively deal with challenges such as HIV and Aids in the classrooms, participatory action research might be the answer for teachers regarding providing the necessary support for the learners. It provides influence of stakeholders and a higher level of support for the implementation of research findings. It can also be seen as an integrated activity that combines social investigation, educational work and actions. Its goal is to work with stakeholders to generate knowledge in order to initiate change. This chapter offers several examples of how researchers and practitioners are using participatory action research to select effective practices and support strategies.

INTRODUCTION AND BACKGROUND

HIV continues to be a major global public health issue. In 2015, an estimated 36.7 million people were living with HIV (including 1.8 million children) – a global HIV prevalence of 0.8%. The vast majority of this number live in low- and middle- income countries. In the same year, 1.1 million people died of AIDS-related illnesses. Since the start of the epidemic, an estimated 78 million people have become infected with HIV and 35 million people have died of AIDS-related illnesses. (UNAIDS 2016)

An estimated 25.5 million people living with HIV live in sub Saharan Africa. The vast majority of them (an estimated 19 million) live in east and southern Africa, which saw 46% of new HIV infections globally in 2015. Around 40% of all people living with HIV do not know that they have the virus. As of

DOI: 10.4018/978-1-5225-2642-1.ch013

December 2015, 17 million people living with HIV were receiving antiretroviral treatment (ART) - up from 15.8 million in June 2015 and 7.5 million in 2010. This means that 46% of all adults and 49% of all children living with HIV are now accessing ART. (UNAIDS 2016).

The disease in a number of ways affects many children: they live with sick parents and relatives in households drained of resources due to the epidemic, and those who have lost parents are less likely to go to school or continue with their education (Risley & Drake 2007). Studies in the regions of Southern Africa and South-East Asia have found that HIV/AIDS negatively impacts both the demand for, and supply of education (Gachui 1999; Risley & Drake 2007). Orphaned children are either pulled out of school or not enrolled at all due to the financial constraints of their affected families, and have to assume responsibilities of heading or providing for households. In this respect, girls are more vulnerable. In Kenya, links were found between parental deaths and children's progress through school. In Tanzania, households that have experienced an adult death have been found to delay the enrolment of younger children in school, but try to keep older children enrolled. In Malawi, it has been shown that the death of an adult encourages children to marry earlier, drop out of school to help support the family, and take on informal labour schemes. In Zambia, it is estimated that more than 7 per cent of approximately two million households, are headed by children (Gachui 1999). The reduction in primary school enrolment has a domino effect on secondary and post-secondary enrolments as well.

The HIV/AIDS pandemic has spread rapidly across the world to affect all nations. Not all groups, however, have been equally affected. HIV/AIDS has been particularly devastating to those sectors of society, which were already marginalised, powerless and vulnerable. It appears that a critical factor in the pandemic is the issue of power relationships. Those most affected are those who have little control over their lives. They are the most vulnerable to infection and the least able to deal with the consequences of the disease (Nelson & Wright 1995; Van der Vliet 1996). Just as the virus depletes the human body of its natural defences, it also depletes the individual and family of their resources and ability to cope (Lyons 1998). There is growing evidence that sustainable solutions for these kinds of problems cannot be found by experts alone. It is necessary for the people to be involved in finding their own solutions (Babbie & Mouton 2001). Thus the motivation for this study is based on the need to engage people living with HIV/AIDS in an empowerment process aimed at finding their own solutions.

In the context that HIV/AIDS is affecting many lives around the globe, especially in sub-Saharan African countries, supporting each and every child in the school seems to be one of the paramount ways to ease the impact of HIV/AIDS. However, there is evidence that HIV/AIDS has flooded education sectors and many children are not accessing education, leaving school before they achieve basic literacy and numeracy skills.

A 1992 World Bank study in Tanzania showed that HIV/AIDS was likely to reduce the number of primary and secondary school children by 22 and 14 per cent respectively (Gachui 1999). The high infant mortality rate due to HIV/AIDS also reduces the number of children in school. The impact of HIV/AIDS on school teachers in terms of mortality, productivity, and costs has negatively affected the quality and availability of education. It is estimated that in 2015, nearly 8000 teachers in the Greater Mekong sub-region would have died of AIDS, and 6000 would be HIV positive (Risley & Drake 2007). In Malawi, the rate of infection among teachers was higher than 30 per cent, with more than four teachers dying per day due to AIDS (Gachui 1999).

A recent study in South Africa on HIV/AIDS in the higher education sector has shown that the mean HIV prevalence for students was 3.4 per cent, while the mean prevalence among the academic staff was about 1.5 per cent (Higher education sector 2010). The prevalence levels are much lower compared to the general population. Since higher education institutions (HEIs) are reservoirs for future leaders and professionals, it is essential that they adopt strategies for lower prevalence, with no new infections.

The economic impact of HIV/AIDS presents huge challenges. While the causality between poverty and HIV is not clear, it is certain that HIV pushes households and individuals into poverty. While many illnesses create catastrophic expenditures which can result in poverty, HIV/AIDS is among the worst because its victims become ill for a prolonged period before they die, and many are the chief household income earners.

In Southern Africa, however, HIV/AIDS is a disease of poverty. Here, it is clear that poor people are most likely to be infected. The first and worst impact of HIV/AIDS on poverty is at the individual and household level. The effect then aggregates, influencing the macro economy through impacts on labour markets, and consequently the allocation of resources by governments (Barnett & Whiteside 2007). In a study on HIV/AIDS in rural Thailand, Michael P. Cameron identifies a vicious poverty-HIV/AIDS cycle in which HIV infected people who are vulnerable to poverty are more likely to engage in high-risk behaviour to cope with the poverty, which in turn makes others susceptible to HIV infection. In a South African national survey, two-thirds of the respondents reported a drop in household income as a result of HIV-related illnesses, including the direct loss of earners. As was also found elsewhere in sub-Saharan Africa, households reported increased expenditure on health, thereby diverting income away from other requirements. Severe illness and death have been found to vary according to the initial economic strength of the households that experience an HIV-related shock. Deaths of young adults from natural causes (including those from HIV/AIDS) had a negative impact on the growth in income of all households over a five-year period, but especially on those households that were initially modestly well off (Risley & Drake 2007).

Lack of education perpetuates the cycle between HIV/AIDS and poverty, often arising from decisions concerning schooling, childbearing, finances and unemployment. A number of studies have measured some of these negative outcomes, which suggest that there are significant intergenerational consequences of households coping strategies (Higher education sector 2010).

While HIV/AIDS has a negative impact on education, it is also through education that appropriate mitigation initiatives can be implemented. The integration of HIV/AIDS education and lifelong skills programmes in the school curricula can be an effective way of influencing behavioural changes in children from an early age. About 40 per cent of the HEIs <<Author: Please use full name with acronym in brackets. In South Africa had different types of units engaged in HIV/AIDS research (Higher education sector 2010). Such units should conduct strategy-relevant research on HIV/AIDS and poverty alleviation, with outputs that can inform policy and other appropriate interventions. HIV/AIDS should also be integrated in the academic curricula. At the same time, the establishment of community-based sustainable development projects to empower communities could enable individuals and households to overcome the poverty impact of the disease (Higher education sector 2010). The author wishes to acknowledge contributions to this article by Professors Julian May and Alan Whiteside of the University of KwaZulu-Natal (Risley & Drake 2007).

HIV/AIDS AND PARTICIPATORY ACTION RESEARCH (PAR)

Participatory action research has been employed extensively in the study of HIV/AIDS. For example, Kesby (2000) and Mabala and Allen (2002) studied how the community could be engaged in the prevention of the spread of HIV/AIDS. Lindsey, Stajduhar and McGuinness (2001) report on a PAR intervention, which engaged the community in developing community-based organisations for caring for people living with HIV/AIDS. However, there appears to be a gap in the literature in the area of individual empowerment in the context of HIV/AIDS and poverty. Some studies were found that focused on empowering individuals to address their personal needs in contexts other than HIV/AIDS, for example, chronic illnesses (Koch & Kralik 2001) and disability (Stewart & Bhagwanjee 1999). These studies bear relevance to the current study as they also deal with chronic conditions and relatively marginalised groups and thus involve similar issues to those raised by the current study.

Central to the goal of the Department of Education's national policy is the prevention of the spread of HIV infection, the reduction of the stigma surrounding HIV/ AIDS, and the support of infected and affected individuals (Department of Education 2000). According to the policy, all schools are expected to formulate an HIV/AIDS school plan, guided by the principles stipulated in the policy document. Schools are, therefore, encouraged to use legislation in developing individualised /AIDS plans, against the backdrop of each school's unique characteristics and needs, as well as those of the communities they serve. As such, an HIV/AIDS school plan may be used to address complex questions in the specific school and community regarding the HIV/AIDS pandemic. Important role players in the school environment and the broader community may thus be involved in the development and drafting of implementation plans for the combating of HIV/AIDS in a school or institution. It is the responsibility of the school principal and the governing body to ensure that such a plan is practical and implemented by all teachers concerned.

With participatory action research, in-depth information is obtained about the manner in which teachers co-develop an HIV and AIDS plan, in discussion with the researchers, and the meanings they attach to the plan they will develop. The purpose of participatory action research is to produce knowledge in an active partnership with those affected by that knowledge, and for the express purpose of improving their social and educational conditions. During participatory action research, the study is guided by the basic approach of involving teachers as co-researchers, signifying a strong partnership between the participating teachers and the university research team. The collaboration formed with the teachers follows the cyclical steps of practice, reflection, research, planning, and action. In this manner, participatory action research allows the participating teachers to obtain insight into their situation as well as their own networks, abilities and resources, leading to their mobilisation of relevant social action.

Participatory action research will be used by teachers and researchers to develop an HIV/AIDS school support plan in the school setting. It is, therefore, very important to build relationships of trust with participating teachers, guiding participants to take ownership and steer action, or dealing with differences and power-related challenges between researchers and participating teachers. The consequences of the HIV pandemic have highlighted the importance of the pastoral role of teachers who are called upon to recognise and deal with the emotional distress that causes problems with children's schoolwork. In addition, children who grow up without parental guidance and support may later develop behavioural problems such as depression, isolation, anxiety, difficulty in making friends, low self-esteem, alcohol and drug abuse. If these emotional problems are not addressed, they may manifest themselves in other destructive ways.

The role of the teacher in this climate is to educate children about the prevention of HIV and encourage openness to promote de-stigmatisation and voluntary testing and counselling. Increasingly, they must also ensure that both infected and affected children can get care, support and treatment, if necessary, by referring them to appropriate agencies as the above factors may affect the children's ability to learn effectively. Teachers have an important role to play in responding to the needs of these children since they act in loco parentis and may be the only responsible adults to whom the children can relate. They are both legally and morally obliged to offer care and support to HIV/AIDS infected and affected learners.

THE ROLE OF EDUCATION IN DEALING WITH HEALTH ISSUES AND HIV/AIDS

Currently, the reach of formal education is immense worldwide and its potential for a positive impact on population health is significant. In a demographic sense, going to school and attending for a considerable number of years is a new, wide-scale change in the behaviour of children, as well as their families and communities without parallel in pre-modern, traditional human society. Over the first few decades of the twentieth century, growth in primary enrolment took off and by 1940, burst into a logarithmic climb. As primary schooling reached large numbers of children, 20 years later enrolment in secondary schooling turned sharply up in the 1960s. And then in the early 1970s, enrolment in higher education began to grow (Meyer et al. 1992). As a function of intensive institutionalisation of formal schooling, 80% of all humans aged 15 or over are able to both read and write a short statement about their life. Most people who are still illiterate and undereducated are living in poor nations, and many are females (UNESCO 2004).

Among the many impoverished nations in Southern Africa, access to basic education has been growing in fits and starts since the middle of the twentieth century to the point that the stage is at least partially set for education to take on the role of a social vaccine. Perhaps nothing illustrates the positive impact formal education can have on the health of whole populations as its association with the much noted demographic transition of modern society, consisting of reduced mortality and fertility, leading to longer life spans. For example, since the early 1950s, low and middle-income countries have seen a rapid increase in life expectancy at birth from 40 years in 1950 to 65 years by 2005, while high-income countries had increased life expectancy at birth to 79 years by 2005 (World Bank 2007).

Although a number of factors are thought to have played a role in this massive change in human life, a consistently identified factor is widespread mass education (Goesling & Baker 2008). In fact, the timing and spread of the rise of mass primary education precedes the timing and spread of the modern demographic transition across the world. There are many trend studies across developing nations showing a correlation between growing mass education and fertility decline, better maternal health, and healthier lifestyles, thus lowering infant and adult mortality and raising life expectancies (Lleras-Muney 2005).

Many of these studies indicate that even when populations in poor nations have received relatively small periods of formal schooling (2–3 years) in basic literacy and numeracy, observable differences arise between schooled and non-schooled populations in terms of risky, unhealthy behaviour and fertility control (de Walque & Wouters 2005). Although macro-demographic studies identify a significant education effect on mortality and fertility, many of the earliest studies failed to control for other factors that also arise with more education (Ross & Wu 1995). Moreover, some argue that any effect of education on health is spurious due to the impact of general social development on both, in other words, increased economic growth, quality of health care, and other factors of modernity often arise simultaneously and could be associated with both mass education and healthier lives. In response, a substantial research literature has arisen that refutes the spurious causality critique, showing that education has a social vaccine effect even after controlling for confounding factors such as relative social status and wealth, and access to health care. For example, consider the education effect on mortality. A recent meta-analysis culled from 33 key journals reporting on social trends in mortality from 1995 to 2005 finds 56 methodologically high-quality, large-sample, multivariate studies from both high-income and low-income countries from all regions of the world.

The predominate finding among these studies is that educational attainment has a significant positive relationship with age at mortality even after controlling for numerous financial resources and social characteristics of individuals—in study after study more educated individuals have lower mortality and longer life expectancy (Baker, Parker, Williams & Clark 2007). The reported effect sizes are moderate to large: less schooled individuals are on average 87% more likely than more schooled ones to die early, even after controlling for many factors such as income, job type, gender, child infirmity, age-cohort and so forth. An illustrative example is Mostafa and van Ginneken's (2000) longitudinal study of a sample of over 10,000 Bangladeshi adults from 1982 to 1999, which reports that after controlling for a host of factors, including socio-economic status, unschooled subjects were 80% more likely to have died over the course of the study than schooled subjects.

A related meta-analysis of 58 studies from 30 public health and medical journals from 1995 to 2005 finds a similar significant and uniform education effect of lowering people's high risk-taking on 14 different types of unhealthy risk-taking behaviour net of economic and social characteristics (Baker et al. 2007). A typical study of education and risk-taking is that of Droomers et al. (2004). This longitudinal study of adults finds that although life stressors such as poverty and unemployment negatively influence alcohol abuse, when individuals with varying levels of education are exposed to the same types of life stressors, less educated individuals are three times more likely to start excessive alcohol drinking than their more educated counterparts. Similar social vaccine effects of education on all types of potential everyday risks such as tobacco use, alcohol and drug abuse, risky sexual and conception practices, over-consumption of food, misuse of safety devices, and even unhealthy criminal activity are reported throughout the literature. Furthermore, recent macro-studies report an education effect on population health. For example, research examining the impact of changes in compulsory schooling laws in the US (Lleras-Muney 2005) for mortality and health, finds significant educational effects. Rogers, Hummer and Nam (2000) identify educational stratification in the US as one significant factor in mortality differences, even after controlling for demographic, social, economic, and behavioural factors in the 1980s and 1990s.

Similar to the shift in education's role in the HIV/ AIDS pandemic in Southern Africa, Walque (2004) finds that earlier in the century, when there was considerable misinformation about supposed positive health outcomes connected with smoking, more educated Americans had a higher propensity to smoke. However, the relationship reversed as more accurate information became available about the negative health risks associated with smoking. All of this empirical literature adds up to the conclusion that formal education has "an enduring, consistent, and growing effect on health" (Mirowski & Ross 2003). And of course, health can beget even more education individually and demographically, so that education as a social vaccine and health as a major characteristic of human capital, go hand in hand (Bloom 2007). The contrast between education's role as a risk factor in the HIV/ AIDS pandemic in SSA and the extensive evidence of its worldwide positive impact on many other aspects of health and risky unhealthy decision-making raises the question whether or not the role of education will shift to a social vaccine against HIV/ AIDS among future generations of youth in these nations.

The role that formal education has played over the course of the HIV/AIDS pandemic is distinctly different from its usual one in population health. In most cases, as individuals increase their educational attainment via formal schooling, their health outcomes are normally seen to be better than their less educated peers (Mirowski & Ross 2003). However, as discussed in this chapter, numerous past research studies identified formal educational attainment as increasing an individual's risk of contracting HIV.

At the same time, new evidence presented here suggests that this relationship may be shifting, with important implications for the region's future educational and public health policies.

Over the past 150 years, mass education has played a significant preventive role in many public health problems associated with higher morbidity and greater mortality, and as a result has been labelled a social vaccine (Kelly 2000; World Bank 2003). As reviewed below, extensive demographic and social epidemiological literatures report that education lessens engagement in risky behaviour and facilitates healthier lifestyles to the point where one could effectively argue that the spread of schooling is a leading cause underlying improvements in global public health. But in the case of HIV/ AIDS, studies report that education has played the unusual role of a risk characteristic—individuals with more educational attainment (particularly males) have been more likely to have multiple sexual partners and engage in unsafe sex, increasing their risk of becoming infected—to the point that many studies routinely refer to formal education and wealth as HIV/AIDS risk factors. Given the contrasting roles education has played in general health versus HIV/AIDS infection, the analysis here examines if over the course of the pandemic, the role of education has shifted from risk factor to social vaccine.

It is widely recognized that schooling can have multiple effects on individuals, ranging from acquisition of facts and cognitive enhancement to greater social status through credentials for better-paying, more prestigious occupations. In the case of HIV/AIDS, the evolving public awareness of the disease was non-existent early in the pandemic. However, even as information about the disease started to become available, it was tempered by social, economic, and political conditions that limited access and even questioned the credibility of information about HIV and its mode of transmission. The lack of information available about the spread of the virus, when coupled with the biology and limited outward symptoms of the HI virus, caused an almost undetected spread of the HI virus in several parts of Africa. We hypothesize that during this period when little or no information and even misinformation was available in the past, the status and wealth-inducing effects of education fuelled the spread of the disease among the highly educated. As accurate information about the disease became available, it is hypothesized that formal education should increasingly assert itself and be associated with lower HIV infection. To foreshadow the results, there is evidence that the effect of formal education is shifting among younger people and may now play a chiefly preventive social vaccine role in respect to HIV infection. A likely factor in this shift is that with accurate information, the central cognitive effects of schooling on reasoning skills become operative and override non-cognitive social effects.

As the disease continued to spread, additional evidence was reported that more educated men spread infection not just to their educated partners, but started infections among uneducated women throughout entire rural regions. In study after study, this pattern was so frequently reported that it is now widely accepted that the spread of HIV into rural sub-Saharan Africa in the 1980s and 1990s was fuelled by infected more educated, relatively wealthier, and geographically mobile men (Barnett & Blaikie 1992; Rushing 1995). For example, by 1999–2000, antenatal clinics in Uganda began to report a shift in new HIV infection rates from more educated to uneducated females (de Walque et al. 2005). Recently, antenatal and voluntary counselling and testing centres in Zambia and Ethiopia have also reported a shift in new HIV infections moving from the more educated to the lesser educated (Bradley et al. 2007). And even though some recent studies of new infections have begun to find some mixed results, by and large education is still considered to be a risk factor throughout (Cogneau & Grimm 2006).

Education everywhere, however, affords higher status and greater income on average, yet as documented below, schooling rarely forms a risk factor in health. In the 1980s and early 1990s, within the brewing pandemic, three additional factors specific to the African context caused education to become a risk factor. HIV and AIDS spread rapidly during the late 1970s and 1980s before public health information was available. However, once the disease was identified, it was narrowly and misleadingly labelled as a disease among homosexual males in the US and Europe. Meanwhile, in Africa the disease was being transmitted via heterosexual transmission as the public there considered this a safe sexual practice. Also, many governments denied the existence of the disease in their countries and did not provide resources or services to address the pandemic until very recently (Hargreaves & Glynn 2002). The cultural practice of transactional sex, and the persistence of multiple sexual partners reflecting an individual's social status, also fuelled the spread of the pandemic before information was even available to the public (Swidler & Watkins 2007). Furthermore, even as accurate medical information became available, heterosexual sex was not identified as a potential mode of transmission until the late 1980s. (World Health Organization/UNAIDS 2000).

DEVELOPING COMPETENT PARTICIPATORY ACTION RESEARCH FOR HIV/AIDS

Participatory action research is defined as "systematic inquiry, with the collaboration of those affected by the issue, for the purposes of education or effecting social change" (George et al. 1999). The participatory action research approach is characterized by six major criteria: (i) it is participatory; (ii) it is cooperative; (iii) it is a co-learning process; (iv) it involves systems development and local capacity building; (v) it is an empowering process for participants; and (vi) it achieves a balance between research and action (Israel et al. 1998). One of the main characteristics of participatory action research is that it eliminates the traditional distinction between the "researchers" and the "researched" (Gaventa 1981). Both HIV/AIDS researchers and participants are partners and responsible for the conceptualization, needs analysis, development, implementation and evaluation of the successful HIV/AIDS ultimate education programme. As a result of this process, participants are empowered with the realization of their own capabilities to be researchers and to induce desired changes within their communities (Hagey 1997). This method seems particularly suitable when one is conducting research on controversial or sensitive issues such as HIV/AIDS.

Most educational efforts in the early phase of the HIV/AIDS pandemic have been concentrated on the high-risk groups, particularly, men having sex with men, and intravenous drug users. Results of the programmatic effort indicated significant improvements in condom use behaviours and reduction of STI among female bar workers (Morisky et al. 2002). It is clear that men still dominate women's sexual behaviour. It is recognized that men are generally more sexually active, have more sexual intercourse than the general female population, have more sexual partners, and generally take risks to maintain their machismo image to the extent that they refuse to practice safe sex. In short, they have a major role in the transmission of sexually transmitted illnesses and HIV/AIDS. Very few studies have assessed multiple intervention programmes targeting a diverse male population using a variety of health education/promotion strategies and a participatory action research component.

Participatory action research has proved to be an important tool for addressing issues of powerlessness. In participatory action research, the participants are seen as partners in the research process and not merely subjects of the study. This means that they are involved in shaping and directing the research and take part in the decision-making process (Cornwall & Jewkes 1995). Participatory action research involves "collective, self-reflective enquiry that is undertaken to improve a situation" (Koch, Selim &

Kralik 2002:110). The process begins with examination and collaborative enquiry into the lived experiences of the participants. The participants are assumed to have essential and valid understandings and experiences, which are vital to comprehending their world and bringing about change. New insights into the situation and the participants' lives are gained through the action reflection cycle, a key component of participatory action research. This cycle involves a circular process of action followed by reflection, followed by further action which is informed by the reflection phase. The goal of this cycle is for the researcher and participants to consciously engage in a process of learning from their actions (Zuber-Skerritt 2001). From the new insights, which the participants gain into their experiences, the researcher helps to facilitate the planning of action to bring about change (Babbie & Mouton 2001). The goal of PAR is to bring about meaningful change in the lives of the participants and the community. Change may come in the form of individual or group empowerment, increased capacity to solve problems, and/or the transformation of organisational structures (Cockbum & Trentham 2002).

Research into HIV/AIDS involves highly personal and sensitive topic areas, and researchers need to exercise respect and circumspection when engaging with participants. This is accentuated because those affected and infected by HIV/AIDS are often the most vulnerable and marginalised social groups (Gray, Lyons & Melton 1995). Writing particularly in the area of participatory action research, Hagey, (1997) states that social justice principles of equity, restitution and procedural justice are important concepts in PAR. Restitution acknowledges institutional responsibility in creating conditions that must now be rectified. The concept of procedural justice values how relationships are lived, how interactions exclude or refrain from including, how particular elite individuals holding office practise dominance and perpetuate systemic disadvantage, how racism hurts and humiliates and is denied, how its perpetrators are unwilling to examine their own practices and how resistance to change is manifested, for example, when institutions have righteous sounding policies that they do not put into daily practice. PAR relies on honesty and veracity both in declaring agendas (reflexivity) and in carrying out the research and implementing its goals. On this foundation, both validity and legitimacy are grounded (for a detailed discussion of PAR and participatory research as well as a review of participatory methods used in researching HIV/AIDS, see School of education and development 2005).

Literature review on HIV/AIDS in education has focused at the macro-level of national education systems within a quantitative research approach. In particular, this body of literature has drawn attention to the destructive impact of the pandemic on teacher numbers, learner attendance, and systemic management. Little attention has been given to the micro-level of analysis of the effects of HIV/AIDS on particular schools and communities, and of the concrete experiences and responses of educators, learners and parents regarding HIV/AIDS. A study by Huber and Gould (2003) suggested that micro-level research, using more qualitative and participatory methods, may elicit very different information and offer valuable insights. It was, therefore, felt that there is a need to complement the quantitative, macro-level studies with qualitative, micro-level research into how participants experience and make meaning of HIV/AIDS at a local level within the education system (School of education, Development and training 2004).

In line with a participatory framework, visual methods and active representations of ideas enable researchers to develop an in-depth understanding of local knowledge and to engage the local people in a process of reflection about their own lives. A range of participatory techniques is used to capture the voices of participants in the study, and to ensure rich qualitative data. These includes individual interviews and focus group interviews with various participatory research techniques such as transect walks, vulnerability matrices, ranking exercises, aerial photograph analysis, photo-voice, social mapping, time lines, and Venn diagrams. However, there are various methodological challenges in the project related to the data production process. A key consideration in selecting researchers for data production is the language issue. The majority of interviews are conducted in the local language

The research process, therefore, had to simultaneously involve planning the research process and data production, and the development of research facilitation skills and reflexive practice. This occurred through a range of training workshops. Although this process developed within a supportive research team that worked as a community of practice, the challenge of addressing inherent power imbalances was constantly reflected on, and it was conceded that these could not be fully resolved. Further tensions related to the context as researchers had to engage with diverse schools and educational settings, in diverse geographical locations, in different communities, and with diversity in terms of race, language, class, political affiliations, etc (Van der Riet, Hough & Killian 2005).

DEALING WITH HIV/AIDS IN THE CLASSROOM

Over the years, participatory action research education programmes that include sexual and reproductive health information have proven to be effective in delaying the onset of sexual intercourse and, among sexually experienced youth, in increasing the use of condoms and decreasing the number of sexual partners. Evaluation shows that participatory action research programmes can contribute to the reproductive and sexual health of young people around the world.

The literature reviewed shows the great potential of participatory action research in helping people to gain control over their lives in disabling environments. However, a number of challenges have also been highlighted. In the context of the great need for sustainable solutions to the many problems caused by the HIV/ AIDS pandemic, it is believed that this study can make a worthwhile contribution by analysing the participatory action research process that was implemented, exploring its impact on empowerment and highlighting the challenges encountered in the process. It is important to note that participatory action research is not a simple method to implement or an easy alternative to other research methods (Cornwall & Jewkes 1995). Gray et al. (2000) point out that the nature of participatory action research and its emphasis on partnership and participant control creates a number of challenges for the process of implementation. These include: problems in transferring control to the participants, conflicting expectations and agendas, problems in motivating people to engage in the research process and the time required for building partnerships.

Participation has been closely linked with empowerment. Research suggests that participation provides a key opportunity for individuals to obtain the skills, knowledge and critical awareness vital for empowerment. It is therefore important to understand what hinders or promotes participation (Prestby, Wandersman, Florin, Rich & Chavis, 1990). Prestby et al. (1990) propose that in general, people decide whether or not to participate by weighing up the perceived benefits of involvement and balancing it against the potential costs. Cornwall and Jewkes (1995) point out; however, that people's motivation to take part tends to fluctuate over time. As circumstances change and/or the participants discover that they are not going to get the benefits they expected to gain from involvement, motivation may dwindle (Cornwall & Jewkes 1995). This is in fact what happened in this research study.

Participatory action research methodology offers an exciting alternative to the more traditional top-down, researcher controlled approach and holds out the possibility of achieving more effective and sustainable outcomes for communities. However, as this research has highlighted, there are many challenges involved in implementing the PAR process.

RECOMMENDATIONS

It is recommended that the participatory action research approach must be an interactive, educational methodology that not only focuses on transmitting knowledge, but also aims at shaping attitudes and developing interpersonal skills. The main goal of the participatory action research approach should be aimed to enhance learners' ability to take responsibility for making healthier choices, resisting negative pressures, and avoiding risk behaviours. Teaching methods must also be learner-centred, gender-sensitive, interactive and participatory. The most common teaching methods that are recommended include working in groups, brainstorming, role-playing, storytelling, debating, and participating in discussions and audio-visual activities.

CONCLUSION

This chapter discussed participatory action research in supporting HIV/AIDS infected and affected learners. In participatory action research, the control over the research process lies in the hands of the participants. In this chapter, it is clear that handing over control to the learners as participants is a vitally important objective of participatory action research as it is through this process, that the participants are empowered and given the opportunity to learn new skills of dealing with HIV/AIDS. However, this is obstructed by a number of problems which include resistance from the participants and their lack of confidence in themselves.

REFERENCES

Babbie, E., & Mouton, J. (2001). *The practice of social research*. Cape Town: Oxford University Press.

Baker, D. W., Parker, R. M., Williams, M. V., & Clark, W. S. (1998). Health literacy and the risk of hospital admission. *Journal of General Internal Medicine*, *13*(12), 791–798. doi:10.1046/j.1525-1497.1998.00242.x PMID:9844076

Bandura, A. (1997). Self-efficacy. *The Harvard Mental Health Letter*, *13*(9), 4–6.

Barnett, T., & Blaikie, P. (n.d.). *AIDS in Africa: its present and future impact*. London: Belhaven Press.

Bradley, H., Bedada, A., Brahmbhatt, H., Kidanu, A., Gillespie, D., & Tsui, A. (2007). Amare Bedada, Heena Brahmbhatt, Aklilu Kidanu, Duff Gillespie, and Amy Tsui. Educational attainment and HIV status among Ethiopian voluntary counseling and testing clients. *AIDS and Behavior*, *11*(5), 736–742. doi:10.1007/s10461-006-9185-4 PMID:17082983

Cockburn, L., & Trentham, B. (2002). Participatory action research: Integrating community occupational therapy practice and research. *Canadian Journal of Occupational Therapy*, 21–31. PMID:11852687

Cogneau, D., & Grimm, M. (2003). *The distribution of AIDS over the population in Africa. Hypothesis building from individual answers to a Demographic and Health Survey with an application to C^ote d'Ivoire. Mimeo*. University of Gottingen.

Cooper, E., Risley, R., Drake, L., Cooper, E., Risley, R., Drake, L., & Bundy, D. (2007). HIV as part of children and youth as life expectancy increases: implications for education. *Journal of International Cooperation in Education.*

Cornwall, A., & Jewkes, R. (1995). What is participatory research?. *Social Science & Medicine*, *41*(12), 1667–1676. doi:10.1016/0277-9536(95)00127-S PMID:8746866

de Walque, D., Nakiyingi-Miiro, J. S., Busingye, J., & Whitworth, J. A. (2005). Changing Association between Schooling Levels and HIV-1 Infection over 11 Years in a Rural Population Cohort in South-west Uganda. *Tropical Medicine & International Health*, *10*(10), 993–1001. doi:10.1111/j.1365-3156.2005.01475.x PMID:16185233

Gachui, D. (1999). *The Impact of HIV/AIDS on Education Systems in the Eastern and Southern African Region.* UNICEF Report.

George, M. A., Daniel, M., & Green, L. W. (1999). Appraising and funding participatory research in health promotion. *International Quarterly of Community Health Education*, *18*(2), 181–197. doi:10.2190/C1B5-7PPE-7TYL-7YN8 PMID:17686723

Gray, R. E., Fitch, M., Davis, C., & Phillips, C. (2000). Challenges of participatory research: Reflections on a study with breast cancer self-help groups. *Health Expectations*, *3*(4), 243–252. doi:10.1046/j.1369-6513.2000.00100.x PMID:11281935

Griffiths, J. (2005). *HIV/AIDS Intervention Programs for Youth in Africa: The role of grass root soccer*. Academic Press.

Hagey, R. S. (1997). The use and abuse of participatory action research. *Chronic Diseases in Canada*, *18*(1), 1–4. PMID:9188513

Hargreaves, J. R., & Glynn, J. R. (2002). Educational attainment and HIV-1 infection in developing countries: A systematic review. *Tropical Medicine & International Health*, *7*(6), 489–498. doi:10.1046/j.1365-3156.2002.00889.x PMID:12031070

Higher Education Sector. (2010). *HIV Prevalence and Related Factors*. Academic Press.

Huber, U., & Gould, W. (2003). *The effect of orphanhood on primary school attendance reconsidered: the power of female-headed households in Tanzania*. HIV/AIDS Impact on Education Clearinghouse. Retrieved from http://iiep.tomoye.com/ev.php

Israel, B.A., Schulz, A.J., Parker, E.A., & Becker, A.B. (1998). Review of community-based research: assessing partnership approaches to improve public health. *Annual Review of Public Health, 19*, 173-202.

Kelly, M. J. (2000). *The encounter between HIV/AIDS and education*. Harare, Zimbabwe: UNESCO, SubRegional Office for Southern Africa.

Kesby, M. (2000). Participatory diagramming as a means to improve communication about sex in rural Zimbabwe: A pilot study. *Social Science & Medicine, 50*(12), 1723–1741. doi:10.1016/S0277-9536(99)00413-X PMID:10798328

Koch, T., Selim, P., & Kralik, D. (2002). Enhancing lives through the development of a community-based, participatory action research programme. *Journal of Clinical Nursing, 11*(1), 109–112. doi:10.1046/j.1365-2702.2002.00563.x PMID:11845746

Lindsey, E., Stajduhar, K., & McGuinness, L. (2001). Examining the process of community development. *Journal of Advanced Nursing, 33*(6), 828–835. doi:10.1046/j.1365-2648.2001.01722.x PMID:11298221

Lleras-Muney, A. (2005). The relationship between education and adult mortality in the United States. *The Review of Economic Studies, 72*(1), 189–221. doi:10.1111/0034-6527.00329

Lyons, M. (1998). *The impact of HIV and AIDS on children, families and communities: risks and realities of childhood during the HTV epidemic*. United Nations Development Programme: HIV and development programme. Issues Paper 30. Retrieved from www.undp.org/hiv/ publications.html

Mabala, R., & Allen, K. (2002). Participatory action research on HIV/AIDS through a popular theatre approach in Tanzania. *Evaluation and Program Planning*.

Mason, J. (2002). *Qualitative researching* (2nd ed.). London: Sage Publications.

Mirowsky, J., & Ross, C.E. (2004, Oct.). Education, social status, and health. *Canadian Journal of Sociology,* 242.

Morisky, D., Snehendu, B.K., Chaudrhry, A.S., Chen, K.R., Shaheen, M., & Chickering, K. (2002). Condom use among female sex workers in China: Role of gatekeepers practices in Pakistan. *Pakistan Journal of Nutrition, 1,* 137–142.

Nelson, N., & Wright, S. (1995). Power and participatory development: Theory and practice. Intermediate Technology Publications.

Rich, R., & Chavis, D. (1990). Benefits, costs, incentive management and participation in voluntary organizations: A means to understanding and promoting empowerment. *American Journal of Community Psychology, 18*(1), 117–149. doi:10.1007/BF00922691

Risley, C. L., & Drake, L. J. (2007). *Impact of HIV/AIDS on education in the Greater Mekong sub-region*. Workshop Presentation. Siem Reap.

Rogers, R. G., Hummer, R., & Nam, C. B. (2000). *Living and dying in the USA: behavioral, health, and social differentials of adult mortality*. San Diego, CA: Academic Press.

Ross, C. E., & Wu, C. (1995). The links between education and health. *American Sociological Review, 60*(5), 719–745. doi:10.2307/2096319

Stewart, R., & Bhagwanjee, A. (1999). Promoting group empowerment and self-reliance through participatory research: A case study of people with physical disability. *Disability and Rehabilitation, 21*(7), 338–345. doi:10.1080/096382899297585 PMID:10471164

Van der Riet, M., Hough, A., & Killian, B. (2005). Mapping HIV/AIDS as a barrier to education: A reflection on the methodological and ethical challenges to child participation. *Journal of Education, 35*, 75–98.

Van Der Vliet, V. (1996). *The politics of AIDS*. London: Bowerdean Publishing Company Ltd.

Watkins, S. C., & Swidler, A. (2007). *Conversations into texts: a method for studying public culture.* Paper originally presented as "hearsay ethnography" at the annual meeting of the Population Association of America, Philadelphia, PA.

World Bank. (2007). *World Bank Participation source book*. Available from http://www.worldbank.org/wbi/sourcebook/sbintro.pdf

World Health Organization/UNAIDS. (2000). *AIDS epidemic update*. Geneva, Switzerland: World Health Organization.

Zuber-Skerritt, O. (2001). *Action learning, action research and process management: theory, practice, praxis*. Brisbane: Action Research Unit, Faculty of Education, Griffith University.

Chapter 14
Participatory Action Research and Learning in Sustainable Local Economic Development

Marius Venter
University of Johannesburg, South Africa

ABSTRACT

Participatory action learning and action research in sustainable local economic development strategies is not a common practice. In this chapter, the author firstly gives an introduction of the concept of sustainable local economic development; the importance of local economic development is outlined, as well as the aims. The author proceeds by discussing participatory action learning and action research as integrated empowering concepts. Lastly, the author describes how participatory action learning and action research were used by the Overstrand Local Economic Development Agency in partnership with the Hawston community to assist them to develop a sustainable neighbourhood development strategy. The steps of the participatory action learning and action research cycle are discussed coupled with the learning and actions taken, that emanated from the reflections during the research.

INTRODUCTION

The purpose of this chapter is to discuss the Hawston Tourism Gateway project as an example of a sustainable local economic development (SLED) strategy and to demonstrate how participatory (P) action learning (AL) and action research (AR), (P)ALAR as integrated concepts can be applied by LED professionals in the (SLED) workplace. According to Zuber-Skerrit (2015):

PALAR is one possible approach to effective, collaborative, creative, innovative, and self-developed community engagement where a shift is needed in mind-sets, paradigms, and skills for social justice and sustainable development (p.5 and p.13)

DOI: 10.4018/978-1-5225-2642-1.ch014

SLED is about local people deciding to take charge of the economic and employment future of their community. The Hawston Gateway Tourism project produced valuable guidelines to LED professionals and communities that could be considered before implementing a neighbourhood development strategy (NDS).

The International Economic Development Council (IEDC) (2011) defines NDS strategies as:

... strategies that facilitate the revitalization or redevelopment of neighbourhoods that are struggling to overcome significant economic, social, and physical distress or devastation (p.2.).

This chapter endeavours to explain the following:

1. What is sustainable local economic development (SLED) as experienced in South Africa?
2. What is (P)ALAR and how can it be applied in the case of a NDS?
3. What were the key lessons learnt; and
4. What are the guidelines that were formulated to assist LED professionals that may to implement NDS in the future to achieve SLED?

BACKGROUND

The aim of this section is to introduce the reader to the concept of SLED in South Africa. It is generally believed that municipalities are the foot-soldiers of local economic development (Venter, 2014). Sections 152 and 153 of the Constitution of South Africa (Act 108 of 1996) (Republic of South Africa, 1996) indicate that municipalities must be structured in such a way that one of their objectives is to promote social and economic development (Rabie, 2011).

The concept of SLED has an interdisciplinary character and application because it evolved over time with influences from a diversity of disciplines, such as geography, urban planning, sociology, public administration, systems thinking, and regional economics.

In 1994, the South African government decided to embrace the principles of a developmental state (Black, Calitz, Steenkamp, & associates, 2005, p.5). This resulted in an effort of the national treasury to allocate additional resources on the national budget to further the government's development functions. Legislation was promulgated to gradually transform local government spending to promote upliftment, empowerment, and poverty relief. Due to the fact that in South Africa municipalities struggle with service delivery, the government introduced the Local Government Turn-Around Strategy (LGTAS) in 2009. In this policy, the local government is described as a

... key part of the reconstruction and development effort in our country. The aims of democratizing our society and growing our economy inclusively can only be realized through a responsive, accountable, effective and efficient local government system that is part of a developmental state. (Department of Local Government, 2009, p. 3.)

In order to intensify and strengthen economic development efforts and to combat poverty and in-equality, the New Growth Path (2010) and the National Development Plan (NDP)–Vision 2030 (NDP,

2011) were conceptualised, and SLED strategies will contribute towards the aims and objectives of these policies, (Draft LED Framework 2013-2018, pp. 21-22).

The unique contribution that SLED can make to national development is through the effective implementation of initiatives at the neighbourhood level. The importance of SLED is confirmed by the fact that it is at the local level where all economic activity takes place and where all jobs are created.

SLED strategies must be incorporated into the Integrated Development Plan (IDP) of a municipality. The IDP of a municipality is its strategic plan for the development of the municipal area. The IDP is a statutory requirement, in terms of the Municipal Systems Act 32 of 2000 (Chapter 5), to guide integrated development planning across the municipality (Draft LED Framework 2013-2018, p. 38).

The Draft LED Framework 2013-2018 (2013) outlines South Africa's Vision for SLED as follows: "Local Economies are inclusive, world-class and dynamic places and brands in which to invest, work, learn, visit and live; and the most successful in creating wealth that is widely shared and benefiting the majority of its local people." (p. 24)

It is about seeking consensus around development issues at a neighbourhood level. SLED is a mindset based upon the simple premise that each community has within, or within its grasp, considerable capacity and opportunities to influence it's economic and employment destiny.

SLED has many objectives. These are, inter alia, the creation of jobs, wealth, or assets and the improvement of the quality of life. SLED can also be defined as a process that influences the growth and restructuring of a local economy to enhance the economic well-being of a community. In recent years, economic diversification has become a primary role for LED professionals who are actively working to leverage job growth in new industries. In tough economic times, these experts have to know how to develop strategies that are resilient in the face of budget cuts and reduced outlays.

According to the IEDC (2011, p.6), in its broadest sense, SLED encompasses three major areas:

- *Policies that governments undertake to meet broad economic objectives, such as price stability, high employment, and sustainable growth. Such efforts include monetary and fiscal policies, the regulation of financial institutions, trade, and tax policies. These policies are national in scope and are referred to as macro-economic policies.*
- *Policies and programmes to provide infrastructure and services, such as building highways, managing nature reserves, and providing medical access to the disadvantaged. Although the primary purpose of these programmes is not SLED, they have implications for SLED.*
- *Policies and programmes explicitly directed at improving the business climate through specific efforts in business finance, marketing, neighbourhood development, small business development, business retention and expansion, technology transfer, and property redevelopment, among others.*

The third category is the primary focus of LED professionals. Claire Patterson (2008) defines SLED as: "Local government committed to working with its citizens and groups within the community to find sustainable ways to meet their social, economic, and material needs, and improve the quality of their lives." (p. 7)

There are several factors that can significantly affect the likelihood of success for SLED strategies, projects and programmes, and the author is of the opinion that it is in this regard that (P)ALAR can play an important role.

The degree to which LED professionals are successful at promoting SLED depends on their capacity to understand the system, identify key objectives, and provide leadership in mobilising stakeholders and resources of the local economy. The proposed method to achieve this is (P)ALAR.

In summary, (P)ALAR in SLED is about creating a platform and environment to engage stakeholders in identifying problems in the local economy, formulating a strategic plan with actions, implementing the plans (strategies), reflecting on the impact, and adjusting and starting the (P)ALAR cycle again. The LED professional as a researcher has a key role to play in the (P)ALAR cycle.

MAIN FOCUS OF THE CHAPTER

In this chapter (P)ALAR (in SLED) are discussed as linked, integrated, and empowering concepts (Zuber-Skerritt & Associates, 2009). According to Zuber-Skerrit (1993, pp. 46-47) AR includes AL, but it is more deliberate and systematic, and it is always made public.

Zuber-Skerrit & Associates (2009, p. 6) define AL as "learning from and through action or concrete experience, as well as taking action as a result of this learning." Similarly, AR "is a cyclical iterative process of action and reflection on and in action including the past, present, and future."

The objectives of using AR in SLED strategies are to improve the practice of learning in SLED, using a systematic approach to encourage modifications to the situation or circumstances under which that practice takes place, and to choose which developments will be effective and appropriate for future SLED strategies.

Albertyn, Bester, Botha, and van der Merwe (2007, p. 10) describe that Swanepoel (1997) "is of the opinion that participative action research (PAR) has a twofold objective namely to produce knowledge and action directly beneficial to a community and to empower people through the process of constructing and using their own knowledge." (p. 153)

In SLED work, the LED professional is confronted with real life issues on a daily basis. These issues can mostly only be solved by facilitating actions to be taken or changes that should be implemented. In order to accomplish this, a methodology is needed where there is a cycle of continuous learning, continuous action, continuous reflection, continuous participation, and continuous improvement. In this section the author will illustrate how (P)ALAR could be used as a tool by LED professionals. In the next section a SLED strategy namely neighbourhood development is discussed with reference to the case of the rural community of Hawston and the key lessons that were learnt.

The Hawston Case Study

For many South Africans and people across the globe, the neighbourhood is the most intimate form of *community*. Where to live is a personal choice (or, in some cases, the personal choice of a previous generation)–and the neighbourhood is the articulation of this choice. The neighbourhood is essentially a collection of people who decide to live near each other and who participate in shared experiences. It is no surprise that people in neighbourhoods sometimes have similar interests and stations in life (retirees, young professionals, and families). Often they share a similar socio-economic status, and may share a cultural background as well. This is no different for the Hawston community.

Hawston is a previously disadvantaged community (neigbourhood) and a former old fishing village named after Charles Haw, who settled as a magistrate in Caledon in June 1859, but frequented the village to spend his vacation. Hawston is located in the Western Cape (South Africa), approximately 99 km from Cape Town, along the R43 a major tourism route towards Hermanus the capital of the Overstrand Municipality. More than 8000 cars with foreign and local people drive past Hawston daily on the R43 coastal road.

The coastal settlements in this Overstrand area are internationally recognized for their whales watching possibilities during spring time. Many tourists visit the area during holidays or drive by on their way along the Garden Route from or to Cape Town.

Hawston has a rich cultural history and is renowned for its abalone divers. This rural town is situated very close to the Kogelberg Nature Reserve, which has been declared as a World Heritage site because of the Fynbos species found in the Reserve.

According to Statistics South Africa (2011), it is estimated that the town has approximately 8200 residents; approximately, 27.7% is in the age range 0-14 and 67. 5% is in the age range 15-64. Hawston has the following community assets: two schools, one clinic, two community halls, one swimming pool, and a caravan camping site. Hawston residents are linked to the fishing culture of the region, but, since the government restricted the fishing and abalone diving licenses because of over fishing about 10 years ago, many people lost their jobs. There is a lack of skills amongst residents and facilities are needed to keep the youth occupied in a constructive manner to counteract the increasing problem of drug abuse. Limited opportunities are provided for the latter group; therefore, this group migrates to more thriving urban areas. This social situation and the poaching of abalone have created a wicked reputation and perception of Hawston over the last years. The town is also often associated with the Western Cape gangster culture, which is a concern to Hawston residents who are well aware of the need to change their image and to open up as a friendly rural town within the Overstrand Municipal Area.

Representatives of the Hawston community approached the Overstrand LED Agency (OLEDA) to assist them to find solutions to their untenable economic situation. OLEDA decided to use (P)ALAR to assist the community with a SLED strategy.

The Establishment Phase

During the establishment phase, a community and stakeholder meeting was organised and, after heated debates, the community and other stakeholders of Hawston nominated representatives to form the Hawston Forum (HF).

The Preplanning Phase

In the preplanning phase, the HF held meetings, developed a constitution, and selected a chairperson, to champion the process. The HF was mandated to coordinate and to initiate the SLED process in partnership with OLEDA. Community and stakeholder meetings were organized. The participants agreed to start a process to understand which problems the community is confronting and to seek innovative solutions to these issues. The difficulties that were identified included unemployment, lack of skills, drug abuse, and crime. A questionnaire was designed and local unemployed youth were trained to conduct the interviews. OLEDA proceeded to conduct surveys to collect baseline information on crime statistics, community income, unemployment, skills, and community assets. The data were analysed and the results presented to the HF to obtain inputs, ideas, and comments.

The Planning Phase

During the planning phase, the strategic plan was developed. The HF used the baseline information to formulate a strategic plan with a new vision for Hawston. The new vision was to establish Hawston as a unique tourism gateway to Hermanus.

The HF invited the community to participate in workshops and brainstorming sessions to develop tourism destination concepts. This culminated into tourism ideas that included the sea economy and the indigenous culture of the community. Land belonging to the Overstrand Municipality was also identified as an area where a tourist destination could be developed. These workshops with members of the community and stakeholders were extremely interactive and produced creative solutions. The most fundamental issue was that the concepts identified should be unique and implementable. The Hawston Gateway Tourism Project was born. The design process started. Artists and other consultants assisted the HF with elemental information in order to support the development of the concept. The design had to incorporate concepts to market Hawston as a small town destination, exploit the natural culture and heritage of the Overstrand Area, and provide a museum which could reflect the residents' interpretation of the region, marine life, geology, the history of the African Khoisan people, and a prime viewing spot for land-based whale watching.

The Action Phase

The next phase was the action phase, during which the strategic plan (SP) was implemented. The SP contained s (specific), m (measurable), a (attainable), r (realistic), t (time bound) objectives with responsible persons, a budget, and timeframes.

The Reflection Phase

The actions were monitored by OLEDA during the observation phase. This was followed by a reflection phase, during which participants gathered to reflect on their experiences from the (P)ALAR cycle. These actions and outcomes were reviewed and the community needs were addressed. The SP was revised for the next phase.

Key Lessons Learnt

Flowing from the reflection phase of (P)ALAR on the Hawston Gateway Tourism project, the following guidelines in terms of a NDS for SLED were formulated:

Lesson 1: Why Should the Community Get Involved?

Mobilizing communities' interest and their involvement in a NDS strategy necessitates processes and actions which empower communities and stakeholder groups to appreciate the importance of implementing initiatives to solve problems relating to employment and economic development. Traditionally, these were left to the government. However, there is growing recognition that a community's economic and employment future depends increasingly on its members' positive initiatives.

A NDS strategy work best when the community takes charge of its own destiny and the sense of responsibility is spread throughout the entire community. However, no community (neighbourhood) should underestimate the potential for opposition to change and initiative and the degree of skepticism often present about a community's ability to instigate and manage change. Any local leadership or neighbourhood action group seeking broad-based community support and participation must use every means to inform, demystify, and stimulate awareness and interest about the NDS project. Examples of actions aimed at motivating the community and stakeholders may include using local media, convening forums and discussion groups, generating interest through competitions, think tanks, study tours, and recruiting respected local people to arouse interest and help to champion interest in the NDS.

Lesson 2: Where Is the Community Now?

The community needs a community analysis (survey) to catalogue its resources, problem areas, opportunities, and options. This is an essential step in the process of identifying and implementing appropriate strategies and actions. Through this analysis, the community and stakeholders may achieve a better understanding of their neighbourhood. The community surveys of Hawston involved the following:

- An examination of current problems, trends, characteristics, and attitudes of the community;
- An analysis of the community's comparative advantages and limitations;
- The community assets;
- The identification of possible opportunities;
- An assessment of the community's interest and commitment to the NDS;
- The collection, analysis, and presentation of relevant data;

There are a variety of techniques and tools that can be used for an effective community analysis:

- The identification of the community's strengths, weaknesses; opportunities and threats (SWOT analysis);
- The preparation of a community profile document, giving an overview of the community's economic, social and environmental features;
- The compilations of a community skills register to record skills and interests of the community.

In the case of Hawston, it was also recognised that the community has structural barriers which must be addressed before strategies could be implemented in order to prepare residents for the jobs of today and tomorrow.

Adult illiteracy was one of the most prevalent challenges. Individuals, who dropped out of the educational system, as well as recent immigrants, struggle with effective workplace communication. Illiteracy prevents people from obtaining well-paying and personally satisfying jobs and can effectively block people from participating in educational and training programmes to develop strongly needed skills. As computer literacy is becoming more and more important in the workplace, individuals who are unable to read are several stages away from being ready for employment.

In addition to lacking basic skills, residents in Hawston also lacked soft skills and life skills. Soft skills include work ethic, integrity, commitment, professionalism, and maintaining interpersonal relationships. Lacking previous and successful employment experiences, some residents may have challenges with showing up to work on time each day, or displaying the appropriate demeanor with bosses and co-workers. Life skills include knowing how to dress appropriately, managing a household, and managing finances. The members of a community (neigbourhood) with limited life skills have difficulties in caring for themselves and meeting work obligations, and may lack the self-esteem that is necessary to be interested in learning these skills.

Hawston residents lack their own means of transportation or and have limited access to transportation services that allow them to get to training or employment opportunities. They do not possess computers or have Internet access. The lack of affordable and accessible child care services is another key barrier.

In Hawston long-term unemployment has been self-exacerbating, not having steady and gainful employment has been the norm, and therefore there is some cultural acceptance of it. The NDS can effectively be utilized to address human capital problems. While this initiative may take several years before the impact is realized, it is anticipated to dramatically alter the lives of the community and break the cycle of multigenerational poverty.

Lesson 3: Where Does the Community Want to Be?

Whilst answering this question the community may want to focus on determining values, goals, and objectives. In essence, the community needs to determine a shared vision of its preferred future. There is the need for situations which enable local residents to articulate what they want their community to be. This vision must be based upon consensus (a unified opinion amongst stakeholders and local leaders as to the mission and scope of the community's NDS activities).

A community vision is how the residents would like to see the neighbourhood in the future, and provides a foundation on which all the NDS activities will be built. Vision statements present an image of the community's future success, based on what is realistic and attainable. Developing a concise statement about the overall vision for the neighbourhood helps to focus efforts and gather stakeholders around a common purpose.

The visioning process should happen as early in the planning process as possible, as it helps setting some of the questions to explore during the process. A vision encompasses the values that stakeholders believe are worthwhile and important. By creating a vision, the community implies that a choice is made among these different values. A shared vision is the first step in bringing stakeholders with different objectives together to work on one effort. A shared vision is, therefore, a call for action, not just an idea. Some find it helpful to draft a vision statement that encompasses the many qualities and characteristics of their ideal community. An alternative method would be for the members of the planning team to identify key words or phrases that express the vision they have for the neighbourhood. Those key words and phrases are then modified and combined to create a single statement on which all can agree. The following are key questions to ask:

- If the neighbourhood continues on its current path, what is likely to happen in the next decade? Is this direction good or bad?
- Do key stakeholders know where the neighbourhood is headed and agree on the direction?
- Do the neighbourhoods' structures, processes, plans, organizations, and information systems support the current direction?
- Are all the relevant stakeholders at the table to participate in developing the vision?
- Is the community being reactive or proactive?

- What will success look like for the community?
- Is the vision realistic?
- Is the vision inspiring?

In the case of Hawston, the visioning process at the neighbourhood level resulted in a physical picture–the community stakeholders found it the easiest way to communicate their desired future in physical terms. New buildings, new community structures, and new levels of coordination among community groups emerged as symbols that encompass a range of desires for improving the quality of life of the Hawston community.

Lesson 4: The Strategic Plan (SP)

The visioning process is the starting point of formulating a SP. In the simplest terms, strategic planning is the art and science of strategic change making where leaders proactively invent their community's future. When the SP is finished, it may lead to a dynamic and highly positive community life-altering experience.

The Hawston case indicated that strategic planning calls for extensive community involvement and involves a creative and practical process whereby public, private, and other stakeholders develop consensus around NDS goals, strategies, and programmes. A SP brings all of the players to the table and allows everyone to share a vision for the area and to be part of the process from concept and design through execution. By involving all stakeholders from the beginning, the foundation is laid for community-wide engagement in the process and outcomes, and there is less chance of surprises popping up, such as non-supportive groups or a part of the plan alienating a key player. Beyond this, there is an exciting potential for the strategic planning efforts and, therefore, the community, to benefit from the expertise and insights of the broad-based citizenry. This sets the stage for a more effective implementation phase. A simple rule to remember is plans do not get things done, people do.

Strategic planning can be used for overall SLED plans or sub-activities. When determining the appropriate size of the project, it is essential to make a realistic assessment of its future demand or use. The adoption of strategic planning for sub-activities should always be done within the broader development plan and vision for the community.

Ultimately, the use of strategic planning for sub-activities should lead to strategies and actions which are:

- Compatible with the community's vision and quality of life;
- Feasible given the community's strengths and weaknesses;
- Implementable given the community's leadership and resources;
- Change-making for addressing the community's most urgent challenges;
- Innovative, enabling the community to achieve its most promising opportunities.

Strategic planning is incredibly hard work and requires significant amounts of time. Energizing the efforts of community leaders and stakeholders in a strategic planning process requires the building and strengthening of trusted relationships, which only happens through repeated interactions. In the Hawston case, the process required each of these key players to volunteer hours out of a 24-hour day which is already highly devoted toward work, family, and other commitments.

Lesson 5: How Does the Community Get to Implement the NDS?

In order to implement the NDS an appropriate organisation or structure is required to oversee and manage the process. In the Hawston case the community and stakeholders formed the HF. The HF partnered with OLEDA to assist with the strategy, using (P)ALAR. Critically reflecting on the Hawston case, the following guidelines were identified:

- Broad-based community support and representation on the management group is needed;
- Endorsement from key community and stakeholder groups including local and provincial government and business groups;
- The formation of task groups;
- Appointing a suitable chairperson, with credibility and knowledge of the community and NDS options;
- The identification and securing of the necessary resources. There is a variety of resources required to implement NDS strategies including: human resources (e.g., skills and expertise, contacts and influence), financial resources, and physical resources (e.g., facilities, equipment and transport, and information resources).
- The establishment of lines of cooperation and communication with the community, municipality, province, and district are essential to keep them informed, demonstrate positive outcomes, and create awareness. This may be achieved through frequent telephone calls, newsletters, user-friendly minutes, press releases, visits to schools and churches, brochures, flyers, community feedback sessions, or by briefing key political, bureaucratic and industry leaders and by participating in workshops, conferences, and training events.
- The ongoing development of local leadership is of utmost importance. Learning from the Hawston example, the local leadership was rejuvenated by participating in community leadership programmes; the identification of local women that can assume leadership roles is a key aspect, and supporting the leadership pool by arranging sponsorships to participate in events of learning and capacity building.

Lesson 6: What Options or Initiatives Are Available to Implement a NDS?

The case of Hawston accentuated that any NDS needs a strategic approach to planning and taking action. There is a need to explore the full range of initiatives available and then choose an appropriate initiative which is feasible and link the comparative advantages to the NDS strategy.

The Hawston case identified seven initiatives that may be incorporated into a NDS:

1. The improvement of the local infrastructure. In terms of this initiative the community may explore ways to maximize local job opportunities and business spin-offs during infrastructural developments, such as the construction of roads, housing, clinics, and commercial and industrial areas. The potential exists that the infrastructural constructions benefit communities through:
    a. The creation of new training and employment opportunities during the construction and maintenance stages;
    b. Special training opportunities for local young people with contractors;
    c. The use of labour intensive methods to maximize local employment;

   d. Assisting with the establishment and consolidation of local enterprises through new local contracting opportunities;
   e. The improvement of access and attractiveness of the town to visitors.
2. The introduction of initiatives to improve the existing local businesses to improve their productivity and market share. Local communities can engage in a series of initiatives that might strengthen existing businesses. The support to these businesses may include:
   a. Business skills training inter alia to improve marketing, merchandising, customer service, management practices, networking, new market opportunities and diversification;
   b. The creation of a local database to link existing businesses with new market and subcontracting opportunities;
   c. The implementation of deregulation measures by local and provincial government;
   d. The marketing of services and products through group branding; and
   e. The development of new tourism products and facilities.
3. The development of initiatives to support new businesses. In order to foster local entrepreneurialism and new business formation, it is important for any community to support them. A new business may mean new jobs as well as expanded trade opportunities for local businesses. There are a range of initiatives that a community may explore, namely:
   a. Idea generation workshops;
   b. The development of business skills;
   c. Business for a day workshops;
   d. Entrepreneurship awareness workshops at schools;
   e. The development of local alternative money; and
   f. The establishment of a walk–in the centre to give business advice or direction in terms of the next steps, work space, access to facilities for example the Internet, e-mail, tender identification and assistance with completing the forms, export advice, research, and networking opportunities.
4. The attraction of new industries, businesses, and investments from outside the community. By attracting new investment to a town, the job opportunities may increase. Place making is an important concept where a town or a community (neighbourhood) can brand itself to attract residents and businesses. Some of the actions that a community may consider are:
   a. Place making campaigns;
   b. Special events;
   c. Database of skills and municipal land available for development;
   d. Lobbying for specific government projects;
   e. Investment incentive packages;
   f. Promotion of lifestyle;
   g. Crime reduction strategies;
   h. Beautification projects; and
   i. Targeting *grey power* (retirees).
5. The implementation of initiatives to keep the flow of money (expenditures and purchases) in the local economy. A rand spent in the local community contributes to the well-being of that community. It is therefore important to encourage local buying by residents and businesses. Local communities can take a variety of steps to limit leakages:
   a. Buy local campaigns and promotions;

   b. The promotion of local businesses and their unique range of goods and services;
   c. The creation of local markets;
   d. The formation and strengthening of local Chambers of Business.
6. The development of capacity to link the skills of unemployed community members to advertised jobs; this initiative may be implemented to assist local persons with job searching, for example:
   a. The creation of a local casual labour exchange that links unemployed persons with casual job opportunities;
   b. Assistance to complete job applications or the development of a curriculum vitae;
   c. The coordination of a local apprenticeship/traineeship scheme where local businesses share a group of apprentices/trainees.
7. The marketing of the community to enhance the image of the town or community (neighbourhood) to potential visitors. The potential of tourism in South Africa is enormous. The case of Hawston emphasized this as a community (neighbourhood) with magnificent potential being bypassed by potential visitors. One of the most important factors to attract visitors to a new destination is broad community commitment to ensure the visitors' safety and security. The actions that a community may consider include:
   a. A focuses on: how to attract visitors or increase the number of visitors, how to increase their length of stay, how to increase the amount each visitor spends and how to increase visitors' satisfaction so that they will return or tell others;
   b. The development of travel break patterns of passing motorists aimed at influencing travellers to visit a neighbourhood, town, or community with attention to attractiveness or uniqueness of the town approaches or entrances, clean toilets, litter free, clear signage, interesting crafts or food, conditioning of roads, accessible and safe parking, and attractive promotional material;
   c. A well-managed Web site;
   d. A strategy promoting cultural events or history tours;
   e. A programme to enhance the quality of service and hospitality at all points of contact with visitors; and
   f. A heritage conservation strategy.

Lesson 7: LED Professional Competencies

In terms of the competencies required by an LED professional, the Hawston case study indicated that, if SLED is defined broadly as a process that affects the wellbeing of a community, region, or province and its residents and if the role of the LED professional is to positively influence that process, the LED professional must take the role of:

- **Analyst:** In this role, the LED professional needs to understand the strengths and weaknesses as well as the comparative advantages of the local economy. The professional also provides quantitative and qualitative information to the decision and policy makers.
- **Catalyst:** In this role, the LED professional provides incentives to leverage the investment or involvement of different public and private sector actors. The professional also invokes enthusiasm and excitement within the community to support initiatives.

- **Gap Filler:** The LED professional provides assistance where the markets and institutions cannot or will not meet the community's needs. The efforts of the public sector should not oust the private sector where the private sector is willing and able to make needed investments.
- **Advocate:** The LED professional speaks out for the wellbeing of the community and protects the interests of businesses. They also move the strategic planning process along by advocating for the good of the entire community.
- **Educator:** The LED professional must educate the general public to that they are adequately informed about SLED and the costs and benefits of such initiatives.
- **Visionary:** The LED professional must be a visionary leader who is able and willing to peer over the economic horizon to see not just what is most likely to happen, but what could be if the community's spirit and resources are focused on catalyst projects that can positively change the community's economic future.
- **Systems Thinker:** The LED professional should embrace systems thinking to understand how different projects, processes, and factors influence one another within a broader system. The professional should play a role of aligning and integrating different related initiatives (and role players) in a way that they together contribute towards LED to benefit the community. At different times and in different contexts, the LED professional will play all of the aforementioned roles.

In summary the Hawston NDS key lessons are an indication that major SLED strategies, projects, and programmes require a long span of awareness and overcoming different challenges with different stakeholders. The LED professional alone cannot improve the economic wellbeing of a community. Comprehensive SLED strategies require the cooperation of various public and private actors, agencies, and organisations and performing (P)ALAR, as in the case of the Hawston Gateway Tourism project, confirmed Zuber-Skerrit' s (1993) findings, namely that:

Action research can be conceived in a similar way of planning an experience or action (including problem analysis and a strategic plan); implementing the strategic plan; observing the action (including an evaluation of the action by appropriate methods and techniques); and reflecting on the results of the evaluation and on the whole action and research process, which may lead to the identification of a new problem or problems and, hence, a new cycle of planning, acting, observing and reflecting. (p. 46)

CONCLUSION

What happens at the neighbourhood level has a profound impact on people's lives. Whether neighbourhoods move forward economically is largely dependent upon the key stakeholders within a neighbourhood, namely the residents, businesses, and property owners. Creating a framework for these stakeholders to not only coexist in a valuable and mutually supportive manner, but to also work together toward creating and sustaining a vibrant and energetic neighbourhood, can be achieved through (P)ALAR as linked, integrated, and empowering concepts.

In this chapter the case of the Hawston Gateway Tourism project, was used to demonstrate how (P) ALAR can improve the practice of SLED strategies by applying it to empower a neighbourhood to find solutions to overcome their challenges and develop guidelines for future NDS.

The case of the Hawston Gateway Tourism project demonstrated that a NDS can take many forms and depends upon the needs of the neighbourhood in question. For neighbourhoods that are declining, LED professionals, in partnership with all stakeholders, must explore questions regarding the root causes of decline and how to best curtail further disinvestment and possible abandonment. For neighbourhoods in growth mode, the stakeholders must weigh how this growth translates into demand for new or expanded spaces and services. In either example, what is happening in the neighbourhood is largely dictated by the surrounding economy, and efforts need to reflect an understanding of both.

NDS employs traditional SLED practices on a micro scale. First and foremost, however, neighbourhoods are residential communities. The reality of neighbourhood development is that there are lots of vested interests. Therefore, a top-down approach does not work and will often be met with great resistance. Rather, building coalitions and working relationships across many interests from the bottom-up produces more sustainable change. Neighbourhoods often have limited resources and limited capacity and, therefore, great challenges. Neighbourhoods are the building blocks for healthy, thriving, and vibrant towns, cities, and regions. Many local governments recognize the benefits of approaching neighbourhood improvement as a distinct task within the larger framework of SLED. Dynamic neighbourhoods are the focus of business, leisure, and daily activities. Residents often gauge the health of the town, city, or region based on the perceived strength of their neighbourhood.

The challenge, of course, is that rural and urban economic forces keep changing and the work of neighbourhood development is never finished. The case of the Hawston Gateway Tourism project as a strategy of SLED where the (Participative) Action Learning and Action Research, (P)ALAR method, was used, is not common to SLED strategies in South Africa.

It is therefore recommended to LED professionals that the (P)ALAR approach form part of any SLED strategy because it will assist in building capacity to make continuous positive change in the face of changing economic and political realities.

REFERENCES

Albertyn, R. M., Bester, A., Botha, M. J., & van der Merwe, M. E. (2007). Entrepreneurial skill development: Participatory action research approach in a rural community. *Journal of Family Ecology and Consumer Sciences*, *35*, 9-16.

Black, P. A., Calitz, E., & Steenkamp, T. J. et al. (2005). *Public Economics* (3rd ed.). Cape Town: Oxford University Press.

Department of Cooperative Governance and Traditional Affairs (COGTA). (2009). *Local Government Turnaround Strategy*. Pretoria: COGTA.

Department of Cooperative Governance and Traditional Affairs (COGTA). (2013). *Draft Local Economic Development Framework (2013-2018)*. Pretoria: COGTA.

International Economic Development Council. (2011a). *Introduction to economic development*. Washington, DC: International Economic Development Council.

International Economic Development Council. (2011b). *Neighbourhood Development Strategies*. Washington, DC: International Economic Development Council.

Patterson, C. (2008). *Country Report Local Economic Development in South Africa' Prepared for the GTZ strengthening local governance programme's LED component in South Africa*. Retrieved September 10, 2016, from http://www2.giz.de/wbf/4tDx9kw63gma/Country_Report_LED_in_South_Africa.pdf

Rabie, B. (2011). *Improving the systematic evaluation of local economic development results in South African local government* (Unpublished PhD thesis). Stellenbosch University, Stellenbosch.

Republic of South Africa. (1996). *Constitution of the Republic of South Africa Act no. 108 of 1996*. Pretoria: Government Gazette.

Republic of South Africa. (2000). *Local Government: Municipal Systems Act no. 32 of 2000*. Pretoria: Government Gazette.

Republic of South Africa. (2011a). *The National Development Plan*. Pretoria: The Presidency- National Planning Commission.

Republic of South Africa. (2011b). *The New Growth Path: Framework*. Pretoria: Economic Development Department.

Statistics South Africa. (2011c). Retrieved September 5, 2016, from www.statssa.gov.za/?page_id=4286&id=171

Swanepoel, H. (1997). *Community Development. Putting plans into action* (3rd ed.). Cape Town: Juta.

Venter, M. (2014). From lip-service to service delivery in Local Economic Development: Guidelines to set up an agency for action. *Journal of Economic and Financial Sciences*, *7*(3), 721–736.

Zuber-Skerrit, O. (2015). Participatory Action Learning and Action Research (PALAR) for Community Engagement: A Theoretical Framework. Educational Research for Social Change, 4(1), 5-25.

Zuber-Skerritt, O. (1993). Improving Learning and Teaching Through Action Learning and Action Research. *Higher Education Research & Development*, *12*(1), 45–58. doi:10.1080/0729436930120105

Zuber-Skerritt, O. et al. (2009). *Action Learning and Action Research Songlines through Interviews*. Rotterdam, The Netherlands: Sense Publishers.

Chapter 15
Conscientize Learners With Littering in Environmental Education by Using Action Research Activities

Tsebo Kgoto Mashiloane
University of South Africa, South Africa

Tome' Awshar Mapotse
University of South Africa, South Africa

ABSTRACT

The aim of this chapter is to provide delegates with a fresh perspective on the littering challenges facing schools and homes in developing countries today, as well as the established skills and awareness intervention strategies necessary to overcome these challenges hence sustaining the teaching of Environmental Education (EE). The extend of Africa's ignorance of EE subject amongst teachers has intensified and reinforced that action research (AR) be regarded as a tool for learner emancipation in the teaching of EE. EE is still seen as a foreign concept by many teachers and a notion that can be down played in school curriculum and home territory. In this chapter EE has find its way into school environment successfully and effectively through engaging learners as co-researchers with action research approach. The study was designed from educational living paradigm and is underpinned by collaborative learning theory. An intervention in the form of action research spiral cycle activities with the learners has managed to make them conscious about littering.

INTRODUCTION

People's environmental behaviour and environmental disasters over the years has worsened. Environmental crises including littering are on the increase caused by modern lifestyle. Availability and accessibility of disposal foam and other packaged items wrapped with plastics contributed to their behaviour littering resulting from unstainable lifestyle.

DOI: 10.4018/978-1-5225-2642-1.ch015

Inevitably, litter is perceived as both an environmental and economic concern, as litter is associated with harmful effects for the environment and high costs (Leijdekkers, Marpaung, Meesters, Naser, Penninx, van Rookhuijzen & Willems, 2015). Littering is unacceptable behaviour (Gossow, 2016: 95) and has become a common phenomenon such that, almost everyone on planet Earth litters in one way or the other (Orock, 2017). On daily basis people intentionally or unintentional do litter the environment owing to lack of proper environmental awareness (EA). In turn, lack of EA increases land degradation both in developed and in developing countries such as South Africa (SA). On the other hand, organisation such as United Nations Educational, Scientific and Cultural Organisation (UNESCO) frequently review educational systems across the countries to response to these environmental issues e.g. littering. Environmental issues constantly are on their agendas.

EA of the individual includes environmental knowledge, environmental values and environmental behaviour (Maravić, Cvjetićanin & Ivković, 2014). Education can be used to mitigate specific environmental issues and respond to their impact, but also to address the behaviour that causes them (UNESCO, 2016). In this instance, environmental education (EE) is a form of education designed to raise people awareness towards environmental issues such as littering among other issues. Therefore, government agencies, faith based organisations, non-profit and community groups, labour organisations and the private sector can all contribute to lifelong EE (UNESCO, 2016). EE is regarded as an important tool in the effort to stimulate a well-informed global response to environmental challenges (Maluleke, 2015).

In response to environmental issues and people environmental behaviour, the 2030 Agenda for Sustainable Development emphasises environmental sustainability issues, the need to transform consumption and production to restore balance to life on land and in water, and the need for urgent action on climate change (UNESCO, 2016). Intrinsically environmental action research (EAR) activities in collaboration with learners as co-researchers were conducted to raise their awareness of littering which adds the voice to the Incheon Declaration for Education 2030 Agenda for Sustainable Development.

Fourteen learners from seven classes served as co-researchers in this study. A sample was determined out of group of volunteers of learner population per class and convenience sampling was used. Researchers selected two learners out of seven classes irrespective of their age and other factors randomly to have manageable population. The study had more females and 15 years of age.

THE ROLE AND CONCEPTUALIZATION OF ENVIRONMENTAL EDUCATION WITHIN SCHOOL AND HOME SETUP

EE is a flexible extra-curricular feature of education that can be taught and practised at home and in all grades levels at school to increase people's awareness on the environment to alleviate environmental challenges. As EE is not bound to the classroom situation only (Maluleke, 2015). Das (2014: 13) corroborate that EE is not restricted to in-class lesson plans only as there are numerous ways learners can learn about the environment in which they live. For example experiential lessons in the schoolyard and field trips to national parks or after-school green clubs and/or school wide sustainability projects, the environment is a topic, which is readily and easily accessible. The theoretical part of learning and hands-on activities of EE in the environment should impart knowledge that leads to action orientated. Accordingly, EAR activities in EE are influential at home and school as these activities offer knowledge and action.

EE given to the learners gradually starts from garbage sorting, rubbish utilisation, hygiene, and to plant treatment (Amini, 2015: 44) to impart environmental knowledge and action to the learners. In this study, co-researchers were engaged in EAR activities in a progressive action research (AR) spiral cycles of exploring the soil, utilising waste materials and other collaborative forms of learning. For co-researchers to develop awareness, knowledge, skills and correct attitudes towards the environment (Maluleke, 2015).

EE is said to be an uninterrupted life-long process that accommodates people from different age groups (Maluleke, 2015). For that reason, EE should continue up to higher institutions of learning and form part of child development at home to raise their awareness towards the environment. The study conducted by Mbatha (2003) indicated that EE should be practiced continuously at every stage of one's life, from youth through to adulthood, so that a continuous sense of EA can be fostered all the way through. Mbatha continues outlining the disadvantage of placing the focus of EE solely on school-based initiatives by stressing that, once learners leave school, they will no longer have access to information that may encourage them to behave in an environmentally-sound way. This approach may perhaps result in learners littering their environment. However, the school is not only institution that can foster EA on learners, so the informal and non-formal education is also responsible to offer EE to children in order to sustain the information gained from school. Both institutions have a role to play on learners' awareness towards the environment and environmental issues including littering. Hence, this study involved co-researchers in EAR activities in a cyclic process of planning, acting, observing and reflecting at home and school. This was to emphasise the importance and linking indoor and outdoor learning activities, and school educational processes and household practices to sustain EE within this institutions.

ACTION LEARNING AS A MEANS TO HELP THE NATIONAL GOVERNMENTS TO ENHANCE ENVIRONMENTAL EDUCATION PROGRAMMES

Learning is a process that includes experiences and other processes. Our habitual action on the environment overtly forms part of learning process. This learning process could be effective or ineffective. For instance throwing garbage inside the litterbin form part of effective action learning. It should be noted that multifaceted relations within ones immediate milieu influence action learning such as habits, peer influence. For that reasons, EE programmes at home and schools should be collaborative not instructive for its effectiveness to occur. As it was established in the EAR activities.

Action learning combines research and learning process through collaborative engagement to raise co-researchers' awareness of littering and improve the environment. AR is about putting ideas into action, not only talking about them (McNiff, 2013). EE programmes incorporating AR will be effective, as every stakeholder is involved in the learning process. Additionally, as AR is collaborative (Atkins & Wallace, 2012; McNiff, 2013; Rose, Spinks & Canhoto, 2015) and practical (McNiff, 2013). Consequently, AR projects enhance learning since are hands-on tasks accordingly EAR activities proved this fact.

EAR activities adds effective programme in EE to improve the current state of the education system struggling to address learners littering behaviour both at home and school. Action learning activities influenced co-researchers in taking action to decrease littering within their close by milieu. As a result, action learning leads to active learning process. Recycling of waste materials for example to make manure accordingly stimulate co-researchers' awareness. This initiative offer solution to curb littering. AR is an environmental tool that empowers co-researchers with knowledge and action orientated to tackle littering. AR has the potential for human, non-human and environmental betterment (McNiff, 2013).

In view of that, action learning adds to well-made curricula and hands-on learning outside of school that could strengthen people's connection with nature (UNESCO, 2016). EE curricular in the education system should infuse knowledge based and action orientated activities such as crocheting to conscientious learners towards environmental problems including littering. The ensuing section outlines EAR activities that relevant stakeholders can use to address the problem of littering both at home and school.

SCHOOL AND HOME RELATED PROJECTS, WHICH REFLECT ENVIRONMENTAL AWARENESS

During the course of the study, co-researchers were engaged in a continuous learning cycle activities to address littering and followed a programme that was designed and implemented to execute EAR activities. These EAR activities raised co-researchers' awareness of littering at their homes and school, littering was reduced. EAR activities were sustained throughout the study. Researchers now outline EAR activities conducted in collaboration with co-researchers both at home and school and the difference it made within these institutions.

Environmental Campaigns

Environmental campaigns were conducted with co-researchers including reciting of poems in the assemblies and picking of litter around the school. Mondays at assembly, one of the co-researchers recites a poem to convey the message to the learners to increase their awareness towards littering. The intention of the campaigns was to encourage school learners to stop littering and respect their immediate environment.

Litter Pick up Duty Roster

Litter pick up duty roster was designed after some of the co-researchers' complaint about the behaviour of learners. The litter pick up duty roster was designed to involve all school members as suggested by co-researchers. Litter pick up duty roster it was further addressing lack of environmental monitoring tool in the school. As it was observed that lack of monitoring influence learners' bad decision of tossing waste in their immediate locale (Msezane, 2014).

School Environmental Policy

School Environmental Policy (SEP) was designed to address environmental issues including littering and to address absence of the environmental policy in the school. The policy was necessary to guide the development of action plans outlined to tackle learners littering behaviour (Simalumba, 2011). The policy was submitted to the School Governing Body (SGB) for its approval and implementation in order to be part of school policies. The school now has the environmental policy, this showed the success of EAR activity of designing environmental mechanism to curb littering, and it is crucial for any AR project to initiate solutions.

Environmental Calendar

Due to absence of environmental day(s) celebration at the school, researchers distributed copies of the environmental calendar to the co-researchers and School Environmental Committee (SEC). Co-researchers filed the calendar in their Social Sciences (SS) book for future references. Their peers as well requested a copy of the calendar. As a result, SEC submitted the copy to the administrative office and the school made A3 copies for all teachers and pasted in their classes and as well in the staff room. This was to allow learners and teachers to have access to the calendar but at the same time to enhance their knowledge about environmental days.

Vegetable Garden

A dumping site was prepared within the schoolyard and converted it into a vegetable garden. The land was prepared using spades, forks and wheelbarrows, and a fence was erected surrounding the garden so that school members could see that the plot is utilised. Grounds staff used to dump garbage and burn it on this plot as they collect waste materials around the school daily. The vegetable garden was meant to raise co-researchers' EA, instil positive attitude and bring co-researchers closer to the environment. Co-researchers brought different seeds from home such as spinach, tomatoes, beetroot, and pumpkin, as parents/guardians of some co-researchers are involved in a community food garden project. The land had to be prepared for almost three months by researchers and co-researchers because there were hazardous items such as bottles found on this land. Researchers and co-researchers buried some of the waste that decays to fertilise the soil and to soften it first before they could start cultivating. After preparing the land, co-researchers started the process of cultivating seeds and seedlings. After cultivation, each co-researcher owned a row to nurture the seedlings and instil care.

In addition, this activity was also conducted at home with co-researchers. Researchers bought three drums and requested sales person to cut the drums so that each family receive half drum and one package of seeds to cultivate. After receiving half cut drum it was filled with the soil and the process of cultivation began. The notion behind this activity was to influence co-researchers that disused objects are useful for environmental activities and instil positive use of three Rs (reduce, recycle and reuse) aspect. As well to make co-researchers aware, that though they do not have enough space within their yards but other avenues are accessible to start a vegetable garden and other environmental activities.

School Registered as Eco-School

The school was registered to Eco-Schools programme that is designed to infuse environmental learning to be part of school ethos. Researchers and co-researchers registered the school as it was not belonging to any environ clubs and to encourage school members to tackle littering inside and outside the school. Learners designed arty tasks in their classes such as mortarboard using waste materials including wool, cardboard boxes and scotch tapes. Their tasks formed part of linking action outside the classroom with the curriculum inside the classroom and raising learners' awareness towards littering.

Litterbins

Researchers bought ten steel water drums and steel poles from scrap yard and researchers and co-researchers requested grounds men to assist us in this regard to design litterbins. Steel water drums were cut on top to remove lids. On these drums at the top outer side grounds men welded pieces of tables/chairs steel to make litterbins to swing. Chairs and tables frames are dumped around the schoolyard resulting from vandalism. Underneath drums, small holes were drilled to release liquid or water during rainy seasons. On these poles grounds men welded steel to make arms to lock the litterbins.

After the whole process of creating litterbins, researchers and co-researchers with the help of grounds men installed two steel poles on the ground and in between, researchers and co-researchers left space in order to hang litterbins to swing. Ten stationed litterbins were installed within the school campus. Two stationed litterbins were installed in each block. A learner entrance/exit gate and at the food court one litterbin was installed. The objective of installing litterbins was to reduce littering in the school and to conscientious learners towards littering.

After the installation of the litterbins, researchers and co-researchers observed that littering behaviour continues and researchers and co-researchers therefore painted litterbins. It was the first time for some of the co-researchers to be involved in such activity. As it was indicated that "Sir, please don't comment about how we painted these litterbins because it is first time doing something like this". Another co-researcher said, "Though it is my first time painting but I am glad to be involved in this activity because of learning". Litterbins were painted with two different colours; the co-researchers suggested this. These colours symbolise school uniform as they wear brown and fawn. Painting of litterbins was to capture learners' attention and influence them to make use of litterbins, and to reduce littering in the school.

In addition, co-researchers have drawn litterbins with the help of Design learners in Grade 10. Co-researchers suggested that the structure should be drawn on paper and trace it on the litterbin. The structure drawn was first drafted on the paper charts to trace it on the litterbin. Co-researcher mentioned, "Tracing this structure on the litterbins will help us to avoid making mistakes and to complete this activity much faster". Litterbins were drawn using two different paint colours comprise of white and black, and with brush. These colours did not have a meaning as compared to previous colours mentioned above. Co-researchers were divided into two members per group to draw litterbins. This EAR activity was to make the litterbin attractive, decorate school and to influence school learners to make use of the litterbins.

Watering Plants

On the 22/04/2016, the Eco-cycle organisation donated trees to the school as part celebrating Earth Day and the trees were planted. During the plantation of trees, it was observed that learners love outdoor activities. The school requested researchers and co-researchers to take care of the trees. Their request was appreciated and each co-researcher adopted a tree and co-researchers watered the trees. The activity of watering trees had two goals. First goal was to teach co-researchers to save water. Co-researcher used buckets to fetch water from the tap to water the trees. Secondly was to instil sense of ownership and respect for the environment.

Installing Boards to Identify Plants

After celebration of Earth Day, trees were put under our care as stated before. Co-researchers were responsible for watering the trees. This activity was intended to influence co-researchers to love trees. They named trees caring words such as gift, hug me and other names. Trees are planted behind the classes at sports field and this area was named baby city. Whenever they come to the area they would say "re lo bona bana ba rona"[1].

The process of designing and writing names on the boards was conducted using paint, barbed wire, sheet boards, steel poles and plier. Co-researchers managed to turn discarded sheet boards into bulletin board to name their plants for identification. A hammer was used to install the bulletin boards with names on it next to their trees. The notion behind this activity was to make co-researchers to love plants and bond with nature in turn to brighten their awareness of littering. Always they were clearing any unwanted items such as plastics wrappers next to their plants. Additionally to instil sense of ownership in their immediate environment starting with these plants.

Crocheting

The co-researchers crochet mats with plastic bags and an instrument called a crotchet. Each co-researcher was involved in cutting the plastic bags into small lengthy pieces. Afterwards researchers and co-researchers joined these pieces together and rolled them to control it when researchers and co-researchers started to crochet. Two co-researchers taught us how to crochet as their parents' crotchet mats and other artistic work at home with plastic bags and wool. As a result, those co-researchers facilitated the whole process of monitoring us to check if researchers and fellow co-researchers crocheted correctly. For some of the co-researchers, it was the first time they were involved in an activity with the researchers. Co-researchers were paired into two members per group with the researchers while two co-researchers facilitated the activity. Co-researcher mentioned, "I never thought I will ever carry out this task because usually I see my mother crocheting. So I thought crocheting is done by women only but I am happy to be part of this team since I am learning new things".

It was a tricky but fun activity, at the end mission was accomplished as co-researchers, and researchers have completed designing plastic mats. This activity intended to teach co-researchers about three Rs of plastic bags and make them aware of littering. Moreover, it empowered co-researchers as they could pursue a career and business venture to design mats and other artistic work with plastics and sell these product to survive in future.

Moreover, co-researchers at home also were engaged in crocheting using plastic bags and using crotchet instrument to design creative tasks. This activity was putting emphasis on three Rs aspect and improves co-researchers' awareness about litter issue and other environmental issues resulting from the incorrect disposal of plastic bags. "I struggled with this activity because I had to ask assistance from my grandmother to teach me how to crochet and I had to request plastic bags from neighbours to complete my mat" co-researcher said.

Class Competition

Class competition was introduced after some of the co-researchers complained about the status of their class. A co-researcher stated, "We don't have cleaning roster and SEP in our class and our class is always dirty. Our class teacher does not care because some of the learners when asked to sweep they refuse". Researchers engaged teachers during staff meeting held on 17/08/2016 to discuss the initiative and it was highly praised. Class managers were requested to inform their classes about the competition. Each Grade level was awarded cleaning material including 5L pine gel, 5L bucket, 2kg floor polish, two brooms, and one mob. This initiative was to encourage school learners and teachers to maintain cleanliness in their class and around the school, and to infuse sense of ownership. After the competition the classes were well organised and some class had charts displayed on the wall as well as SEP was pasted on the wall. This showed improvement from the time when class competition commenced.

Celebrating Arbour Day

In this activity, co-researchers celebrated Arbour Day by planting flowers, a day on which schools and other organisations in SA plant flowers, trees and so forth. It was the first time that some of these co-researchers were engaged in such an activity. Some of the co-researchers mentioned that they had never celebrated Arbour Day before, and thanked researchers for giving them the opportunity to do so.

One co-researcher mentioned, "The school should encourage learners to bring flowers on this day to beautify the school and this could teach us to know about environmental days". In this school, flowers are only planted at the administrative block. At this school, it was evident that the focus is on indoor learning, but the outdoor activities seemed to heighten the learners' awareness of environmental issues, among them the scourge of litter. The celebration of Arbour Day at the school was intended to raise learners' awareness towards plants and littering and to link indoor and outdoor learning processes.

Researchers also gave co-researchers flowers as part of celebrating Arbour Day at home on the 01/09/2016. The aim of planting flowers at home was to link school and home activities and to heighten their awareness towards the environment. Flowers were irrigated with used water as a way of saving water and to stress three Rs.

Making Manure From Waste Materials

Co-researchers on daily basis carry boxes and buckets containing vegetable waste materials collected from the school's kitchen and are throw down within the garden yard to make manure for the garden. Comments such as "Besides learning how to make manure I'm also learning ways of saving money as we don't buy manure from the shop but we utilise garbage materials". Rubbish is free and a powerful visual tool (McEwen, 2015) to influence learners to stop littering. Another co-researcher stated, "By collecting these waste materials we are also teaching ourselves and food handlers' ways of disposing vegetable waste in an environmentally friendly manner". It was observed that food handlers throw waste materials from the kitchen to the dumping site in the schoolyard. A study conducted by Sibanyoni, Tshabalala and Tabit (2017) found that food handlers lacked the knowledge, awareness and attitude on many important microbial food safety hazards. In turn, this perhaps translated into their negative behaviour of disposing waste incorrectly.

Because of their actions, researchers and co-researchers informed grounds staff and food handlers to keep waste materials safe. Co-researchers collected waste every day and dropped these items within the garden yard. Then the process of making manure started. After throwing the waste in the garden yard, researchers and co-researchers placed soil on top of it so that they can decompose. Co-researchers executed the process of mixing soil with decomposed waste. After this process co-researchers applied manure on seedlings. This environmental activity was directed to teach co-researchers about compost making and the three Rs, showing them how to reduce litter by re-using organic waste.

After researchers and co-researchers made manure, researchers and co-researchers shared the compost with other co-researchers so that they fertilise their vegetable garden at home. Co-researcher indicated, "Thank you guys for sharing manure my garden now is blossoming". Another co-researcher mentioned, "My mother was surprised when I arrived at home holding a container containing mulch and I told her this is compost to fertilise the garden". Moreover, co-researchers made their manure at home, as they took residuals in the washbasin and poured inside bottles to decay and then fertilised their garden. This was to encourage them to use waste materials at home to make manure and in turn to increase their awareness of littering.

Emblems Attached on School Walls

This activity involved the process of designing the emblem using lids removed from the drums that were turned into litterbins. Reasons behind using lids was to teach co-researchers about three Rs. Co-researchers washed off oil and dust on the lids using a brush and afterwards co-researchers started painting the lids with a white colour and drawn with black colour. "Love your school, stop littering" message was written on the bulletin board and with hands holding the Earth at the centre of the emblem. Two hands holding the Earth on the emblem represent that co-researchers should respect environment because is delicate. Emblems were put strategically next to learners' toilets entrance and taps so that learners read the message right through. Co-researcher indicated, "Since we have started with this project I have noticed that we have been utilising discarded objects. I never thought lids removed from the drums could be turned into emblems. Thank you sir for opening our eyes". The aim of this emblem was to draw an everlasting impression and instil the motive behind the preservation of the environment. Msezane (2014: 69) reiterates that visibility of posters campaigns in the school could be important to reduce the spread of littering.

In summation of the above, cyclic process of conducting EAR activities with co-researchers was completed successfully. Continuous learning circles were conducted in a spiral cycles. The process involved four steps consist of planning, action, observation and reflection. Researchers and co-researchers conducted EAR activities to enhance learners' awareness of littering in the school. These EAR activities covered vast psychological factors that might have influenced learners to litter the school grounds. For instance, the litterbin is far and ugly. Study conducted by Mathe (2014) found the condition of the litterbins as another factor that contributes to untidy environment. Researchers and co-researchers therefore painted and drawn litterbins addressing some of the psychological aspects. All EAR activities were hands-on and envisioned to address littering and develop learners' awareness. As it was noted that learners' attention could be caught via practices that involve environmental problems, they face or may face in their region (Simsekli, 2015) to foster good environmental habit and practice.

Co-researchers were engaged in vegetable garden throughout the cycles. From preparing the land for cultivation, fertilising the soil, cultivating and watering. This activity reached its objectives of bringing co-researchers closer to the environment based on their comments and behaviour. As well their willingness of setting up the garden. Co-researchers took ownership of the seedlings in each row and they were competing among themselves, which made them to love the garden more than before.

A study conducted by Rakotomamonjy, Jones, Razafimanahaka, Ramamonjisoa and Williams, (2015) suggested that a single day EE programme can have a lasting influence. Indeed this study was conducted for a period of nine months and proved that a prolonged EE programme thus conscientious co-researchers towards littering and heighten their awareness. If approached through AR in a collaborative learning process.

SHARING INNOVATIVE STRATEGIES THAT CAN BRING CHANGE AROUND ENVIRONMENTAL EDUCATION PERCEPTION

This section discusses innovative strategies that EAR activities employed to address littering in collaboration with co-researchers to widen EE curricular at school and home. Co-researchers implemented these EAR activities through progressive AR spiral cycles. Progressive AR spiral cycles conducted with co-researchers intended to boost their awareness of littering. A number of cyclic activities were embarked on with the co-researchers through collaborative learning. EAR activities used different strategies such as exploration of soil, wastes, litterbins and environmental monitoring tools.

At School

Co-researchers undertook through four progressive cyclic activities. These activities discussed different aspects such as exploration of soil, wastes, litterbins, and environmental monitoring tools.

On the first progressive cycle, co-researchers were taken through the process of exploring the soil in preparation for vegetable garden and celebration of Arbour Day. The school did not have a vegetable garden and co-researchers identified dumping site within the schoolyard to start the garden. The land was prepared for months to remove hazardous items and buried other items to fertilise the soil. Afterwards co-researchers irrigated the land in preparation for cultivation. Successfully co-researchers turned dumpsite into vegetable garden. Co-researchers once more celebrated Arbour Day at school by planting flowers. In turn thus, beautify the school and improve their awareness of the importance of soil and environmental days. Vegetable garden proved to be environmental avenue both at home and school to raise brighten awareness towards the environment and littering. From preparing the dumpsite all the way through. This activity reached its objectives of bringing co-researchers closer to the environment based on their annotations and manner, and their keenness to start-up the garden. Competition among co-researchers of nurturing their seedlings in their own rows instilled ownership and made them to love the garden greatly.

Second strategy of progressive AR spiral cycle activity to increase co-researchers' awareness was utilisation of wastes. Co-researchers collected vegetable waste materials from school kitchen to make manure for the garden. Plastic bags were used to crochet mats plastic. In addition, co-researchers managed to turn discarded metal sheet boards into bulletin board to name their plants for identification. Additionally, learners used discarded items such as cardboard boxes to design creative tasks including mortarboard as part of promoting indoor and outdoor learning activities. The activity improved co-researchers' awareness towards wastes with reference to the success of the activities.

On the third cycle, co-researchers observed that the school has shortages of litterbins. Iron drums containers were bought and turned these drums into litterbins and installed on the school ground. Then afterwards these litterbins were painted and drawn to catch learners' attention to use litterbins. Then again, lids removed from the drums co-researchers created emblems and attached it on the school wall to convey a message to learners to stop littering. This activity showed improvement since the school now has sufficient litterbins around the campus.

Cycle four, co-researchers introduced environmental monitoring tools which included litter pick up activities to encourage their peers to keep the school clean. Class competition was introduced to encourage learners and teachers to keep their classrooms clean. Subsequently, co-researchers designed SEP since the school did not have one. The policy envisioned to put into action of tackling littering in the school.

In addition, teachers pasted environmental calendar inside their classrooms for learners to be aware of environmental days and to supplement their environmental knowledge. For the purpose of sustainability of environmental activities, consequently researchers and co-researchers registered the school to Eco-Schools programmes anticipated to encourage all school members to tackle environmental issues including littering. The school was successfully registered and received bronze certificate.

When researchers reflect how they have started the journey of incorporating EAR activities with the co-researchers at both their homes and school, littering was experienced daily resulting from their day-to-day activities and learners were not engage in environmental activities at the school. The co-researchers were lacking EA, which contributed towards littering on these earmarked research sites.

At Home

Two progressive AR spiral cycles were conducted with co-researchers to raise their awareness of littering. Cycles cover aspects such as exploration of soil and wastes, which they have to gain experience from the school activities.

The first cycle co-researchers were engaged in exploration of the soil at home. Co-researchers received empty half drum and filled the drum with soil in preparation for the vegetable garden. Co-researchers watered the soil before starting to cultivate. Co-researchers further prepared the soil to plant flowers as part of celebrating Arbour Day at home. In general, co-researchers successfully carried out the activity of vegetable garden and planting of flowers at home. Families now have gardens and decorated their yards with flowers.

Co-researchers went through the process of turning wastes into useful resource on cycle two. Co-researchers used plastic bags to design plastic mats and hat using an instrument called crotchet. Additionally, co-researchers used waste residuals and vegetable waste to make manure for their garden. Manure was used to fertilise the soil filled in the drum to nurture their seedlings.

ACTION LEARNING AND ENVIRONMENTAL EDUCATION: TWINS FOR POSITIVE LEARNING AND HEALTHY STAY

EAR activities revealed approaches and/or strategies that parents and teachers can use to teach children about, for and in the environment. EAR activities promotes hands-on within EE to curb littering infusing knowledge based and action orientated. EAR activities made action learning to be fun simultaneously raising co-researchers' awareness for the benefit of the environment and their health. In view of that, the health of the environment is directly related to the health of people (Kuwahara, 2013). Positive environmental behaviour towards the environment will reduce land filth and subsequently reduce environmental diseases such as bilharzia emanating from contaminated water. Littered items can be turned into artistic work such as crocheting mats, which promotes action learning within EE and in turn will decrease environmental disease and issues.

Besides EAR activities reducing littering and conscientising co-researchers towards littering. EAR activities are instrumental in EE curriculum for example vegetable gardens could be used as teaching resources to enlighten learners about the process of photosynthesis, the effects of climate change, absorption, and other phenomena (Maluleke, 2015). EAR activities broaden EE curricular through educational living paradigm and collaborative learning theory.

EAR activities as well address EE obstacles faced by teachers in South African schools and parents at home to infuse environmental activities in their practices. For instance, co-researchers used waste materials to design artistic work that sort out lack of resources in schools. Consequently, it is essential that children be empowered to act for the environment. Hands-on, active education is a way in which children can learn about the environment, and more importantly, it can teach children how to practice positive and repetitive environmental actions (McEwen, 2015).

LEARNERS' VOICES: SUCCESS STORIES OF PARTAKING IN ENVIRONMENTAL EDUCATION ACTIVITIES

This section share co-researchers voice during the research journey. The voices of the co-researchers thus indicate the success of EAR activities that were conducted in collaborative manner. When the individual works collaboratively with more capable peers, the potential level of development will be increased. In other words, with the help of an expert, the individual can do more things, and this is referred to as potential development (Lin, 2015).

Some co-researchers said, "All teachers and learners must join us since this garden will contribute more vegetable especially in the school kitchen". Another co-researcher mentioned, "This garden besides producing vegetable but also makes us to respect the environment and learn how to plant seedlings and seeds because at home some of us don't have vegetable garden". Cross (2013: 2) states one-way, to connect young people to the environment may be to create a school garden, where the learners will be responsible for designing, planting, maintaining and harvesting vegetables. Exposing children to gardening at school could be an avenue for cultivating future environmental stewards. Co-researcher said, "My mother was so happy when I brought the spinach that we have harvested yesterday because she thought I was lying when I told her that we have started the garden at school". Ray, Fisher and Fisher-Maltese (2016: 392) reaffirm that school gardens may enhance healthy eating and increase both physical activity and civic participation among learners and their families.

Co-researcher indicated, "It was nice doing this activity because usually I make drawings inside my exercise book but now drawing this on the litterbins I was overwhelmed and when I look at these litterbins I still can't believe we produced such beautiful drawings". This activity explored co-researchers creativity. "I used to throw papers on the ground but now I stopped littering because we have litterbins which are closer" co-researcher said. A co-researcher indicated, "Installing these litterbins we are leaving the legacy in the school since they will be used for the next coming decades". As this addressed shortage of litterbins in the school.

A learner said, "It was refreshing to engage in this activity and being away from the classroom. I wish the school could continue to celebrate environmental days". In addition, some learners were even hesitant to go back to their classes after plantation of the trees as this task was carried out in the morning. This was indication that outdoor learning has an effect on the learners and enhances their awareness towards the environment. Outdoor activities provide learners with valuable experience in getting closer to nature, which sensitises them on environmental issues and enhances their knowledge in a systematic manner (Simalumba, 2011).

"I used to throw plastic bags away not knowing that they can be used for environmental activities. From the time when we have started with this activity I now stopped discarding plastic, I have started practicing to crochet at home" co-researcher said. This shows the interlinking of EAR activities at school escalated to co-researchers home. The activity of crocheting reduces littering on the ground at home and school.

EAR activities at home and school were successfully conducted and had an impact on the co-researchers because awareness campaign are often most successful when they are targeted at specific groups because information can be tailored to the activities, needs and challenges of the group (Verma, 2016).

CONCLUSION

Stated EAR activities evinces that intervention programme had positive effect in bringing EA among the learners (D'Souza, Sharma, & Singh, 2013). The execution of EAR activities proven that implemented programmes in collaboration with co-researchers produced positive results as they managed to heighten co-researchers awareness and reduced littering. Moreover, it was found that an educational intervention boosts children's awareness, perceptions of consequences and self-reported action (Hartley, Thompson & Pahl, 2015).

Experience is a central tenant in the action learning of EE from early childhood through to adult education (Cutter-Mackenzie & Edwards, 2013). Experience gained during the course of EAR activities served as lifelong learning process. Co-researchers learned to crochet, cultivate seeds and seedlings, drawn structure on litterbins and other activities. These activities heightened co-researchers' awareness of littering and empowered co-researchers to convert vegetable waste materials into compost. Raising the individuals' awareness of the environmental problems and improving environmental sensitivity will make great contributions to the creation of sustainable environments (Juahir, Toriman, Azid, Abdullah, Aziz & Kamarudin, 2014).

In conclusion, EAR activities if approached through progressive AR spiral cycles will improve learners' awareness and reduce littering within the school and home setup. Subsequently EAR activities will fill gaps within education system to address impeding factors that obstruct teachers to deliver EE effectively in schools and parents to engage their children in environmental activities daily.

REFERENCES

Amini, R. (2015). Outdoor based environmental education learning and its effect in caring attitude toward environment. *Indonesian Journal of Science Education*, *4*(1), 3–47.

Atkins, L., & Wallace, S. (Eds.). (2012). *Qualitative research in education.* Los Angeles, CA: Sage. doi:10.4135/9781473957602

Cross, B. A. (2013). *The impact of a school garden and environmental education on the environmental awareness of fifth graders* (MA dissertation). Southeastern Louisiana Hammond University.

Cutter-Mackenzie, A., & Edwards, S. (2013). Toward a model for early childhood environmental education: Foregrounding, developing, and connecting knowledge through play-based learning. *The Journal of Environmental Education*, *44*(3), 195–213. doi:10.1080/00958964.2012.751892

D'Souza, D. A., Sharma, D., & Singh, R. P. (2013). Effectiveness of selected instructional approaches in developing environmental awareness among IX grade students In Patna district, India. *Researchers World*, *4*(2), 109–114.

Das, M. (2014). Secondary school students' awareness on plastic Pollution and environmental education: A study. *Global Online Electronic International Interdisciplinary Research Journal*, *2*(5), 10–25.

Gossow, C. (2016). *An exploration of the daily environmentally sustainable practices by South African youths living in Urban Gauteng* (D.Ed. thesis). University of South Africa.

Hartley, B. L., Thompson, R. C., & Pahl, S. (2015). Marine litter education boosts childrens understanding and self-reported actions. *Marine Pollution Bulletin*, *90*(1), 209–217. doi:10.1016/j.marpolbul.2014.10.049 PMID:25467869

Juahir, H., Toriman, M. E., Azid, A., Abdullah, A. M., Aziz, N. A. A., & Kamarudin, M. K. A. (2014). Young Environmental Scientist (YES) program to enhance the environmental awareness among the Malaysian school students. In *INTED2014 Proceedings* (pp. 7122-7128). IATED.

Kuwahara, J. L. (2013). Impacts of a place-based science curriculum on student place attachment in Hawaiian and Western cultural institutions at an urban high school in Hawaii. *International Journal of Science and Mathematics Education*, *11*(1), 191–212. doi:10.1007/s10763-012-9387-3

Leijdekkers, S., Marpaung, Y. M., Meesters, M., Naser, A. K., Penninx, M., van Rookhuijzen, M., & Willems, M. (2015). *Effective Interventions on littering behaviour of youngsters. What are the ingredients?*. MSc programme.

Lin, L. (2015). Exploring collaborative learning: Theoretical and conceptual perspectives. In Investigating Chinese HE EFL Classrooms (pp. 11-28). Heidelberg, Germany: Springer.

Maluleke, H. M. (2015). *Curriculum policy implementation in the South African context, with reference to environmental education within the natural sciences* (D.Ed. thesis). University of South Africa.

Maravić, M., Cvjetićanin, S., & Ivković, S. (2014). Level of environmental awareness of students in republic of Serbia. *World Journal of Education*, *4*(3), 13–18. doi:10.5430/wje.v4n3p13

Mathe, R. (2014). *Effects of the Biographic factors and Religious Convictions on Littering to Enhance Waste Management in Pretoria City* (MSc dissertation). University of South Africa.

Mbatha, N. M. (2003). *Impact of Environmental Education on Environmental awareness levels of grade 9 learners in Soweto* (MA dissertation). Rand Afrikaans University.

McEwen, K. E. (2015). *The primary school factors that shape the environmental Knowledge, attitudes, and behaviours of children* (MSc dissertation). University of Waikato.

McNiff, J. (2013). *Action research: principles and practice* (3rd ed.). Oxon, UK: Routledge.

Msezane, S. B. (2014). *An exploration of the impact of environmental education innovation on students in Sustaining land resources: A case of Mkhondo village* (MEd dissertation). University of South Africa.

Orock, S. O. (2017). *The Effects of Littering on Tourism in Limbe, Southwest Region of Cameroon*. BSc course.

Rakotomamonjy, S. N., Jones, J. P. G., Razafimanahaka, J. H., Ramamonjisoa, B., & Williams, S. J. (2015). The effects of environmental education on childrens and parents knowledge and attitudes towards lemurs in rural Madagascar. *Animal Conservation*, *18*(2), 157–166. doi:10.1111/acv.12153

Ray, R., Fisher, D. R., & Fisher-Maltese, C. (2016). School gardens in the city: Does environmental equity help close the achievement gap?. *Du Bois Review: Social Science Research on Race*, *13*(2), 379–395. doi:10.1017/S1742058X16000229

Rose, S., Spinks, N., & Canhoto, A. I. (2015). *Management research: Applying the principles*. New York: Routledge.

Sibanyoni, J. J., Tshabalala, P. A., & Tabit, F. T. (2017). Food safety knowledge and awareness of food handlers in school feeding programmes in Mpumalanga, South Africa. *Food Control*, *73*, 1397–1406. doi:10.1016/j.foodcont.2016.11.001

Simalumba, P. M. (2011). *The implementation of environmental learning in grades 8-10 geography in the Caprivi region. Namibia* (MEd dissertation). University of South Africa.

Simsekli, Y. (2015). An implementation to raise environmental awareness of elementary education students. *Procedia: Social and Behavioral Sciences*, *191*, 222–226. doi:10.1016/j.sbspro.2015.04.449

United Nations Educational, Scientific and Cultural Organization (UNESCO). (2016). *Education for people and planet: Creating sustainable futures for all. Global education monitoring report*. UNESCO.

Verma, S. K. (2016). Environmental awareness: A need of nature. *Journal of Global Resources*, *2*, 182–185.

ENDNOTE

1 We are going to see our kids.

Chapter 16
Diversity, Equity, and Inclusion for Learners With Diverse Needs:
An Action Research Approach

Lloyd Dan Tlale
University of South Africa, South Africa

ABSTRACT

Diversity, inclusion and equity are complex, interdependent and interlinked concepts and embedded in social justice. Diversity summarises numerous differences that exist among people. It means not only people of differing race and culture but also differing languages, religions, values, abilities, socioeconomic status, gender and any other aspect that makes people different from one another. It, therefore, recognises that each person is different even if they belong to the same ethnic group, community or family. Diversity is an evolving concept in inclusive education. While it once referred to children from different ethnic backgrounds, it now encapsulates every kind of individual difference that exists for children, families and professionals. Diversity is the celebration of difference and the acknowledgement that difference is complex and that it exists in some way for every child. Diversity refers to attributes that contribute to each person's uniqueness, such as social class, gender, occupational status, income, sexual orientation, ability, disability, religion, and education.

INTRODUCTION

When we talk about diversity in inclusive education, we refer to the right of every child to be socially, academically and physically included in all facets of life (Mastropieri & Scruggs, 2013). Inclusivity does not mean uniformity among all people, but a society which makes room for diversity and nurtures collaboration and cooperation (Dressel & Hodge, 2013). To achieve social integration, the voice of all people articulating their needs and concerns needs to be heard. All members of society from whichever background must have a say and a stake (Schwartz, Weinberg, Hagenbunch & Scott, 2008). Inclusivity

DOI: 10.4018/978-1-5225-2642-1.ch016

creates and maintains stability as well as a readiness to embrace change where and whenever necessary. In an inclusive society all its members not only have a right to education but also take part in it. The commitment of each and every member in the process is most significant in creating inclusive settings (Deloitte, 2011).

Like inclusion, diversity is also inseparably associated with equity. The concept of equity is principally concerned with providing each and every child with access to fair, just and non-discriminatory education. In inclusive settings equity means providing high-quality education to diverse learners (Ontario Ministry of Education, 2014). Each child has a unique learning and development path that may require individualised or additional support for him or her to be fully engaged in quality education (Dressel & Hodge, 2013). Equity does not mean providing the same experiences for every child, but requires teachers and caregivers to identify and address barriers to learning. It is clear that every child has a different learning context that can result in barriers to learning (Mastropieri, & Scruggs, 2013). Challenges such as poverty, gender, language, culture, race, ability, disability and living context mean that each child requires individualised, differentiated support to reach his or her full potential. Individualised and diverse support is especially important for the meaningful inclusion of children with diverse needs (Page, 2007). Equity and diversity are based on every child's right to be a valued member of the learning community and to be supported to engage in all learning settings. The relationship between diversity, inclusion and equity is that in conjunction these values ensure that learners from various backgrounds are all accepted and treated both fairly and uniquely (Novak & Bartelheim, 2012).

Today's teachers are challenged to expand their list of teaching strategies to meet the needs and strengths of learners from a great diversity of circumstances and cultures (Williams, 2013). These learners either face societal discrimination or live in poverty, or both. The ways in which we teach these learners exert a powerful influence on their linguistic, social, cognitive, and general educational development. Many effective instructional approaches build on learners' backgrounds to further the development of their abilities. It is critically important to recognise that the use of effective instructional practices will improve achievement for all learners, including those who are at risk. The implementation of sound, action research-based strategies that recognise the benefits of diversity, inclusion and equity can build a better future for all learners (Williams, 2013).

The differences in achievement observed among learners of culturally and ethnically diverse circumstances are not necessarily the result of differences in their ability to learn. These are rather the result of differences in the quality of the teaching and learning these learners received in school. Educational programmes must not only take into account their academic needs, but also consider learners as whole persons with social, emotional, and physical needs and strengths and that they come from a particular family context (Samuels, 2014). In addition, children who experience cultural incoherence between home and school may develop a negative self-concept. Teachers are expected to be the primary resources when learners experience different and sometimes conflicting cultural contexts. Teachers can play a vital role as learners move toward fluid identities and start steering within and across spaces of cultural differences through action research in their classrooms (Schwartz, Weinberg, Hagenbunch & Scott, 2008).

Embracing and Dealing With Diversity, Inclusion, and Equity

Studies have indicated a need for improving teachers' abilities to understand, respect, and respond to diverse situations when they interact with learners who bring a host of home cultures and values to the classroom. There is a need for going beyond the recognition of differences and questioning the ways in which classroom practices and processes honour equity and diversities (Madhlangobe & Gordon, 2012). Though some teachers recognise that learners' behaviours are regulated by culturally-appropriate practices and rules, most teachers are not likely to understand cultural differences by means of interaction with learners from various cultures (Maloney, 2012). This is possible when the teacher reflects on his/her classroom practice through action research.

Apart from not engaging in culturally appropriate or responsive ways of interacting with and including all learners, teachers' responses and views in classrooms will most likely be at odds with those of learners who had not been assimilated into the dominant culture. Although cultural similarities between learner and teacher are likely to make the learner see more affection and emotional support from the teacher and less difficulty in the relationship (Saft and Pianta 2001). This match of backgrounds is often not a reality in classrooms because teachers and learners usually occupy different socioeconomic, cultural, linguistic, and racial spaces. If teachers try to level the ground by socialising learners into their own cultural contexts and expect learners to act in certain ways, learners are likely to perceive themselves as inferior and lose the desire to maintain their parents' culture and language (Espinosa 2005; Rhedding-Jones 2002). To avoid these consequences, teachers need to enhance their knowledge of inclusion, equity and cultural diversity and to consider cultural aspects that affect how learners learn in school (Bernhard 1995; Buysse et al. 2005; Espinosa 2005; Lahman & Park 2004; Landerholm et al. 2004).

Teachers need a thorough knowledge of each child's home culture and family values, as this might be helpful in accurately interpreting the meaning of a child's behaviour, honouring and respecting each child's culture, as well as minimising the incidents of cultural conflict and enhancing equity in the classroom (Bernhard 1995; Buysse et al. 2005; Espinosa 2005; Lahman & Park 2004; Landerholm et al. 2004). Studies have highlighted the importance of a strong relationship with parents and families, as they are critical resources for culturally-receptive teaching (Gay 2000; Pappamihiel 2004). Seeking to move towards culturally-reactive teaching (Gay 2000) in a diverse classroom, one teacher may take it upon him-/herself to understand the learners' home cultures, family values, and parental expectations. S/he could work diligently to build collaborative partnerships with parents based on reciprocal trust (Espinosa 2005). Seeing the mismatch between his/her cultural experiences and those of learners in his/her classroom, the teacher could embark on a continuous action research mission aimed at refining the ways in which equity and cultural practices of all learners could be honoured in his/her classroom (Page, 2007).

The broad range of experiences and perspectives brought to school by culturally, linguistically, and ethnically diverse learners offer a powerful resource for everyone to learn in different ways, in new environments, and with different types of people. Every single learner in this extremely diverse and ever-changing world has the power to serve as an invaluable resource for other learners, teachers, and the community as a whole (Johnson, 2012). Rather than constituting a problem for learners and teachers, the increasingly diversified classroom demands and encourages the development and use of diverse teaching strategies designed to respond to each learner as a unique individual. Foreign learners also bring opportunities to be explored and are valuable gems to be appreciated, and they help to challenge the status quo (Johnson, 2012).

Adopting a truly global perspective allows us to view culturally and linguistically diverse learners and their parents as resources who provide supreme opportunities for enrichment. However, there is a need for greater variety in approaches to teaching and learning so as to cope with varied styles of learning (Ainscow, 2015). Teachers and learners alike must cultivate interpersonal skills, equity and respect for other cultures. Indeed, a diverse classroom is an ideal research laboratory in which to learn the diverse perspectives required by a global society and to put to good use information concerning diverse cultural patterns (Brown, 2015). Learners who learn to work and play with classmates from various other cultures are better prepared for the global village and its diversity. Teaching and learning strategies that draw on the social history and the everyday lives of learners and their cultures can only assist in this learning process.

Teachers need to promote critical thinking when they make the rules of the classroom culture clear and enable learners to compare and contrast them with other cultures. Learners can develop cross-cultural skills in culturally and linguistically diverse classrooms. For such learning to take place, however, teachers must have the attitudes, knowledge, and skills to turn their classrooms into inclusive and effective learning environments for all learners. Given the opportunity, learners can participate in learning communities within their schools and neighbourhoods and can be ready to assume constructive roles as responsible citizens in a global society (Brown, 2015).

Including Diverse Learners

When teachers believe that learners can learn, they communicate this expectation explicitly and encourage learners. They also spend more time creating challenging activities. Furthermore, they ask higher-order questions that require not only identification and categorisation, but also comprehension and analysis, application to other situations, synthesis, and value judgments. They include learners in the practical aspects such as fundraising, and promote involvement, responsibility, and the learning of mathematical skills (Ainscow, 2015). Learners acquire social skills along with communication and performance skills. When doing such inclusive, collaborative work, the diversity of skills is seen as a resource that benefits the entire group. Everyone brings something different to the table. When journal writing forms part of learners' group responsibilities, for action research they reflect on what they are learning, practice their writing skills, and keep the teacher informed of their individual progress and well-being (Breslow, Crowell, Francis & Gordon, 2015).

Classroom instruction in a multicultural context is enhanced when learners learn more about themselves through oral history projects; for example, in which they involve their parents, grandparents, and other older adults who supply information about their family background and history. When shared in the classroom, such information becomes a powerful tool both for identifying similarities among learners and for highlighting how they differ from one another in positive rather than negative ways (Cooke, 2014).

Teaching the multicultural aspects of learners cultivates a school culture that celebrates inclusivity and diversity, supports equity, mutual acceptance of, respect for, and understanding of all human differences. It promotes a balanced viewpoint on key issues involved in such teaching (Brown, 2015). Furthermore, it provides learners with a global, international perspective of the world they live in. It seeks to eliminate racial, ethnic, cultural, and gender stereotypes and to solve problems associated with prejudice and biased attitudes. It stresses the importance of teaching ethics, values, and citizenship in a democratic culture (Breslow, Crowell, Francis & Gordon, 2015).

Culturally and Linguistically Diverse Learners

It is not surprising that many learners who do not speak English fall behind in their studies from early on because they are not taught in their native language. When they eventually learn English, they have lost so much ground in their schoolwork that they find it difficult and even impossible to catch up with their peers. In far too many cases these learners become discouraged and drop out of school (Evans, Thurton & Usinger, 2012). The number of bilingual teachers is totally insufficient, and the use of existing bilingual teachers is far from adequate (Overton, Wrench & Garrett, 2017). Furthermore, schools do not use bilingual teachers to their best advantage (Warikoo & de Novais, 2015). Learners who speak a language other than English need to be taught content, for a time, in their natural language, while, at the same time, they are also given intensive training in English. When they rejoin their English-speaking peers, they will be up to speed in their studies (Fredette, 2012).

Nowadays teachers prefer that learners develop linguistic competence in both English and their home language, rather than learn English alone to the detriment of their home language only to relearn it as a second language later on. Learners' cognitive development proceeds more readily in their home language and they learn content far more easily in it while, at the same time, they learn English as a second language (Fredette, 2012). An equitable and interdisciplinary approach to curriculum after decades of separation between the various disciplines is a powerful tool in teaching culturally and linguistically diverse learners. For instance, instead of teaching reading as a separate subject, teachers must now view reading as a means of learning new concepts and exploring subjects and their connections (Ainscow, 2015). Cooperative learning groups and peer tutoring work well in conjunction with computer-mediated language learning. Moreover, parents are partners in their children's schooling and serve as resources for teachers who want to understand learners' cultural patterns of communication and interaction (Cooke, 2014).

Learners who come to school with a home language other than English, learn more from programmes in which their home language is one of the languages of instruction. By continuing to learn subject content in their home language, these learners do not fall behind in their academic subjects while acquiring English (Garces & Cogburn, 2015). When potentially bilingual learners continue to learn in their home language while learning English, they continue to develop cognitively and acquire skills, such as reading, that can later be transferred to English (Fredette, 2012). Once they have acquired a wider English vocabulary, they can comprehend what they interpret. Learning is difficult if instruction is entirely in a learner's second language. Learners taught solely in the second language also risk losing the opportunity to become bilingual. A school that respects the languages and cultures of its learners develops educational situations that maximise the resources these learners bring to school (Jaeger, 2013). Instead of being confused and distressed by trying to cope in a language they only partially understand, learners continue to learn content and skills and develop a feeling of self-worth as well as belonging to their school. If the school does not provide for linguistic and cultural diversity, learners are more likely to feel excluded, isolated and confused (Evans, Thurton & Usinger, 2012).

Teachers can enhance learning a second language by structuring informal situations in which learners who are proficient in a language other than English, act as peer tutors for monolingual English-speaking learners learning a second language and vice versa. Second-language learning for both groups is enhanced when they can communicate informally at certain times during the school day in their second language. This alternative social organisation of learning a second language does not rely solely on the teacher as the locus of teaching (Fishman, Penuel, Allen, Cheng & Sabelli, 2013). Learners become teachers and resources for one another. Second-language learning is reciprocal. The learning of language cannot be separated from what is being learned. Too often learners with limited proficiency in English are required to learn the abstract or grammatical aspects of language as opposed to the functional and communicative aspects. These more important functional skills are best developed in conjunction with the learning of content and learning makes sense and is more interesting. Learners also benefit by learning cross-cultural skills (Breslow, Crowell, Francis & Gordon, 2015).

Learning to greet others in a second language as well as the respectful behaviour associated with that language enable learners to communicate more easily in a new culture (Breslow, Crowell, Francis & Gordon, 2015). To ask for food requires basic grammar and good-mannered behaviour. Learners can go on to discuss how courteous behaviour differs in different cultures and what being courteous means in the classroom among friends, in a restaurant, and in the school cafeteria. In addition, instead of removing learners from a content lesson in mathematics because they are not yet proficient in English, the teacher can pair bilingual and monolingual learners in small groups and provide mathematics related tasks within these groups. Bilingual learners will assist the monolingual learners in completing these tasks while providing natural models of language development within the content area (Fredette, 2012).

Cultural Relevance in the Classroom

Learners' self-esteem is strengthened when they see and read about the contributions made by their own ethnic groups to the history and culture of their country. Whenever possible, teachers should adapt the curriculum to focus lessons on topics that are meaningful to learners. This kind of focus allows learners to practice language, thinking, reading and writing skills in real, meaningful and interactive situations. Learners also realise that teachers value and appreciate each learner's culture and language. This way, learners feel more comfortable in school and participate more actively in learning situations (Jaeger, 2013).

Learners who care at home for their siblings have a foundation for cooperative learning and peer teaching. They can excel in this if they are given the opportunity and support of the teacher. If learners are accustomed to caring for their physical environment at home, they often feel comfortable in caring for and managing the school environment as well. The benefits of involving the parents will now be discussed (Cooke, 2014).

Parental Involvement

For parents who do not speak English, the language barrier can pose another challenging obstacle. Parents are encouraged to be an active presence at school. Social activities bring families and school staff together and parents start to trust the school. Similar activities are reported to have contributed to lower dropout rates amongst other benefits (Messiou, Ainscow, Echeita, Goldrick, Hope, Paes, Sandoval, Simon & Vitorino, 2016). Parents can be directly involved in the work of the school by assisting teachers and learners with lessons; supervising class trips; participating in classroom activities; aiding administration officials, teachers, and school staff in the school cafeteria, library, laboratories, and workshops; organising parent groups in fund-raising, community relations, political awareness, and programme development; attending learner assemblies, sports events, and special presentations; and participating in workshops, discussion groups, and training sessions. Parents can involve themselves in learning activities at home by developing a child's social and personal skills and by contributing to basic-skills education, the development of advanced skills, and enrichment (UNESCO, 2015).

Teachers must understand and respect the countless ways of being a parent and empathise with parents' concern about the education of their children. Parental involvement is linked to learners' academic achievement (Epstein & Sheldon, 2006). Parent-teacher associations can hold meetings at times convenient for parents to attend, and can provide translators for those who do not speak English. A room in the school can be set aside for parents to meet in and discuss issues concerning their children's education or the school community. Teachers can visit parents at their homes, or they can use parent-teacher meetings to discuss homework and discipline.

Parents who are welcomed into the school in ways that are culturally appropriate for them become more accessible both as resources and as learners. Parents who are bilingual may be asked to translate for those who have not yet achieved fluency in a new language. When learners see that their parents are respected by the school, there may be less conflict between their home and school cultures that can cause a breakdown of discipline (Jaeger, 2013).

Furthermore, teachers can assist parents to understand the expectations of the school and their classroom as they learn from parents what their expectations of teachers and learners are. Teachers also can suggest ways in which parents might converse more often with their children in order to prepare them for communication in the classroom (Miles & Ainscow, 2011). The benefits of family involvement are not confined to primary schooling; involving parents throughout high school can offer good results. Parents can help a great deal regardless of their level of formal education (Pickett & Vanderbloemen, 2015). We cannot regard the school and the home as separate locations. We must take note of how they connect with each other and with the world at large. The importance of action research for diverse learners will be discussed in the next section (UNESCO, 2015).

The Importance of Action Research for Diverse Learners

The teacher has to understand that truly recognising diverse learners is a continuous process. Whenever a teacher positions himself/herself as a learner while seeking to address issues in his/her own classroom, the teacher is definitely engaged in action research (AR). This section of the chapter will explain the process whereby the teacher negotiates action research as a way to move towards more inclusive and equitable, relevant practices for the diverse needs of learners (Ladson-Billings 1995) in his/her classroom.

Action research can be conducted by teachers who are fully aware of the day-to-day concerns in their teaching environments and who decide to undertake systematic inquiry into the identified concerns. Teachers who apply this approach are those who wish to promote a better understanding of their practice, support the learning of their learners in the classroom, or have been invited to do so by decision-makers who are aware of a problem that requires some action research (Stark, 2014).

Action research missions may be conducted by one individual in one classroom, by a team of two or several teachers, by an entire school, by an entire school district, or in collaboration with a university or other agency. Although action research is usually focused on specific issues in specific classrooms, there are many benefits to collaboration during and after the process. Conducting action research in teams allows shared support for the resolution of common concerns. According to Greenwood and Levin (1998, p. 4), teams can work together to "define the problems to be examined, co-generate relevant knowledge about them, learn and execute social research techniques, take actions, and interpret the results of actions based on what they have learned." It is helpful to find colleagues and work with them because it breaks the exclusion often caused when a teacher is the only adult in the classroom. When teachers collaborate on an action research venture, they know that they are not alone in finding answers to challenging situations. Action research promotes professional conversations as teachers work together in pairs or teams to address related anxieties (Brozo, 2011).

The idea of action research is that educational problems and challenges are best identified and investigated where the action is: at classroom and school level. By integrating research into these settings and engaging those who work at this platform in research activities, findings can be applied immediately and problems solved more quickly (Guskey, 2000). Good teachers are those who are also lifelong learners who seek to improve their knowledge and practice throughout their careers. Action research adds to their knowledge by addressing gaps and expanding on what they know. Action research improves practice by helping teachers to gain new ideas for their teaching and new insights into their approaches, and connect with other teachers. Action research allows teachers to weigh different perspectives on concerns and challenges and make informed decisions (Brozo, 2011). Action research provides a means of focusing instruction on issues directly related to the classroom or the school. It positions teachers as learners who want to narrow the gap between their practice and their vision of quality education (Harding, 2015). The action research itself and the implementation of the resulting information provide a form of professional development directly related to and supportive of teachers' and learners' educational needs. This has great potential for creating long-lasting school transformation. These modifications can happen in everyday practice rather than being a theoretical generalisation addressed to a wide-ranging audience (Castro, 2014). Each researcher discovers something new and applicable to his or her own classroom because the question he or she asks is unique. Action research provides teachers with the opportunity to gain knowledge and skills in research methods and to become more aware of the options and possibilities for change (Jacobs, Yamamura, Guerra & Nelson, 2013). It gives them a rich source of data for improving their classrooms and schools. They have the opportunity to try new teaching practices and reflect on the changes they introduce in their classrooms and learners (Solis, 2015).

Reflection Within Action Research

The difference between simply teaching and becoming a great teacher is reflective practice. The action research report is an opportunity for a teacher to reflect on his/her research data and results. Action research means trying out and reflecting on ideas in practice as a technique for development and as a means of gathering knowledge (Jacobs, Yamamura, Guerra & Nelson, 2013). A cycle of action and reflection is at the heart of action research. Once the teacher has collected the data and analysed it, s/he will need to act on the information. Did his/her plan achieve the desired results? Does s/he need to modify the plan to make another change and examine that change? If so, the teacher needs to alter only one element because if too many changes are made at once, it will be hard to know which action is responsible for the results. With the plan modified, the teacher will continue to collect and analyse data. It is important to remember that action research proceeds through moments of action and reflection. It does not end with a single planning and acting cycle, but continues on a coil of action and reflection (Solis, 2015).

Reflective practice is not always easy. What if a project does not deliver the results the teacher expected? Key to effective reflection is that the teacher reflects on his/her own practice. Reflection may be ongoing or a reaction to a specific event or unexpected occurrence (Adams, 2016). In case of action research, the teacher reflects on a question that arises from his/her concern. The immediate aim of reflective practice is professional or self-development. Reflective practice is enhanced when the reflection is shared and discussed with colleagues who support the teacher and share his/her concern. The objective of reflection may not be the resolution of an issue, but a better understanding of it (Solis, 2015).

There is a need to observe and record what happens as a result of the first action stage. Collaboration, talking with other teachers or asking them to observe you and your classroom and give feedback, will add an additional perspective to one's research. Design specific tasks for learners that will show changes in their work. Interview selected learners and interview learners who do well and who do not in a focus group or individually for more clarity. The teacher can use an audio- or a videotape to record the interviews so that s/he can review them and analyse the data provided by the interviews (Brozo, 2011). After identifying a research concern, gathering additional information, and refining the focus of the research, the plan of action needs to be developed and implemented. The action plan involves specification on the participants, strategies, available resources, evaluation of procedures, and timeframes (Harding, 2015).

Developing an Action Plan

The teacher needs to develop an action plan that involves parents and capitalises on their knowledge which consist of cultural relics and the knowledge present in the family circle which underlie family practices (Moll 2000). Certain cultural resources are important in a cultural and family network. In addition, there are knowledge-based and authentic resources. Cultural knowledge varies from one family to another and from one community to another. By developing a plan and inviting parents to share their cultural practices, the teacher allows cultural practices to enter the classroom. In addition, s/he will gain insight into how to genuinely value diverse cultural practices in the classroom (Howell & Saye, 2016).

As the teacher engages in action research for data analysis, his/her practice starts to identify and value the significance of inclusion, equity and diversity in his/her classroom and there may even be a change in his/her attitude and standpoint (Howell & Saye, 2016). As change happens, there may an alignment in his/her storylines that specifies common stances that s/he embraces as s/he seeks to promote inclusivity, equity and diversity in the classroom. This is particularly interesting as there is often a belief that change must happen within the learners. In this instance, a change in a teacher's standpoint will transform the atmosphere of the classroom, making it respectful of differences and embracing of diversity (Jacobs, Yamamura, Guerra & Nelson, 2013). Because he or she respects and seeks to respect learners' backgrounds and cultural contexts, the teacher will adopt a humble stance. S/he will accommodate and embrace learners' cultures and narratives in the classroom and allow them to take the focal point. When s/he adopts such a stance, the teacher becomes the model of respect for diversity, equity, inclusion, for cultural difference and for multiple perspectives in the classroom, the school and the community as a whole. This is illustrated by the supportive role s/he plays when parents and families come to share their cultural knowledge and are regarded as experts in the classroom. S/he respects and values all learners and their families as human beings and starts to know them on a personal level. S/he listens to their narratives and encourages an environment of approval and respect in his/her classroom, fostering a setting that nurtures diversity, inclusion and equity (Howell & Saye, 2016). By concealing the roles of teacher and learner, the teacher expresses the position of an action researcher. In these encounters all participants learn about inclusion, equity and diverse world views. They realise that their cultural background and values are neither superior nor inferior; they are just different and unique. While this undertaking is important, it is only a tool to get learners to think and a way of challenging their suppositions. This profoundly considerate stance adopted by the teacher comes from narratives that take place in the classroom (Harding, 2015).

It is imperative for the teacher to step outside his/her comfort zone as s/he immerses himself/herself in the homes and communities of his/her learners. Additionally, it is important to socialise learners to respect others who talk as experts on their cultural practices. This should be done often and in a voluntary way in the classroom so that learners feel comfortable to describe their own cultural practices. Learners should not only represent their ethnic group, but also should report on what happens in their homes and communities (Hargreaves & Ainscow, 2015).

This new standpoint allows both teachers and learners to understand prejudice and act on it, even if initially only in the classroom. Although it might make teaching more interesting, such an approach allows for an explicit link to the learners' circumstances and everyday lives. It further allows for inclusivity, equity and diversity to be truly nurtured (Howell & Saye, 2016). When inclusivity, equity and diversities are nurtured, teachers gain the potential to reach learners from a range of socio-cultural backgrounds while valuing and knowing the learners for who they are. After years of reflective teaching and action research the teacher will realise that a few themes constantly recur which, if appropriately triangulated, could shed some light on other classrooms in the school (Jacobs, Yamamura, Guerra & Nelson, 2013).

The result of an action plan is the presence of a specific and evident approach to testing out fresh ideas as a means to solve the initial problem (Creswell, 2005). The action plan is a proposed strategy for implementing the results of the action research. As the action plan is implemented, its effectiveness must be examined, appraised, and reviewed constantly to continue the cyclical nature of action research. The action plan may be planned by an individual teacher or a classroom, or by a group of teachers, or by a school or even district. It may sometimes be necessary to draft an official document outlining the action plan and procedures for executing the resolutions. In action research it is prudent to have all information documented and to never rely on collective memories for the implementation of solutions (Stark, 2014).

Discussing Findings With Fellow Teachers and Peers

Teachers need to be open to learning and sharing information with others in order to improve their teaching. With years of experience teachers have excellent ideas and strategies to approach teaching and learning in a diverse classroom (Locke, Alcorn & O'Neill, 2013). They also give good advice on dealing with diverse learners. Therefore, discussing their findings with fellow teachers and peers will help teachers to solve problems which emerge from practice (Howell & Saye, 2016).

In addition, careful planning helps the teacher to master specific techniques and processes designed for the education of learners with diverse needs. The action plan has to be applied in practice in order to be of value. Regular reflection on practice enables the teacher to find out what policies have worked well and what areas still need improvement. Therefore, in order to improve teaching and learning for learners with diverse needs, the framework of planning, practice, and reflection should consistently be applied (Jensen & Sonnemann, 2014).

CONCLUSION

The purpose of this chapter was to examine diversity, equity and inclusion in the classroom with a particular focus on learners with diverse needs. This chapter looked at aspects of reflecting on practices and the process of action research, as well as how an action plan is developed to improve school teaching practice.

This chapter offered inferences of how a teacher can honour learners' multiple voices and cultural practices in the classroom. The teacher's voyage highlights practices and processes that are situated in social, historical, and cultural contexts. Moreover, the teacher's action research leads the process of reflecting on and practising equity, inclusion and diversity in the classroom as an ongoing dialogue and a continuous process. The teacher's experience is representative of a larger phenomenon and sheds light on other contexts. It seeks to inspire other teachers to engage in teacher action research and enact culturally-responsive teaching within their classrooms by honouring and valuing differences and affirming learner diversity. Solutions include ideas on how teachers, parents, and the curriculum can nurture a more diverse classroom that honours and respects diversity, equity and inclusion in the classroom.

REFERENCES

Adams, M. (2016). Pedagogical Foundations for Social Justice Education. In Teaching for Diversity and Social Justice (3rd ed.). New York: Routledge.

Ainscow, M. (2015). *Towards Self-Improving School Systems: Lessons From a City Challenge*. London: Routledge.

Atherton, T. J., Barthelemy, R. S., Deconinck, W., Falk, M. L., Garmon, S., Long, E., & Reeves, K. et al. (2016). *LGBT Climate in Physics: Building an Inclusive Community*. College Park, MD: American Physical Society.

Bell, P., Lewestein, B., Shouse, A. W., & Feder, M. A. (Eds.). (2009). *Learning science in informal environments: People, places and pursuits*. Washington, DC: National Academies Press.

Bernhard, J. K. (1995). Child development, cultural diversity, and the professional training of early childhood educators. *Canadian Journal of Education*, *20*(4), 415–436. doi:10.2307/1495080

Bockting, W. O., Miner, M. H., & Romine, R. E. (2013). Stigma, mental health, and resilience in an online sample of the US transgender population. *American Journal of Public Health*, *103*(5), 943–951. doi:10.2105/AJPH.2013.301241 PMID:23488522

Breslow, K., Crowell, L., Francis, L, & Gordon, S.P. (2015). Initial efforts to coordinate appreciative inquiry: Facilitators' experiences and perceptions. I.*E.: Inquiry in Education, 6*(1).

Brinkworth, C. (2016). *From Chilly Climate to Warm Reception: Experiences and Good Practices for Supporting LGBTQ Students in STEM*. CGU Theses & Dissertations. Paper 97. Retrieved from http://scholarship.claremont.edu/cgu_etd/97

Brown, T. C. (2015). *Collaborative autobiography: A process for addressing educators' stress and awareness in 21st century public education* (Doctoral dissertation).

Brozo, W. G. (2011). *Action Research* [PowerPoint slides]. Newark, DE: International Reading Association.

Buysse, V., Castro, D. C., West, T., & Skinner, M. (2005). Addressing the needs of Latino children: A national survey of state administrators of early children programs. *Early Childhood Research Quarterly*, *20*(2), 146–163. doi:10.1016/j.ecresq.2005.04.005

Castro, E. L. (2014). Underprepared and at-risk: Disrupting deficit discourses in undergraduate STEM recruitment and retention programming. *Journal of Student Affairs Research and Practice*, *51*(4), 407–419. doi:10.1515/jsarp-2014-0041

Cech, E. A., & Waidzunas, T. J. (2011). Navigating the heteronormativity of engineering: The experiences of lesbian, gay, and bisexual students. *English Studies*, *3*(1), 1–24.

Claeys, A., Kempton, J., & Paterson, C. (2014). *Regional Challenges: A Collaborative Approach to Improving Education*. Centre Forum, London, UK.

Cooke, N. A. (2014). Creating Opportunities for Empathy and Cultural Competence in the LIS Curriculum. *SRRT Newsletter, 187*. Retrieved from http://libr.org/srrt/news/srrt187.php#9

Deloitte. (2011). *Only skin deep? Re-examining the business case for diversity*. Deloitte Point of View Human Capital Australia.

Dressel, P., & Hodge, G. (2013). *Analysis of Policies, Practices, and Programs for Advancing Diversity, Equity and Inclusion: Full Report*. D5 Coalition. Retrieved on February 10, 2014 from http://www.d5coalition.org/wp-content/uploads/2013/09/ PPP-Full-Report-11.14.13.pdf

Epstein, J. L., & Sheldon, S. B. (2006). Moving forward: Ideas for research on school, family, and community partnerships. In C. F. Conrad & R. Serlin (Eds.), *SAGE handbook for research in education: Engaging ideas and enriching inquiry* (pp. 117–138). Thousand Oaks, CA: Sage. doi:10.4135/9781412976039.n7

Espinosa, L. M. (2005). Curriculum and assessment considerations for young children from culturally, linguistically, and economically diverse backgrounds. *Psychology in the Schools*, *42*(8), 837–853. doi:10.1002/pits.20115

Evans, L., Thurton, B., & Usinger, J. (2012). Theoretical frameworks to guide school improvement. *NASSP Bulletin*, *96*(2), 154–171. doi:10.1177/0192636512444714

Fishman, B., Penuel, W. R., Allen, A., Cheng, B. H., & Sabelli, N. (2013). Design-based implementation research: An emerging model for transforming the relationship of research and practice. *National Society for the Study of Education Yearbook*, *112*(2), 136–156.

Fredette, C. (2012). *Leadership Diversity in the Non-profit Sector: Baby Steps, Big Strides and Bold Stances. A DiverseCity Counts Research Project*. Toronto: DiverseCity.

Freeman, R. B., & Wei, H. (2014). Collaboration: Strength in diversity. *Nature*, *513*(7518), 305. doi:10.1038/513305a PMID:25230634

Fullan, M., Rincon-Gallardo, S. & Hargreaves, A. (2015). Professional capital as accountability. *Education Policy Analysis Archives, 23*(15), 1-17. 10.14507/epaa.v23.1998

Garces, L. M., & Cogburn, C. D. (2015). Beyond declines in student body diversity how campus-level administrators understand a prohibition on race-conscious postsecondary admissions policies. *American Educational Research Journal*, *52*(5), 828–860. doi:10.3102/0002831215594878

Gay, G. (2000). *Culturally responsive teaching: Theory, research, & practice*. New York: Teachers College Press.

Greaves, E., Macmillan, L., & Sibieta, L. (2014). *Lessons from London schools for attainment gaps and social mobility*. London: The Social Mobility and Child Poverty Commission.

Greenwood, D. J., & Levin, M. (1998). *Introduction to Action Research: Social Research for Social Change*. London: Sage Publications.

Guskey, T. R. (2000). *Evaluating professional development*. Thousand Oaks, CA: Corwin Press.

Harding, S. (2015). *Objectivity and Diversity: Another Logic of Scientific Research*. Chicago, IL: University of Chicago Press. doi:10.7208/chicago/9780226241531.001.0001

Hargreaves, A., & Ainscow, M. (2015). The top and bottom of leadership and change. *Phi Delta Kappan*, *97*(3), 42–48. doi:10.1177/0031721715614828

Howell, J. B., & Saye, J. W. (2016). Using lesson study to develop a shared professional teaching knowledge culture among 4th grade Social Studies teachers. *Journal of Social Studies Research*, *40*(1), 25–37. doi:10.1016/j.jssr.2015.03.001

Jacobs, J., Yamamura, E., Guerra, P. L., & Nelson, S. W. (2013). Emerging leaders for social justice: Negotiating the journey through action research. *Journal of School Leadership*, *23*, 91–121.

Jaeger, P. T. (2013). Internet Justice: Reconceptualizing the Legal Rights of Persons with Disabilities to Promote Equal Access in the Age of Rapid Technological Change. *Review of Disability Studies*, *9*(1), 39–59.

Jensen, B., & Sonnemann, J. (2014). *Turning Around Schools: It Can Be Done*. Melbourne: Grattan Institute.

Johnson, D. R. (2012). Campus racial climate perceptions and overall sense of belonging among racially diverse women in STEM majors. *Journal of College Student Development*, *53*(2), 336–346. doi:10.1353/csd.2012.0028

Ladson-Billings, G. (1998). Teaching in dangerous times: Culturally relevant approaches to teacher assessment. *The Journal of Negro Education*, *67*(3), 255–267. doi:10.2307/2668194

Lahman, M., & Park, S. (2004). Understanding children from diverse cultures: Bridging perspectives of parents and teachers. *International Journal of Early Years Education*, *12*(2), 131–142. doi:10.1080/0966976042000225525

Landerholm, E., Gehrie, C., & Hao, Y. (2004). Educating early childhood teachers for the global world. *Early Child Development and Care*, *174*(7–8), 593–606. doi:10.1080/0300443042000187095

Lee, K. (2015). *Culture of STEM Education*. Presentation at the SOWA 2015 Bridge conference. Retrieved from https://bridge2015.sched.org/event/4AGR/culture-of-stem-education

Locke, T., Alcorn, N., & ONeill, J. (2013). Ethical issues in collaborative action research. *Educational Action Research*, *21*(1), 107–123. doi:10.1080/09650792.2013.763448

Madhlangobe, L., & Gordon, S. P. (2012). Culturally responsive leadership in a diverse school: A case study of a high school leader. *NASSP Bulletin*, *96*(3), 177–202. doi:10.1177/0192636512450909

Maloney, M. M. (2012). Cultivating Community, Promoting Inclusivity: Collections as Fulcrum for Targeted Outreach. *New Library World*, *113*(5–6), 281–289. doi:10.1108/03074801211226364

Mastropieri, M. A., & Scruggs, T. E. (2013). Inclusive classroom: Strategies for effective differentiated instruction (5th ed.). Prentice Hall.

McIsaac, E., Park, S., & Toupin, L. (2013). *Shaping the Future: Leadership in Ontario's Nonprofit Labour Force*. Toronto: ONN and The Mowat Centre.

Mendelsohn, M., Mulholland, E., & Shamshiri, N. (2011). *Strengthening the Third Pillar of the Canadian Union*. Toronto: The Mowat Centre.

Messiou, K., Ainscow, M., Echeita, G., Goldrick, S., Hope, M., Paes, I., & Vitorino, T. et al. (2016). Learning from differences: A strategy for teacher development in respect to student diversity. *School Effectiveness and School Improvement*, *27*(1), 45–61. doi:10.1080/09243453.2014.966726

Miles, S., & Ainscow, M. (2011). *Responding To Diversity in Schools*. London: Routledge.

Moll, L. (2000). *Exploring biliteracy and beyond: Developing a case example of educational sovereignty*. Retrieved May 4, 2009 from http://www.gened.arizona.edu/tmiller/comlit/moll.htm

Novak, A. D., & Bartelheim, F. J. (2012). General Education Students' Changing Perceptions of Students with Special Needs. *Current Issues in Education*, *15*(2), 1–10.

Ontario Ministry of Education. (2014). *Achieving Excellence: Ensuring Equity*. Queen's Printer for Ontario. Retrieved from http://www.edu.gov.on.ca/eng/about/ equity.html

Overton, H., Wrench, A., & Garrett, R. (2017). Pedagogies for inclusion of junior primary students with disabilities in PE. *Physical Education and Sport Pedagogy*, *22*(4), 414–426. doi:10.1080/17408989.2016.1176134

Page, S. E. (2007). *The Difference: How the Power of Diversity Creates Better Groups, Firms, Schools and Societies*. Princeton University Press.

Pappamihiel, N. E. (2004). Hugs and smiles: Demonstrating caring in a multicultural early childhood classroom. *Early Child Development and Care*, *174*(6), 539–548. doi:10.1080/0300443042000187059

Pickett, K., & Vanderbloemen, L. (2015). *Mind the gap: tackling social and educational inequality*. Cambridge Primary Review Trust.

Rhedding-Jones, J. (2002). An undoing of documents and other texts: Toward a critical multi-culturalism in early childhood education. *Contemporary Issues in Early Childhood*, *3*(1), 90–116. doi:10.2304/ciec.2002.3.1.10

Royal Bank of Canada. (2012). *RBC Diversity Blueprint 2012-2015*. Retrieved from http://www.rbc.com/ diversity/pdf/rbc-diversity-blueprint.pdf

Saft, E. W., & Pianta, R. C. (2001). Teachers perceptions of their relationships with students: Effects of child age, gender, and ethnicity of teachers and children. *School Psychology Quarterly*, *16*(2), 125–141. doi:10.1521/scpq.16.2.125.18698

Samuels, D. (2014). *The Culturally Inclusive Educator: Preparing for a Multicultural World*. Teachers College Press, Columbia University.

Schwartz, R., Weinberg, J., Hagenbunch, D., & Scott, A. (2008). *The Voice of Nonprofit Talent: Perceptions of Diversity in the Workplace*. Retrieved on February 10, 2014 from: http://www.cgcareers.org/diversityreport.pdf

Smith, W. A., Mustaffa, J. B., Jones, C. M., Curry, T. J., & Allen, W. R. (2016). You make me wanna holler and throw up both my hands! Campus culture, Black misandric microaggressions, and racial battle fatigue. *International Journal of Qualitative Studies in Education*, *29*(9), 1189–1209. doi:10.1080/09518398.2016.1214296

Solis, R. (2015). *Teachers' experiences of learning through a reflective inquiry process focused on the relationship between teaching beliefs and behaviors* (Doctoral dissertation).

Stark, J. L. (2014). The potential of Deweyan-inspired action research. *Education and Culture*, *30*(2), 87–101. doi:10.1353/eac.2014.0013

UNESCO. (2015). *Education for all 2000-2015: achievements and challenges*. Paris: UNESCO.

Warikoo, N. K., & de Novais, J. (2015). Colour-blindness and diversity: Race frames and their consequences for White undergraduates at elite US universities. *Ethnic and Racial Studies*, *38*(6), 860–876. doi:10.1080/01419870.2014.964281

Williams, D. A. (2013). *Strategic Diversity Leadership: Activating Change and Transformation in Higher Education*. Sterling, VA: Stylus Publishing, LLC.

Compilation of References

(1993). Townsend (1993) EA. Muriel Driver Memorial Lecture: Occupational therapys social vision. *Canadian Journal of Occupational Therapy*, *60*, 174–184. doi:10.1177/000841749306000403

Abbott, G. N. (2006). *Exploring evidence-based executive coaching as an intervention to facilitate expatriate acculturation: Fifteen case studies*. Australian National University.

Academic Partner. (2017). *A teacher friendly summary of Action Learning principles*. Unpublished report.

ACARA. (2015, Aug 5). *MySchool*. Retrieved from: https://www.myschool.edu.au/

Acker-Hocevar, M., Pisapia, J., & Coukos-Semmel, E. (2002, April). *Bridging the abyss: adding value and validity to leadership development through action learning—cases-in-point*. Paper presented at the Meeting of the American Educational Research Association, New Orleans, LA.

Ackerman, P. L., & Lohman, D. F. (2006). Individual differences in cognitive functions. In P. R. Alexander & P. H. Winne (Eds.), *Handbook of educational psychology* (2nd ed.; pp. 139–161). Mahwah, NJ: Lawrence Erlbaum Associates.

Action Plan 2014: Towards the Realisation of Schooling 2025. (2012). *The goals dealing with the how of improving schooling*. Problem Statement. Pretoria: Department of Basic Education.

Adams, M. (2016). Pedagogical Foundations for Social Justice Education. In Teaching for Diversity and Social Justice (3rd ed.). New York: Routledge.

Adams, P. (2010). Action Research. In N. J. Salkind (Ed.), Encyclopedia of Research Design (Vol. 1, pp. 4 - 9). Thousand Oaks, CA: SAGE Publications, Inc. doi:10.4135/9781412961288.n3

Agarwal, B. (1988). Patriarchy and the 'modernizing' state: An introduction. *Structures of Patriarchy*, 1-27.

Agarwal, P. (2015). Private deemed universities in India. *International Higher Education*, (49).

Agarwal, R., & Gautam, A. (2011). Effect of Constructivist methods of teaching on academic achievement of primary school students. *GYANODAYA*: *The Journal of Progressive Education*, *4*(1).

Ainscow, M. (2015). *Towards Self-Improving School Systems: Lessons From a City Challenge*. London: Routledge.

AITSL. (2012). *Australian charter for the professional learning of teachers and school leaders*. Australian Institute for Teaching and School Leadership. Retrieved from: http://www.aitsl.edu.au/docs/default-source/default-document library/australian_charter_for_the_professional_learning_of_teachers_and_school_leaders

Albareda Tiana, S., & Alférez Villarreal, A. (2016). A collaborative programme in sustainability and social responsibility. *International Journal of Sustainability in Higher Education*, *17*(5), 719–736. doi:10.1108/IJSHE-07-2016-0134

Albertyn, R. M., Bester, A., Botha, M. J., & van der Merwe, M. E. (2007). Entrepreneurial skill development: Participatory action research approach in a rural community. *Journal of Family Ecology and Consumer Sciences*, *35*, 9-16.

Al-Daami, K. K., & Wallace, G. (2007). Curriculum reform in a global context: A Study of teachers in Jordan. *Journal of Curriculum Studies*, *39*(3), 339–360. doi:10.1080/00220270601057790

Allen, T. D., Eby, L. T., & Lentz, E. (2006). Mentorship behaviors and mentorship quality associated with formal mentoring programs: Closing the gap between research and practice. *The Journal of Applied Psychology*, *91*(3), 567–578. doi:10.1037/0021-9010.91.3.567 PMID:16737355

Allison, A. (2008). *Critical theory and pre-service art education: one art teacher's journey of equipping art teachers for inclusion*. Austin, TX: University of North Texas.

Alsop, R. (2008). *The trophy kids grow up: How the millennial generation is shaking up the workplace*. John Wiley & Sons.

Altbach, P. G., & Knight, J. (2007). The internationalization of higher education: Motivations and realities. *Journal of Studies in International Education*, *11*(3-4), 290–305. doi:10.1177/1028315307303542

Altrichter, H., Posch, P., & Somekh, B. (1993). *Teachers investigate their work: An introduction to the methods of action research*. London: Routledge.

Aluko, R., & Hendrikz, J. (2012). Supporting distance education students: The pilot study of a tutorial model and its impact on student's performance. *Progressio*, *34*(2), 68–83.

Amechi, E. P. (2009). Poverty, Socio-Political Factors and Degradation of the Environment in Sub-Saharan Africa: The Need for a Holistic Approach to the Protection of the Environment and Realisation of the Right to Environment. *Law, Environment & Development Journal*, *5*(2), 107–129.

Amini, R. (2015). Outdoor based environmental education learning and its effect in caring attitude toward environment. *Indonesian Journal of Science Education*, *4*(1), 3–47.

Amir, A., Mandler, D., Hauptman, S., & Gorev, D. (2017). Discomfort as a means of pre-service teachers' professional development–an action research as part of the 'Research Literacy' course. *European Journal of Teacher Education*, 1–15.

Aniekwe, C. C., Hayman, R., Mdee, A., Akuni, J., Lall, P., & Stevens, D. (2012). *Academic-NGO collaboration in international development research: A reflection on the issues*. A Working Paper.

Ankiewicz, A. (2003). *Technology education at school: Illusion or reality? Inaugural address*. Johannesburg, South Africa: University of Johannesburg.

Appleton, K. (2003). *Pathways in professional development in primary science: extending science pedagogical content knowledge*. Paper presented at the annual conference of the Australian Science Education Research Association, Melbourne, Australia.

Appleton, K. (2006). Science pedagogical content knowledge and elementary school teachers. In K. Appleton (Ed.), Elementary science teacher education: International perspectives on contemporary issues and practice (pp. 31–54). Lawrence Erlbaum in association with the Association for Science Teacher Education (ASTE).

Argyris, C., & Schön, D. A. (1974). *Theory in practice: Increasing professional effectiveness*. San Francisco: Jossey-Bass.

Argyris, C., & Schön, D. A. (1978). *Organizational learning: A theory of action perspective* (Vol. 173). Reading, MA: Addison-Wesley.

Ashman, M. R., & Puri, G. (2002). *Essential Soil Science*. Oxford, UK: Blackwell Publishing.

Astier, M., Masera, O., & López-Ridaura, S. (1999). *Sustentabilidad y manejo de recursos naturales* (M. P. México, Ed.). SA de CV.

Asunda, P. A., & Hill, R. (2007). Critical features of Engineering Design in Technology Education. *Journal of Industrial Teacher Education*, *44*(1), 25–48.

Atherton, T. J., Barthelemy, R. S., Deconinck, W., Falk, M. L., Garmon, S., Long, E., & Reeves, K. et al. (2016). *LGBT Climate in Physics: Building an Inclusive Community*. College Park, MD: American Physical Society.

Atkins, L., & Wallace, S. (Eds.). (2012). *Qualitative research in education*. Los Angeles, CA: Sage. doi:10.4135/9781473957602

Atlantic Provinces Education Foundation. (2001). *Foundation for the Atlantic Canada Arts Education Curriculum*. Halifax, Canada: Atlantic Provinces Education Foundation.

Aubusson, P., Brady, L., & Dinham, S. (2005). *Action learning: what works*. NSW Department of Education and Training. Retrieved from: http://sydney.edu.au/education_social_work/professional_learning/resources/papers/AGQTP_reports/2005_Action_Learning-What_Works.pdf

Auger, W., & Wideman, R. (2000). Using action research to open the door to life-long professional learning. *Education*, *12*(1), 120–127.

Avison, D., Lau, F., Myers, M., & Nielsen, P. A. (1999, January). Action Research. *Communications of the ACM*, *42*(1), 94–97. doi:10.1145/291469.291479

Avriel Avni, N., Avni, Y., Pney-Gil, D., & Ginat, H. (2016). *Formulation of management principles of runoff and soil erosion in agricultural projects in upper Zin basin*. Interim Report to Dead Sea Drainage Authority.

Axelrod, B., Handfield-Jones, H., & Welsh, T. (2001). War for talent, part two. *The McKinsey Quarterly*, *2*, 9–11.

Azapagic, A. (2010). Life cycle assessment as a tool for sustainable management of ecosystem services. *Issues in Environmental Science & Technology*, *30*, 140–168. doi:10.1039/9781849731058-00140

Babbie, E., & Mouton, J. (2001). *The practice of social research*. Cape Town: Oxford University Press.

Bacon, T. R., & Spear, K. I. (2003). *Adaptive Coaching: The art and practice of a client-centered approach to performance improvement*. Palo Alto, CA: Davis-Black.

Badawy, M. (2011). Is open innovation a field of study or a communication barrier to theory development?: A perspective. *Technovation*, *31*(1), 65–67. doi:10.1016/j.technovation.2010.09.006

Baker Collins, S. (2005). An understanding of poverty from those who are poor. *Action Research*, *3*(1), 9–31. doi:10.1177/1476750305047983

Baker, D. W., Parker, R. M., Williams, M. V., & Clark, W. S. (1998). Health literacy and the risk of hospital admission. *Journal of General Internal Medicine*, *13*(12), 791–798. doi:10.1046/j.1525-1497.1998.00242.x PMID:9844076

Balakrishnan, V., & Conforth, S. (2013). Using working agreements in Participatory Action Research: Working through moral problems with Malaysian students. *Educational Action Research*, *21*(4), 582–602. doi:10.1080/09650792.2013.832347

Bandura, A. (1997). Self-efficacy. *The Harvard Mental Health Letter*, *13*(9), 4–6.

Banga Chhokar, K. (2010). Higher education and curriculum innovation for sustainable development in India. *International Journal of Sustainability in Higher Education*, *11*(2), 141–152. doi:10.1108/14676371011031865

Banning, M. (2005). Approaches to teaching: Current opinions and related research. *Nurse Education Today*, *25*(7), 502–508. doi:10.1016/j.nedt.2005.03.007 PMID:16140425

Bantanur, S., Mukherjee, M., & Shankar, R. (2015). Emerging dimensions of sustainability in institutes of higher education in India. *International Journal of Sustainable Built Environment*, *4*(2), 323–329. doi:10.1016/j.ijsbe.2015.03.004

Barbara, M. G., Gayle, B. M., Cortez, D., & Preiss, R. W. (2013). Safe Spaces, Difficult Dialogues, and Critical Thinking. *International Journal for the Scholarship of Teaching and Learning*, *7*(2), 5.

Barnett, T., & Blaikie, P. (n.d.). *AIDS in Africa: its present and future impact.* London: Belhaven Press.

Barnett, R. (2004). Learning for an unknown future. *Higher Education Research & Development*, *23*(3), 247–260. doi:10.1080/0729436042000235382

Baron, L., & Morin, L. (2009). The coach-coachee relationship in executive coaching: A field study. *Human Resource Development Quarterly*, *20*(1), 85–106. doi:10.1002/hrdq.20009

Barr, R. B., & Tagg, J. (1995). From teaching to learning – A new paradigm for undergraduate education. *Change*, *27*(6), 12–25. doi:10.1080/00091383.1995.10544672

Barth, M., Godemann, J., Rieckmann, M., & Stoltenberg, U. (2007). Developing key competencies for sustainable development in higher education. *International Journal of Sustainability in Higher Education*, *8*(4), 416–430. doi:10.1108/14676370710823582

Basu, M. (2015). A class apart at IIT-Bombay - Some students of IIT Bombay takes an unconventional career decision. *Indian Express*. Retrieved Sept 8, 2016 from http://indianexpress.com/article/cities/mumbai/a-class-apart-at-iit-bombay/

Bayar, A. (2014). The Components of Effective Professional Development Activities in Terms of Teachers' Perspective. *Online Submission*, *6*(2), 319–327.

Beard, C. M., & Wilson, J. P. (2006). *Experiential learning: A best practice handbook for educators and trainers*. London, UK: Kogan Page Publishers.

Becher, T., & Trowler, P. R. (2001). *Academic Tribes and Territories: Intellectual enquiry and the culture of disciplines* (2nd ed.). Buckingham, UK: Society for Research into Higher Education and Open University Press.

Becker, H. S. (1958). Problems of inference and proof in participant observation. *American Sociological Review*, *23*(6), 652–660. doi:10.2307/2089053

Beckett, D., & Hager, P. (2002). *Life, Work And Learning: Practice in Postmodernity*. London: Routledge.

Bejoy, T. (2010). Participation in the knowledge society: The free and open source software (FOSS) movement compared with participatory development. *Development in Practice*, *20*(2), 270–276. doi:10.1080/09614520903566509

Bell, E., & Bryman, A. (2007). The ethics of management research: An exploratory content analysis. *British Journal of Management*, *18*(1), 63–77. doi:10.1111/j.1467-8551.2006.00487.x

Bell, P., Lewestein, B., Shouse, A. W., & Feder, M. A. (Eds.). (2009). *Learning science in informal environments: People, places and pursuits*. Washington, DC: National Academies Press.

Bennet, A., & Bennet, D. (2008). A new change model: Factors for initiating and implementing personal action learning. *Vine*, *38*(4), 378–387. doi:10.1108/03055720810917633

Berg, B. L., & Lune, H. (2011). *Qualitative research methods for the social sciences* (8th ed.). Upper Saddle River, NJ: Pearson.

Berger, P. L., & Luckmann, T. (1966). *The social construction of reality*. New York, NY: Doubleday.

Berkes, F., Folke, C., & Colding, J. (2000). *Linking social and ecological systems: management practices and social mechanisms for building resilience*. Cambridge University Press.

Bernhard, J. K. (1995). Child development, cultural diversity, and the professional training of early childhood educators. *Canadian Journal of Education*, *20*(4), 415–436. doi:10.2307/1495080

Bettman, M. (2012). *500 School Project – making schools better.* Pretoria: UNISA Press.

Beveridge, L. (2010). Implementing pedagogical reform through action learning: Emerging issues from the local experience. *ALAR: Action Learning and Action Research Journal*, *16*(1), 39.

Beveridge, L., & Hinde McLeod, J. (2009). Action learning through Indigenous literature. *Intercultural Education*, *20*(2), 187–197. doi:10.1080/14675980902922218

Bhattacharya, S., & Bhattacharya, S. (2014). Indian Renaissance in the Twenty-First Century: Management Education, Thoughts and Practices for the Betterment of Business, Nation and Society. *Journal of Human Values*, *20*(1), 45–58. doi:10.1177/0971685813515602

Bialek, T. K. (2014). *An examination of how participation in a cohort-based leadership development program for high-potential employees contributes to the development of leaders at a major professional services firm in the United States* (Unpublished doctoral dissertation). University of Minnesota.

Biggs, J. (1996). Enhancing teaching through constructive alignment. *Higher Education*, *32*(3), 347–364. doi:10.1007/BF00138871

Biggs, J., & Tang, C. (2007). *Teaching for quality learning at university* (3rd ed.). Buckingham, UK: Society for Research into Higher Education and Open University Press.

Black, P. A., Calitz, E., & Steenkamp, T. J. et al. (2005). *Public Economics* (3rd ed.). Cape Town: Oxford University Press.

Blake, E., Tuckery, W., & Glaser, M. (2014, October). Towards communication and information access for Deaf people. *South African Computer Journal*, *54*, 10–19. doi:10.18489/sacj.v54i0.236

Blume Dahl, K. K. (2014). From worse to better: How Kenyan student-teachers can use Participatory Action Research in health education. *Educational Action Research*, *22*(2), 159–177. doi:10.1080/09650792.2013.859089

Blumer, H. (1969). *Symbolic interaction: Perspective and method.* Englewood Cliffs, NJ: Prentice-Hall, Inc.

Bockheim, J. G., Gennadiyev, A. N., Hammer, R. D., & Tandarich, J. P. (2005). Historical development of key concepts in Pedology. *Geoderma*, *124*(1-2), 23–36. doi:10.1016/j.geoderma.2004.03.004

Bockting, W. O., Miner, M. H., & Romine, R. E. (2013). Stigma, mental health, and resilience in an online sample of the US transgender population. *American Journal of Public Health*, *103*(5), 943–951. doi:10.2105/AJPH.2013.301241 PMID:23488522

Bødker, S., Ehn, P., Kammersgaard, J., Kyng, M., & Sundblad, Y. (1987). A utopian experience: On design of powerful computer-based tools for skilled graphic workers. In G. Bjerknes, P. Ehn & M. Kyng (Eds.), *Proceedings of the Aarhus Conference on Computers and Democracy—A Scandinavian challenge* (pp. 251-278). Academic Press.

Bolhuis, S. (2003). Towards process-oriented teaching for self-directed lifelong learning: A multidimensional perspective. *Learning and Instruction*, *13*(3), 327–347. doi:10.1016/S0959-4752(02)00008-7

Bordo, S. R. (1987). *The flight to objectivity: Essays on Cartesianism and culture*. Albany, NY: State University of New York Press.

Boshuizen, H. P., Bromme, R., & Gruber, H. (2004). *Professional learning: Gaps and transitions on the way from novice to expert*. Dordrecht: Kluwer Academic. doi:10.1007/1-4020-2094-5

Boshyk, Y., & Dilworth, R. (2010). *Action learning: History and evolution*. Springer. doi:10.1057/9780230250734

Bothamley, J. (2004). *Dictionary of Theories*. New York: Barnes & Noble Books.

Boudreau, J. (2004). Talentship: A decision science for HR. *Strategic HR Review*, *3*, 28–31.

Bowles, S., & Gintis, H. (1976). *Schooling in capitalist America* (Vol. 57). New York: Basic Books.

Braa, K., Bratteteig, T. & Øgrim, L. (1996). Organizing the redesign process in system development. *Journal of Systems and Software*, *33*(2), 133–140.

Braa, J., Monteiro, E., & Sahay, S. (2004). Networks of Action: Sustainable Health Information Systems across Developing Countries. *Management Information Systems Quarterly*, *28*(3), 337–362.

Braa, K., & Nielsen, P. (2015). Sustainable Action Research: the networks of actions approach. In *Proceedings of the 13th International Conference on Social Implications of Computers in Developing Countries* (pp. 331 – 343). IFIP.

Bradbury, L. U., & Koballa, T. R. Jr. (2007). Borders to cross: Identifying sources of tension in mentor-intern relationships. *Teaching and Teacher Education*, *24*(8), 2132–2145. doi:10.1016/j.tate.2008.03.002

Bradley, H., Bedada, A., Brahmbhatt, H., Kidanu, A., Gillespie, D., & Tsui, A. (2007). Amare Bedada, Heena Brahmbhatt, Aklilu Kidanu, Duff Gillespie, and Amy Tsui. Educational attainment and HIV status among Ethiopian voluntary counseling and testing clients. *AIDS and Behavior*, *11*(5), 736–742. doi:10.1007/s10461-006-9185-4 PMID:17082983

Braimoh, D. (2015). Open and Distance Learning in Sub-Saharan Africa: promises and threats. In A. Ravhudzulo & S. Ndlangamandla (Eds.), *Practical and Critical Issues in Open Distance Learning: 2016* (pp. 69–78). Pretoria: Unisa Press.

Bransford, J. (2000). *How people learn: Brain, mind, experience, and school*. Washington, DC: National Academy Press.

Bransford, J., Pellegrino, J. W., & Donovan, S. (1999). *How people learn: Bridging research and practice*. Washington, DC: National Academy Press.

Bråten, S. (1973). Model Monopoly and Communications: Systems Theoretical Notes on Democratization. *Acta Sociologica*, *2*(16), 98–107. doi:10.1177/000169937301600202

Breslow, K., Crowell, L., Francis, L, & Gordon, S.P. (2015). Initial efforts to coordinate appreciative inquiry: Facilitators' experiences and perceptions. I.*E.: Inquiry in Education*, *6*(1).

Brinkworth, C. (2016). *From Chilly Climate to Warm Reception: Experiences and Good Practices for Supporting LGBTQ Students in STEM*. CGU Theses & Dissertations. Paper 97. Retrieved from http://scholarship.claremont.edu/cgu_etd/97

Brook, C., Pedler, M., Abbott, C., & Burgoyne, J. (2016). On stopping doing those things that are not getting us to where we want to be: Unlearning, wicked problems and critical action learning. *Human Relations*, *69*(2), 369–389. doi:10.1177/0018726715586243

Brook, C., Pedler, M., & Burgoyne, J. (2012). Some debates and challenges in the literature on action learning: The state of the art since Revans. *Human Resource Development International*, *15*(3), 269–282. doi:10.1080/13678868.2012.687926

Brown, T. C. (2015). *Collaborative autobiography: A process for addressing educators' stress and awareness in 21st century public education* (Doctoral dissertation).

Brown, M., & Macatangay, A. (2002). The Impact of Action Research for Professional Development: Case studies in two Manchester schools. *Westminster Studies in Education*, *25*(1), 35–45. doi:10.1080/0140672020250104

Brown, P., Morello-Frosch, R., Brody, J. G., Gasior Altman, R., Rudel, R. A., Senier, L., & Simpson, R. et al. (2010). Institutional Review Board challenges related to Community-based Participatory Research on human exposure to environmental toxins: A case study. *Environmental Health*, *9*(1), 39–50. doi:10.1186/1476-069X-9-39 PMID:20637068

Brundtland, G., Khalid, M., Agnelli, S., Al-Athel, S., Chidzero, B., Fadika, L., …. Singh, M. (1987). *Our common future*. Brundtland Report.

Brundtland, G., Khalid, M., Agnelli, S., Al-Athel, S., Chidzero, B., Fadika, L.,… & Singh, M., (1987). *Our common future*. Brundtland Report.

Bruner, J. S. (2009). *The process of education*. Harvard University Press.

Brunning, R. H., Schraw, G. J., & Ronning, R. R. (1999). Cognitive psychology and instruction (3rd ed.). Prentice-Hall.

Bruns, D., & Worsley, S. (2015). *Navigating complexity in international development. Facilitating sustainable change at scale*. Rugby, UK: Practical Action Publishing. doi:10.3362/9781780448510

Brydon-Miller, M., Greenwood, D., & Maguire, P. (2003). Why Action Research? *Action Research*, *1*(1), 9–28. doi:10.1177/14767503030011002

Bryman, A. (2003). *Quantity and quality in social research*. Routledge.

Buchanan, I. (2010). *Oxford Dictionary of Critical Theory: The Invaluable Multidisciplinary guide to Theory*. New York: Oxford University Press.

Buchanan, L., & Connell, O, A. (2006). A brief history of decision making. *Harvard Business Review*, *84*(1), 32. PMID:16447367

Buysse, V., Castro, D. C., West, T., & Skinner, M. (2005). Addressing the needs of Latino children: A national survey of state administrators of early children programs. *Early Childhood Research Quarterly*, *20*(2), 146–163. doi:10.1016/j.ecresq.2005.04.005

Caldwell, J. E. (2007). Clickers in the Large Classroom: Current Research and Best-Practice Tips. *CBE Life Sciences Education*, *6*(1), 9–20. doi:10.1187/cbe.06-12-0205 PMID:17339389

Camara, G., & Fonseca, F. (2007). Information Policies and Open Source Software in Developing Countries. *Journal of the American Society for Information Science and Technology*, *58*(1), 121–132. doi:10.1002/asi.20444

Carlson, J. A. (2010). Avoiding traps in member checking. *Qualitative Report*, *15*(5), 1102.

Carroll, J. B. (1989). The Carroll Model: A 25-Year Retrospective and Prospective View. *Educational Researcher*, *18*(1), 26–31. doi:10.3102/0013189X018001026

Carr, W., & Kemmis, S. (1986). *Becoming Critical: Education, Knowledge, and Action Research*. London: Falmer Press.

Carter, A. D. (2015). *Building novice capacity through action learning coaching* (Doctoral dissertation). University of Georgia, Athens, GA.

Castro, E. L. (2014). Underprepared and at-risk: Disrupting deficit discourses in undergraduate STEM recruitment and retention programming. *Journal of Student Affairs Research and Practice*, *51*(4), 407–419. doi:10.1515/jsarp-2014-0041

Cavaye, J. (2001). *'Community engagement framework project: Scoping and review paper.'* Cavaye Community Development/CEO Committee on Land Resources''.

Cecchini, S., & Scott, C. (2003). Can information and communications technology applications contribute to poverty reduction? Lessons from rural India. *Information Technology for Development*, *10*(2), 73–84. doi:10.1002/itdj.1590100203

Cech, E. A., & Waidzunas, T. J. (2011). Navigating the heteronormativity of engineering: The experiences of lesbian, gay, and bisexual students. *English Studies*, *3*(1), 1–24.

Chambers, R. (1997). *Whose reality counts? Putting the first last. Intermediate Technology Publications Ltd.* doi:10.3362/9781780440453

Chambers, R. (2008). 'PRA, PLA and pluralism: practice and theory. In P. Reason & H. Bradbury (Eds.), *The Sage handbook of action research: participative inquiry and practice* (2nd ed.; pp. 297–318). London, UK: Sage. doi:10.4135/9781848607934.n28

Chapman, J. (2002). *Monitoring and Evaluating Advocacy, PLA Notes No. 43*. London: IIED.

Charalambous, C. Y., & Hill, H. C. (2012). Teacher knowledge, curriculum materials, and quality of instruction: Unpacking a complex relationship. *Journal of Curriculum Studies*, *44*(4), 443–466. doi:10.1080/00220272.2011.650215

Charles, L., & Ward, N. (2007). *Generating change through research: Action research and its implications*. Retrieved September 15, 2016 from: www.ncl.ac.uk/cre/publish/discussionpapers/pdfs/dp10.pdf

Checkland, P., & Holwell, S. (1998). Action Research: Its Nature and Validity. *Systemic Practice and Action Research*, *11*(1), 9–21. doi:10.1023/A:1022908820784

Checkland, P., & Poulter, J. (2010). Soft Systems Methodology. In M. Reynolds & S. Holwell (Eds.), *Systems Approaches to Managing Change: A Practical Guide* (pp. 191–242). Springer-Verlag London Limited. doi:10.1007/978-1-84882-809-4_5

Chesbrough, H. (2003). *Open Innovation: The New Imperative for Creating and Profiting from Technology*. Harvard Business School Press.

Chesbrough, H., & Crowther, A. K. (2006). Beyond high tech: Early adopters of open innovation in other industries. *R & D Management*, *36*(3), 229–236. doi:10.1111/j.1467-9310.2006.00428.x

Chetty, R., & Lubben, F. (2010). The scholarship of research in teacher education in a higher institution in transition: Issues of identity. *Teaching and Teacher Education*, *26*(4), 813–820. doi:10.1016/j.tate.2009.10.018

Chiaroni, D., Chiesa, V., & Frattini, F. (2011). The Open Innovation Journey: How firms dynamically implement the emerging innovation management paradigm. *Technovation*, *31*(1), 34–43. doi:10.1016/j.technovation.2009.08.007

Chikasanda, V. K. M., Otrel-Cass, K., Williams, J., & Jones, A. (2012). Enhancing teachers' technological pedagogical knowledge and practices: A professional development model for Technology teachers in Malawi. *International Journal of Technology and Design Education*. doi:10.1007/s10798-012-9206-8

Chokwe, J. M. (2013). Factors Impacting Academic Writing Skills of English Second Language Students. *Mediterranean Journal of Social Sciences*, *4*(14), 377–383.

Cho, Y. (2013). What is Action Learning? Components, Types, Process, Issues, and Research Agendas. *Learning and Performance Quarterly*, *1*(4), 1–11.

Cho, Y., & Egan, T. M. (2009). Action learning research: A systematic review and conceptual framework. *Human Resource Development Review*, *8*(4), 431–462. doi:10.1177/1534484309345656

Claeys, A., Kempton, J., & Paterson, C. (2014). *Regional Challenges: A Collaborative Approach to Improving Education*. Centre Forum, London, UK.

Clark, I. (2008). Assessment is for Learning: Formative Assessment and Positive Learning Interactions. *Florida Journal of Educational Administration & Policy*, *2*(1), 1–16.

Clark, S. G. (2011). *The policy process: A practical guide for natural resource professionals*. New Haven, CT: Yale University Press.

Clutterbuck, M. D., Megginson, M. D., & David, M. S. (2013). *Beyond goals: Effective strategies for coaching and mentoring*. New York: Routledge.

Cockburn, L., & Trentham, B. (2002). Participatory action research: Integrating community occupational therapy practice and research. *Canadian Journal of Occupational Therapy*, 21–31. PMID:11852687

Coetzee, E. (2012). Community Engagement by Higher Education Institutions – a practical model and guidelines. *Africa Education Review*, *9*(3), 501–517. doi:10.1080/18146627.2012.742669

Coghlan, D. & Coughlan, P. (2007). *Action learning and action research (ALAR): A methodological integration in an inter-organizational setting*. Springer Science+Business Media, LLC.

Cogneau, D., & Grimm, M. (2003). *The distribution of AIDS over the population in Africa. Hypothesis building from individual answers to a Demographic and Health Survey with an application to C^ote d'Ivoire. Mimeo*. University of Gottingen.

Cohen, J. (2006). Social, emotional, ethical, and academic education: Creating a climate for learning, participation in democracy, and well-being. *Harvard Educational Review*, *76*(2), 201–237. doi:10.17763/haer.76.2.j44854x1524644vn

Cohen, L., Manion, L., & Morrison, K. (2007). *Research methods in education* (6th ed.). London: Routledge.

Cohen, W. M., & Levinthal, D. (1990). Absorptive Capacity: A New Perspective on Learning and Innovation. *Administrative Science Quarterly*, *35*(1), 128–152. doi:10.2307/2393553

College of Education. (2012). *College of Education brochure Unisa*. Retrieved from www.unisa.ac.za

Collins, A., Joseph, D., & Bielazcyc, K. (2004). Design research: Theoretical and methodological issues. *Journal of the Learning Sciences*, *13*(1), 15–42. doi:10.1207/s15327809jls1301_2

Collins, S. L., Carpenter, S. R., Swinton, S. M., Orenstein, D. E., Childers, D. L., Gragson, T. L., & Knapp, A. K. et al. (2011). An integrated conceptual framework for long term social–ecological research. *Frontiers in Ecology and the Environment*, *9*(6), 351–357. doi:10.1890/100068

Cooke, N. A. (2014). Creating Opportunities for Empathy and Cultural Competence in the LIS Curriculum. *SRRT Newsletter, 187*. Retrieved from http://libr.org/srrt/news/srrt187.php#9

Cook, T. (1998). The importance of mess in Action Research. *Educational Action Research*, *6*(1), 93–109. doi:10.1080/09650799800200047

Cooper, E., Risley, R., Drake, L., Cooper, E., Risley, R., Drake, L., & Bundy, D. (2007). HIV as part of children and youth as life expectancy increases: implications for education. *Journal of International Cooperation in Education.*

Cope, F. (1998). Current practices: Current issues in selecting high potentials. *Human Resource Planning*, *21*(3), 15–16.

Cope, J., & Watts, G. (2000). Learning by doing-An exploration of experience, critical incidents and reflection in entrepreneurial learning. *International Journal of Entrepreneurial Behavior & Research*, *6*(3), 104–124. doi:10.1108/13552550010346208

Corey, S. M. (1960). Action Research in Education. *Guidance Readings for Counselors*, 603.

Corey, S. M. (1949). Curriculum development through action research. *Educational Leadership*, *7*(3), 147–153.

Cornbleth, C. (1984). Beyond Hidden Curriculum?. *Journal of Curriculum Studies*, *16*(1), 29–36. doi:10.1080/0022027840160105

Cornwall, A., & Jewkes, R. (1995). What is participatory research?. *Social Science & Medicine*, *41*(12), 1667–1676. doi:10.1016/0277-9536(95)00127-S PMID:8746866

Cosby, B. (2012). *About Technology Education*. Academic Press.

Cowan, C. A. (2013). *An exploration of significant leadership development factors in action learning: A comparison of three action learning programs* (Doctoral dissertation). The George Washington University.

Cox, E. (2013). *Coaching Understood: A pragmatic enquiry into the coaching process*. London: SAGE.

Creswell, J. W. (2009). *Research Design: Qualitative, Quantitative, and Mixed Methods Approaches* (3rd ed.). Thousand Oaks, CA: Sage.

Creswell, J. W. (2014). *Research Design: Qualitative, Quantitative, & Mixed Methods Approaches* (4th ed.). London: SAGE.

Creswell, J. W., & Miller, D. L. (2000). Determining validity in qualitative inquiry. *Theory into Practice*, *39*(3), 124–130. doi:10.1207/s15430421tip3903_2

Cross, B. A. (2013). *The impact of a school garden and environmental education on the environmental awareness of fifth graders* (MA dissertation). Southeastern Louisiana Hammond University.

Cullen, J., Hadjivassiliou, K., Hamilton, E., Kelleher, J., Sommerlad, E., & Stern, E. (2002). *Review of current pedagogic research and practice in the fields of post-compulsory education and lifelong learning*. London: The Tavistock Institute.

Custer, R. L. (1995). Examining the dimensions of technology. *International Journal of Technology and Design Education*, *5*(3), 219–244. doi:10.1007/BF00769905

Cutter-Mackenzie, A., & Edwards, S. (2013). Toward a model for early childhood environmental education: Foregrounding, developing, and connecting knowledge through play-based learning. *The Journal of Environmental Education*, *44*(3), 195–213. doi:10.1080/00958964.2012.751892

D'Souza, D. A., Sharma, D., & Singh, R. P. (2013). Effectiveness of selected instructional approaches in developing environmental awareness among IX grade students In Patna district, India. *Researchers World*, *4*(2), 109–114.

Daley, B. J. (1999). Novice to expert: An exploration of how professionals learn. *Adult Education Quarterly*, *49*(4), 133–147. doi:10.1177/074171369904900401

Darling-Hammond, L., Wei, R. C., Andree, A., Richardson, N., & Orphanos, S. (2009). State of the Profession: Study Measures Status of Professional Development. *Journal of Staff Development*, *30*(2), 42.

Das, M. (2014). Secondary school students' awareness on plastic Pollution and environmental education: A study. *Global Online Electronic International Interdisciplinary Research Journal*, *2*(5), 10–25.

Davey, C., Powell, J. A., Powell, J. E., & Cooper, I. (2002). Action learning in a medium-sized construction company. *Building Research and Information*, *30*(I), 5–15. doi:10.1080/09613210110090403

David, S., Clutterbuck, D., & Megginson, D. (2014, August). Goal orientation in coaching differs according to region, experience, and education. *International Journal of Evidence Based Coaching and Mentoring*, *12*(2).

Davids, M. N. (2014). Traditional tutorial system-fit for purpose or past its sell-by date? University student's pedagogical experiences. *South African Journal of Higher Education, 28*(2), 338–354.

Davies, D. (1997). The relationship between science and technology in the primary curriculum: Alternative perspectives. *The Journal of Design and Technology Education, 2*(2), 101–111.

Davies, P. (1999). What is evidence-based education? *British Journal of Educational Studies, 4*(2), 108–121. doi:10.1111/1467-8527.00106

Davison, R. M. (1998). Research Methodology. In *An Action Research Perspective of Group Support Systems: How to Improve Meetings in Hong Kong. PhD thesis* (pp. 3-1 - 3-20). Hong Kong: City University of Hong Kong. Retrieved September 13, 2016, from http://www.is.cityu.edu.hk/staff/isrobert/phd/ch3.pdf

Davison, R., Martinsons, M. G., & Kock, N. (2004). Principles of canonical action research. *Information Systems Journal, 14*(1), 65–86. doi:10.1111/j.1365-2575.2004.00162.x

de Haan, E., & de Ridder, L. (2006). Action learning in practice: How do participants learn?. *Consulting Psychology Journal: Practice and Theory, 58*(4), 16- 31.

de Haan, E., Bertie, C., Day, A., & Sills, C. (2010). Clients Critical Moments of Coaching: Toward a Client Model of Executive Coaching. *Academy of Management Learning & Education, 9*(4), 607–621. doi:10.5465/AMLE.2010.56659879

De Haan, G. (2006). The BLK 21programme in Germany: A Gestaltungskompetenz-based model for Education for Sustainable Development. *Environmental Education Research, 12*(1), 19–32. doi:10.1080/13504620500526362

De Jager, M. L. E. (2011). *Teachers' understanding and implementation of the design process as it relates to teaching Technology Education in the intermediate and senior phases (Grades 4-9) in the General Education and Training Band in KwaZulu-Natal* (Unpublished PhD thesis). Durban University of Technology, Durban, South Africa.

de Negri, B., Thomas, E., & Ilinigumugabo, A. (1998). Community-based program development.: Vol. 1. *Trainer's Manual and Participant's Handbook.* Nairobi: Academic Press.

de Villiers, C. (1998). South African Environmental Reporting: What It Is, What It Should Be. *APIRA Conference Proceedings.*

De Villiers, M. R. (2005). Three approaches as pillars for interpretive information systems research: development research, action research and grounded theory. *Proceedings of the 2005 annual research conference of SAICSIT on IT research in developing countries,* 142-151.

De Vries, M. J. (1996). Technology Education: Beyond the Technology is applied science paradigm. *Journal for Technology Education, 8*(1), 7–15. doi:10.21061/jte.v8i1.a.1

de Walque, D., Nakiyingi-Miiro, J. S., Busingye, J., & Whitworth, J. A. (2005). Changing Association between Schooling Levels and HIV-1 Infection over 11 Years in a Rural Population Cohort in South-west Uganda. *Tropical Medicine & International Health, 10*(10), 993–1001. doi:10.1111/j.1365-3156.2005.01475.x PMID:16185233

Dean, J. W., Brandes, P., & Dharwadkar, R. (1998). Organizational cynicism. *Academy of Management Review, 23*(2), 341–352.

Deloitte. (2011). *Only skin deep? Re-examining the business case for diversity.* Deloitte Point of View Human Capital Australia.

Department of Basic Education (DBE). (2011). *Integrated Strategic Planning Framework for Teacher Education and Development in South Africa, 2011–2025: Technical Report.* Pretoria: Government Printers.

Department of Basic Education. (2011). Curriculum and assessment policy statement grades 7-9: technology. Pretoria, South Africa: Government Printers.

Department of Basic Education. (2011). *Education in South Africa*. Available at: http://www.southafrica.info/about/education/education.htm

Department of Cooperative Governance and Traditional Affairs (COGTA). (2009). *Local Government Turnaround Strategy*. Pretoria: COGTA.

Department of Cooperative Governance and Traditional Affairs (COGTA). (2013). *Draft Local Economic Development Framework (2013-2018)*. Pretoria: COGTA.

Department of Education. (2003). *Revised National Curriculum Statement Grades R-9 (Schools): Teacher's guide for developing learning programmes Technology*. Pretoria: Government Printers.

Department of Education. (2004, September 2). White Paper on e-Education: Transforming Learning and Teaching through Information and Communication Technologies (ICTs). *Government Gazette,* (26734), 3 - 46.

Deshpande, A., Mathur, A., & Krishnamurthy, S. (2016). Application of Internet of Things in Healthcare Sector for Bottom of Pyramid in India. IJEAST. *International Journal of Engineering Applied Sciences and Technology*, *1*(11), 28–33.

Design-Based Research Collective. (2003). Design-based research: An emerging paradigm for educational inquiry. *Educational Researcher*, *32*(1), 5–8. doi:10.3102/0013189X032001005

Desimone, L. M. (2009). Improving impact studies of teachers professional development: Toward better conceptualizations and measures. *Educational Researcher*, *38*(3), 181–199. doi:10.3102/0013189X08331140

DeWalt, K. M., & DeWalt, B. R. (2002). *Participant observation: A guide for fieldworkers*. Walnut Creek, CA: Altamira Press.

Dewar, N., & Shippey, K. (2002). Meeting the challenge of change in tertiary level environmental education in South Africa. *International Journal of Sustainability in Higher Education*, *3*(4), 324–334. doi:10.1108/14676370210442370

Dewey, J. (1910). *How we think*. Boston: D. C. Heath. doi:10.1037/10903-000

DFE. (2016). *Educational Excellence Everywhere*. Available at https://www.gov.uk/government/publications/educational-excellence-everywhere

Dick, B. (1997). *Action learning and action research*. Retrieved from http://www.scu.edu.au/schools/gcm/ar/arp/actlearn.html

Dick, B. (1997). *Action Learning and Action Research*. Retrieved from: http://www.aral.com.au/resources/actlearn.html

Dick, B. (2010). *Action Research thesis. You want to do an action research thesis? How to conduct and report action research*. Retrieved from http://www.scu.edu.ac/schools/gcm/ar/art/artesis.html

Dick, B. (2011). Organizational survive and improvement through resilience. In M. J. Celinski & K. Gow (Eds.), *Continuity Versus Creative Response to Challenge: The Primacy of Resilience and Resourcefulness in Life and Therapy*. Nova Science Publishers.

Dickens, L., & Watkins, K. (1999). Action research: Rethinking Lewin. *Management Learning*, *30*(2), 127–140. doi:10.1177/1350507699302002

Dick, R. (1990). *Convergent Interviewing*. Brisbane, Australia: Inter-change.

Dilworth, R. L., & Boshyk, Y. (Eds.). (2010). *Action learning and its applications*. Basingstoke, UK: Palgrave Macmillan. doi:10.1057/9780230250741

Dinham, S., Aubusson, P., & Brady, L. (2006). *Distributed leadership through action learning, keynote address*. Paper presented at the Fifth International Conference on Educational Leadership, Australian Centre for Educational Leadership, University of Wollongong, NSW.

Dmochowski, J. E., Garofalo, D., Fisher, S., Greene, A., & Gambogi, D. (2016). Integrating sustainability across the university curriculum. *International Journal of Sustainability in Higher Education*, *17*(5), 652–670. doi:10.1108/IJSHE-10-2014-0154

Dochy, F. J. R. C., Segers, M., & Sluijsmans, D. (1999). The use of self-, peer and co-assessment in higher education: A review. *Studies in Higher Education*, *24*(3), 331–350. doi:10.1080/03075079912331379935

Domask, J. J. (2007). Achieving goals in higher education: An experiential approach to sustainability studies. *International Journal of Sustainability in Higher Education*, *8*(1), 53–68. doi:10.1108/14676370710717599

Doyle, A., Seery, N., & Canty, D. (2016). *Exploring the Relationship between Technology Teachers Orientations towards Teaching and their Associated Professional Life Phases*. Paper presented at PATT–32, Technology Education for 21st Century Skills, Netherlands.

Dressel, P., & Hodge, G. (2013). *Analysis of Policies, Practices, and Programs for Advancing Diversity, Equity and Inclusion: Full Report*. D5 Coalition. Retrieved on February 10, 2014 from http://www.d5coalition.org/wp-content/uploads/2013/09/ PPP-Full-Report-11.14.13.pdf

Dreyfus, H. L. (2008). *On the internet*. Routledge.

Drucker, P. F. (1993). *Post-Capitalist Society*. New York, NY: HarperCollins.

Du Chatenier, E., Verstegen, J., Biemans, H., Mulder, M., & Omta, O. (2009). The Challenges of Collaborative Knowledge Creation in Open Innovation Teams. *Human Resource Development Review*, *8*(3), 350–381. doi:10.1177/1534484309338265

Dugger, W. (2008). The perspective of technology education. *Proceedings of the Japanese International Symposium of Educational Cooperation for Industrial Technology Education*.

Dugger, W. E. Jr, & Naik, N. (2001, Sept.). Clarifying misconceptions between Technology Education and Educational Technology. *Technology Teacher*, 31–35.

Dulewicz, S. V., & Higgs, M. J. (1998). *Emotional intelligence: can it be measured reliably and validly using competency data?* Competency Journal.

Duncombe, R., & Heeks, R. (2005). *Information & Communication Technologies (ICTs), Poverty Reduction and Micro, Small & Medium-Scale Enterprises (MSMEs)*. Vienna: United Nations Industrial Development Organization. Retrieved from http://www.unido.org/file-storage/download/?file_id=83230

Edge, J. (Ed.). (2001). *Action research: Case studies in TESOL practice*. Alexandria, VA: TESOL Inc.

Elder, Z. (2012). *The MLG Tool: Micro Action Research*. Accessed online on 25 August 2016 at: https://marginallearninggains.com/methodology-micro-action-research/

Elkington, J. (2004). Enter the triple bottom line. *The Triple Bottom Line: Does It All Add Up*, 1-16.

Ellinger, A. D., Hamlin, R. G., & Beattie, R. S. (2008). Behavioural indicators of ineffective managerial coaching: A cross-national study. *Journal of European Industrial Training*, *32*(4), 240–257. doi:10.1108/03090590810871360

Ellinger, A. D., & Kim, S. (2014). Coaching and human resource development examining relevant theories, coaching genres, and scales to advance research and practice. *Advances in Developing Human Resources*, *16*(2), 127–138. doi:10.1177/1523422313520472

Elzen, B., Geels, F. W., & Green, K. (Eds.). (2004). *System innovation and the transition to sustainability: theory, evidence and policy*. Edward Elgar Publishing. doi:10.4337/9781845423421

Enkel, E., Gassmann, O., & Chesbrough, H. (2009). Open R&D and open innovation: Exploring the phenomenon. *R & D Management*, *39*(4), 311–316. doi:10.1111/j.1467-9310.2009.00570.x

Epstein, J. L., & Sheldon, S. B. (2006). Moving forward: Ideas for research on school, family, and community partnerships. In C. F. Conrad & R. Serlin (Eds.), *SAGE handbook for research in education: Engaging ideas and enriching inquiry* (pp. 117–138). Thousand Oaks, CA: Sage. doi:10.4135/9781412976039.n7

Ertmer, P. A., & Newby, T. J. (1993). Behaviorism, Cognitivism, Constructivism: Comparing Critical Features from an Instructional Design Perspective. *Performance Improvement Quarterly*, *6*(4), 50–72. doi:10.1111/j.1937-8327.1993.tb00605.x

Espinosa, L. M. (2005). Curriculum and assessment considerations for young children from culturally, linguistically, and economically diverse backgrounds. *Psychology in the Schools*, *42*(8), 837–853. doi:10.1002/pits.20115

European Union. (2015). *Growing a Digital Social Innovation Ecosystem for Europe*. Retrieved from https://www.nesta.org.uk/sites/default/files/dsireport.pdf

Evans, L., Thurton, B., & Usinger, J. (2012). Theoretical frameworks to guide school improvement. *NASSP Bulletin*, *96*(2), 154–171. doi:10.1177/0192636512444714

Evenari, M., Shanan, L., & Tadmor, N. (1982). *The Negev: the challenge of a desert*. Harvard University Press. doi:10.4159/harvard.9780674419254

Ewing, R., Groundwater-Smith, S., Mockler, N., Loughland, T., Simpson, A., Smith, D., & Books, D. et al. (2010). *Meta analysis of quality teaching action learning project*. Sydney, NSW: University of Sydney.

Ewing, R., Groundwater-Smith, S., Mockler, N., Loughland, T., Simpson, A., Smith, D., & Books, D. et al. (2010). *Meta-analysis of quality teaching action learning project*. Sydney, Australia: University of Sydney.

Fals Borda, O. (1979). *El problema de como investigar la realidad para transformarla: Por la praxis*. Bogotá, Colombia: Ediciones Tercer Mundo.

Fals Borda, O. (1988). *Knowledge and people's power: Lessons with peasants in Nicaragua, Mexico and Colombia*. New York, NY: New Horizons Press.

Fawcett, S.B., Paine-Andrews, A., & Francisco, V.T. (1995). Using empowerment theory in collaborative partnerships for community health and development. *Community Psychology*, *23*(5), 677–697.

Feldman, A. (1999). The role of conversation in collaborative action research. *Educational Action Research*, *7*(1), 125–147. doi:10.1080/09650799900200076

Ferrance, E. (2000). *Action research*. Retrieved September 14, 2016 from: www.google.co.za/?gws_rd=ss1#q=themes+in+education+action+research+by+eileen+ferrance

Ferrance, E. (2000). *Action research*. Brown University.

Ferreira, R. (2016). *Action research as research design*. Pretoria, South Africa: University of Pretoria.

Fey, M. V. (2010). A short guide to the soils of South Africa, their distribution and correlation with World Reference Base soil groups. *Proceedings of: 19th World Congress of Soil Science, Soil Solutions for a Changing World.*

Fiol, C. M., & Lyles, M. A. (1985). Organizational learning. *Academy of Management Review*, *10*(4), 803–813.

Fischer, G. (2013). Learning, social creativity, and cultures of participation. In A. Sannino & V. Ellis (Eds.), *Learning and Collective Creativity: Activity-Theoretical and Sociocultural Studies* (pp. 198–213). Routledge.

Fishman, B., Penuel, W. R., Allen, A., Cheng, B. H., & Sabelli, N. (2013). Design-based implementation research: An emerging model for transforming the relationship of research and practice. *National Society for the Study of Education Yearbook*, *112*(2), 136–156.

Flick, E. (2011). *Creating a Culture of Avid Readers: The Western School Literacy Project* (Unpublished honors dissertation). University of Wollongong, NSW, Australia.

Flood, R. L. (2010). The relationship of systems thinking to action research. *Systemic Practice and Action Research*, *23*(4), 269–284. doi:10.1007/s11213-010-9169-1

Flowers, J. C. (1994). Applied philosophy Technology Education. *Technology Teacher*, *54*(4), 6–7.

Folke, C. (2006). Resilience: The emergence of a perspective for social–ecological systems analyses. *Global Environmental Change*, *16*(3), 253–267. doi:10.1016/j.gloenvcha.2006.04.002

Fosnot, C. T. (1996). *Constructivism: Theory, Perspectives, and Practice*. New York, NY: Teachers College Press.

Fourie, I., & Krauss, K. (2010, December). Information literacy training for teachers in a developing South African context: Suggestions for a multi-disciplinary planning approach. *Innovation (Abingdon)*, *41*, 107–122.

Fox, R. (2001). Constructivism examined. *Oxford Review of Education*, *27*(1), 23–35. doi:10.1080/03054980125310

Fredette, C. (2012). *Leadership Diversity in the Non-profit Sector: Baby Steps, Big Strides and Bold Stances. A DiverseCity Counts Research Project*. Toronto: DiverseCity.

Freeman, R. B., & Wei, H. (2014). Collaboration: Strength in diversity. *Nature*, *513*(7518), 305. doi:10.1038/513305a PMID:25230634

Fullan, M., Rincon-Gallardo, S. & Hargreaves, A. (2015). Professional capital as accountability. *Education Policy Analysis Archives, 23*(15), 1-17. 10.14507/epaa.v23.1998

Fullan, M. (1993). *Change forces: probing the depth of education reform*. Bristol, PA: Falmer Press.

Fullan, M. (1993). *Change forces: Probing the depth of education reform*. Bristol, PA: Falmer Press.

Fullan, M. (2004). *Leadership & sustainability: System thinkers in action*. Corwin Press.

Funtowicz, S. O., & Ravetz, J. R. (1993). *The emergence of post-normal science. Science, politics and morality* (pp. 85–123). Springer. doi:10.1007/978-94-015-8143-1_6

Gachui, D. (1999). *The Impact of HIV/AIDS on Education Systems in the Eastern and Southern African Region*. UNICEF Report.

Gajardo, M. (2007). *Country profile prepared for the Education for All Global Monitoring Report 2008 Education for All by 2015: Will we make it? Dominican Republic case study*. Retrieved May 3, 2015, from http://unesdoc.unesco. org/images/0015 /001555/155539e.pdf

Garces, L. M., & Cogburn, C. D. (2015). Beyond declines in student body diversity how campus-level administrators understand a prohibition on race-conscious postsecondary admissions policies. *American Educational Research Journal*, *52*(5), 828–860. doi:10.3102/0002831215594878

Gardner, H. (1995). Reflections on multiple intelligences: Myths and messages. *Phi Delta Kappan*, *77*(3), 200.

Garet, M. S., Porter, A. C., Desimone, L., Birman, B. F., & Yoon, K. S. (2001). What makes professional development effective? Results from a national sample of teachers. *American Educational Research Journal*, *38*(4), 915–945. doi:10.3102/00028312038004915

Garmendia, E., & Stagl, S. (2010). Public participation for sustainability and social learning: Concepts and lessons from three case studies in Europe. *Ecological Economics*, *69*(8), 1712–1722. doi:10.1016/j.ecolecon.2010.03.027

Garratt, B. (1991). The power of action learning. In Action learning in practice (pp. 15-30). Gower Publishing, Ltd.

Garrison, G. (2012). Clever vs. wise. *Journal of Healthcare Contracting*. Retrieved from http://www.jhconline.com/wisdom-vs-cleverness.html

Gassmann, O. (2006). Editorial: Opening up the innovation process: Towards an agenda. *R & D Management*, *36*(3), 223–228. doi:10.1111/j.1467-9310.2006.00437.x

Gauteng Department of Education (GDE). (2010). *Gauteng Mathematics, Science and Technology Education improvement strategy: 2009–2014*. Johannesburg: Government Printers.

Gay, G. (2000). *Culturally responsive teaching: Theory, research, & practice*. New York: Teachers College Press.

Gearty, M. R., Bradbury-Huang, H., & Reason, P. (2015). Learning history in an open system: Creating histories for sustainable futures. *Management Learning*, *46*(1), 44–66. doi:10.1177/1350507613501735

George, M. A., Daniel, M., & Green, L. W. (1999). Appraising and funding participatory research in health promotion. *International Quarterly of Community Health Education*, *18*(2), 181–197. doi:10.2190/C1B5-7PPE-7TYL-7YN8 PMID:17686723

Georgescu-Roegen, N. (2002). Four challenges of sustainability. *Conservation Biology*, *16*(6), 1457–1460. doi:10.1046/j.1523-1739.2002.01668.x

Gergen, K. J., & Gergen, M. M. (1991). Toward reflexive methodologies. In F. Steier (Ed.), *Research and Reflexivity* (pp. 76–95). Newbury Park, CA: SAGE Publications, Inc.

Gibson, R. (1983). *Critical Theory and Explanation*. London: Hodder & Stodder.

Gil-pe'rez, D., Vilches, A., Fernana'ndez, I., Cachapuz, A., Praia, J., Valde, P., & Salinas, J. (2005). Technology as 'applied science': A serious misconception that reinforces distorted and impoverished views of science. *Science and Education*, *14*, 309–320.

Giroux, H. A. (2003). Public pedagogy and politics of resistance: Notes on critical theory of educational struggle. *Educational Philosophy and Theory*, *35*(1), 5–16. doi:10.1111/1469-5812.00002

Gizaw, A., Bygstad, B., & Nielsen, P. (in press). Open Generification. *Information Systems Journal*.

Glasser, W. (1986). *Control theory in the classroom*. Perennial Library/Harper & Row Publishers.

Goel, A., Gupta, A., & Bande-Vilela, B. (2014). It's all in interpersonal interactions: Empowerment, organizational cynicism, and perception of organizational politics. In Values in Shock - The role of contrasting management, economic, and religious paradigms in the workplace. Academic Press.

Goel, A., Gupta, A., & Bande-Vilela, B. (2014). It's all in interpersonal interactions: Empowerment, organizational cynicism, and perception of organizational politics. Values in Shock - The role of contrasting management, economic, and religious paradigms in the workplace, 418.

GOI. (1956). *Government of India, Sixth Five Year Plan*. Planning Commission, Government of India.

Goleman, D. (2001). Emotional intelligence: Issues in paradigm building. *The Emotionally Intelligent Workplace, 13*, 26.

Goleman, D. (2002). *The New Leaders: Transforming the art of leadership into the science of results*. London: Little Brown.

Goleman, D., Boyatzis, R. E., & McKee, A. (2002). *The new leaders: Transforming the art of leadership into the science of results*. London: Little, Brown.

Gomez, R., & Pather, S. (2011). ICT Evaluation: Are we asking the right questions?. *The Electronic Journal of Information Systems in Developing Countries, 50*(5), 1 - 14. Retrieved January 20, 2014, from http://www.ejisdc.org

Goosen, L. (2015). Educational Technologies for Growing Innovative e-Schools in the 21st Century: A Community Engagement Project. In D. Nwaozuzu, & S. Mnisi (Ed.), *Proceedings of the South Africa International Conference on Educational Technologies* (pp. 49 - 61). Pretoria: African Academic Research Forum.

Goosen, L. (2016a, February 18). *"We don't need no education"? Yes, they DO want e-learning in Basic and Higher Education!*. Retrieved from http://uir.unisa.ac.za/handle/10500/20999

Goosen, L. (2016b). Policy Issues in ICT Education towards Effective Teaching and Meaningful Learning. In J. Kriek, B. Bantwini, C. Ochonogor, J. J. Dhlamini, & L. Goosen (Ed.), *Proceedings of the Institute of Science and Technology Education (ISTE) Conference On Mathematics, Science And Technology Education* (pp. 667 - 679). Mopani Camp, Kruger National Park, South Africa: UNISA Press.

Goosen, L., & Van der Merwe, R. (2015). e-Learners, Teachers and Managers at e-Schools in South Africa. In C. Watson (Ed.), *Proceedings of the 10th International Conference on e-Learning (ICEL)* (pp. 127 - 134). Nassau: Academic Conferences and Publishing International.

Goosen, L., & Gouws, P. (2016). Inspiring and Sustaining Learners' and Their Communities' Interest in Science, Engineering and Technology. In W. Mwakapenda, T. Sedumedi, & M. Makgato (Ed.), *Proceedings of the 24th Annual Conference of the Southern African Association for Research in Mathematics, Science and Technology Education (SAARMSTE)* (pp. 530 - 541). Pretoria: Tshwane University of Technology.

Gossow, C. (2016). *An exploration of the daily environmentally sustainable practices by South African youths living in Urban Gauteng* (D.Ed. thesis). University of South Africa.

Gottlieb, H. (2007). Community Engagement Step-by-Step Action Kit. Tucson, AZ: Renaissance Press.

Goyal, K. A. (2006). Impact of Globalization on Developing Countries (with special reference to India). *International Research Journal of Finance and Economics*, *5*, 166–171.

Grant, A. (2005). From a Distance: Tutors as a Community of Practice. *Education As Change*, *9*(1), 80–100. doi:10.1080/16823200509487104

Gray, R. E., Fitch, M., Davis, C., & Phillips, C. (2000). Challenges of participatory research: Reflections on a study with breast cancer self-help groups. *Health Expectations*, *3*(4), 243–252. doi:10.1046/j.1369-6513.2000.00100.x PMID:11281935

Greaves, E., Macmillan, L., & Sibieta, L. (2014). *Lessons from London schools for attainment gaps and social mobility*. London: The Social Mobility and Child Poverty Commission.

Greenwood, D. J. (2002). Action research: Unfulfilled promises and unmet challenges. Concepts and Transformation. *Concepts and Transformation*, *7*(2), 117–139. doi:10.1075/cat.7.2.02gre

Greenwood, D. J., Foote Whyte, W., & Karkavy, I. (1993). Participatory Action Research as a process and as a goal. *Human Relations*, *46*(2), 175–192. doi:10.1177/001872679304600203

Greenwood, D. J., & Levin, M. (1998). *Introduction to Action Research: Social Research for Social Change*. London: Sage Publications.

Greenwood, D. J., & Levin, M. (1998). *Introduction to Action Research: Social research for social change*. Thousand Oaks, CA: SAGE Publications.

Greenwood, D. J., & Levin, M. (1998). *Introduction to action research: Social research for social change*. Thousand Oaks, CA: Sage.

Griffiths, J. (2005). *HIV/AIDS Intervention Programs for Youth in Africa: The role of grass root soccer*. Academic Press.

Griggs, D., Stafford-Smith, M., Gaffney, O., Rockström, J., Öhman, M. C., Shyamsundar, P., & Noble, I. et al. (2013). Policy: Sustainable development goals for people and planet. *Nature*, *495*(7441), 305–307. doi:10.1038/495305a PMID:23518546

Guba, E. G., & Lincoln, Y. S. (1994). *Competing paradigms in qualitative research*. Newbury Park, CA: SAGE.

Gumbo, M. T. (2010). *A critical reflection on curriculum review of Technology Education in South Africa*. Paper presented at the 4th Annual Teacher Education at a Distance Conference, Lanseria, South Africa.

Gumbo, M. T. (2016b). The question of Technology Education: A pertinent issue. Inaugural Lecture of Professor Mishack T Gumbo, Pretoria, South Africa.

Gumbo, M. (2017). An Indigenous Perspective on Technology Education. In P. Ngulube (Ed.), *Social, Cultural, and Educational Consideration of Indigenous Knowledge in Developing Countries*. IGI Global. doi:10.4018/978-1-5225-0838-0.ch008

Gumbo, M. T. (2015). Indigenous technology in Technology Education curricula and teaching. In P. J. Williams, A. Jones, & C. Buntting (Eds.), *Contemporary issues in Technology Education: The future of Technology Education* (pp. 57–75). Singapore: Springer. doi:10.1007/978-981-287-170-1_4

Gumbo, M. T. (2016a). Pedagogical principles in Technology Education: An indigenous perspective. In G. Emeagwali & E. Shizha (Eds.), *African Indigenous Knowledge and the Sciences: Journeys into the past and present* (pp. 13–32). Rotterdam, The Netherlands: Sense. doi:10.1007/978-94-6300-515-9_2

Gumbo, M., Makgato, M., & Müller, H. (2012). The impact if In-service Technology Training Programme on Technology teachers. *The Journal of Technology Studies*, *38*(2), 23–33.

Gunther, F., & Chatoney, F. (2016). *Technology Education in France: efficiency of tools from functional analysis in learning process for describing objects. Pupils Attitude towards Technology – PATT 32 Conference*.

Guskey, T. R. (1991). Enhancing the effectiveness of professional development programs. *Journal of Educational & Psychological Consultation*, *2*(3), 239–247. doi:10.1207/s1532768xjepc0203_3

Guskey, T. R. (2000). *Evaluating professional development*. Thousand Oaks, CA: Corwin Press.

Gustavsen, B. (2003). New Forms of Knowledge Production and the Role of Action Research. *Action Research*, *1*(2), 153–164. doi:10.1177/14767503030012003

Haberl, H., Winiwarter, V., Andersson, K., Ayres, R. U., Boone, C., Castillo, A., & Furman, E. et al. (2006). From LTER to LTSER : Conceptualizing the Socioeconomic Dimension of Long-term Socioecological Research. *Ecology and Society*, *11*(2), 13. doi:10.5751/ES-01786-110213

Habermas, J. (1970). Towards a theory of communicative competence. *Inquiry: An Interdisciplinary Journal of Philosophy*, *13*(1-4), 360–375. doi:10.1080/00201747008601597

Hacker, K. (2013). *Community-based Participatory Research*. New York, NY: SAGE Publications, Inc.

Hagey, R. S. (1997). The use and abuse of participatory action research. *Chronic Diseases in Canada*, *18*(1), 1–4. PMID:9188513

Hallström, J., Hultén, M., & Lövheim, D. (2014). The study of technology as a field of knowledge in general education: Historical insights and methodological considerations from a Swedish case study, 18422010. *International Journal of Technology and Design Education*, *24*(2), 121–139. doi:10.1007/s10798-013-9252-x

Hammel, A. M. (2007). Professional development research in general education. *Journal of Music Teacher Education*, *17*(1), 22–32. doi:10.1177/10570837070170010106

Hammond, J. S. (1980). *Learning by the case method*. Boston: Harvard Business School.

Hamovitch, B. A. (1997). *Staying after school: At-risk students in a compensatory education program*. Westport, CT: Praeger Publishers.

Hannaway, D. (2016). *Learners' profiles when experiencing Technology-based learning in the Foundation phase*. South African International Conference on Educational Technologies (SAICET), Pretoria, South Africa.

Harding, S. (2015). *Objectivity and Diversity: Another Logic of Scientific Research*. Chicago, IL: University of Chicago Press. doi:10.7208/chicago/9780226241531.001.0001

Hargreaves, A., & Ainscow, M. (2015). The top and bottom of leadership and change. *Phi Delta Kappan*, *97*(3), 42–48. doi:10.1177/0031721715614828

Hargreaves, J. R., & Glynn, J. R. (2002). Educational attainment and HIV-1 infection in developing countries: A systematic review. *Tropical Medicine & International Health*, *7*(6), 489–498. doi:10.1046/j.1365-3156.2002.00889.x PMID:12031070

Harland, T. (2012). *University Teaching*. London: Routledge.

Harris, A. (2003). Teacher leadership as distributed leadership: Heresy, fantasy or possibility?. *School Leadership & Management*, *23*(3), 313–324. doi:10.1080/1363243032000112801

Harrison, R. T., Leitch, C. M., & Chia, R. (2007). Developing paradigmatic awareness in university business schools: The challenge for executive education. *Academy of Management Learning & Education*, *6*(3), 332–343. doi:10.5465/AMLE.2007.26361624

Hartley, B. L., Thompson, R. C., & Pahl, S. (2015). Marine litter education boosts childrens understanding and self-reported actions. *Marine Pollution Bulletin*, *90*(1), 209–217. doi:10.1016/j.marpolbul.2014.10.049 PMID:25467869

Harvard. (2016). *The Harvard business school case method*. Retrieved from www.hbs.edu/mba/academic-experience/Pages/the-hbs-case-method.aspx

Harvey, M., Ambler, T., & Cahir, J. (2017). Spectrum Approach to Mentoring: An evidence-based approach to mentoring for academics working in higher education. *Teacher Development*, *21*(1), 160–174. doi:10.1080/13664530.2016.1210537

Hashweh, M. Z. (2005). Teacher pedagogical constructions: a reconfiguration of pedagogical content knowledge. Teachers and Teaching: Theory and Practice, 11(3), 273–292. doi:10.1080/13450600500105502

Haski-Leventhal, D., & Bargal, D. (2008). The volunteer stages and transitions model: Organizational socialization of volunteers. *Human Relations*, *61*(1), 67–102. doi:10.1177/0018726707085946

Hawkins, P. (2008). The coaching profession: Some key challenges. *Coaching: An International Journal of Theory, Research and Practice*, *1*(1), 28–38.

Heeks, R. (2010). Do information and communication technologies (ICTs) contribute to development? *Journal of International Development*, *22*(5), 625–640. doi:10.1002/jid.1716

Heidegger, M. (1977). *The question concerning technology and other essays*. New York: Harper & Row.

Henning, E., Van Ransburg, W., & Smit, B. (2004). *Finding your way in a Qualitative Research*. Pretoria: Van Schaik.

Herr, K., & Anderson, G. L. (2005). *The action research dissertation*. Thousand Oaks, CA: Sage.

Higgs, P. (2016). The African Renaissance and the decolonisation of the curriculum. In V. Msila & M. T. Gumbo (Eds.), *Africanising the Curriculum: Indigenous Perspective and Theories*. African Sun Media Press.

Higher Education Sector. (2010). *HIV Prevalence and Related Factors*. Academic Press.

Hill, H. C. (2002). *Learning: A survey of psychological interpretations* (7th ed.). Boston: Allyn & Bacon.

Hirsch, B. J., Deutsch, N. L., & DuBois, D. L. (2011). *After-school centers and youth development: Case studies of success and failure*. New York, NY: Cambridge University Press. doi:10.1017/CBO9780511994913

HISP UIO Strategy 2014-2016. (n.d.). Retrieved from http://www.mn.uio.no/ifi/english/research/networks/hisp

Hmelo-Silver, C. E., Duncan, R. G., & Chinn, C. A. (2007). Scaffolding and achievement in problem-based and inquiry learning: A response to Kirschner, Sweller, and Clark (2006). *Educational Psychologist*, *42*(2), 99–107. doi:10.1080/00461520701263368

Hofman, N. G., & Rosing, H. (2006). *Pedagogies of Praxis: Course-based action research in the social sciences*. Anker Publishing Company, Incorporated.

Hofstee, E. (2006). *Constructing a good dissertation: a practical guide to finishing a master's, MBA or PhD on schedule*. Sandton: EPE.

Holwell, S. (2004). Themes, iteration and recoverability in action research. In B. Kaplan (Eds.), *Action Research* (pp. 353–362). Kluwer Academic Publishers. doi:10.1007/1-4020-8095-6_20

Household Survey. (2015). Household Survey by student volunteers, Household Survey conducted by the 'One Million SoUL' project in Palghar District of Maharashtra from 15th to 17th of November 2014, with help of Symbiosis Centre for Information Technology [SCIT]. Retrieved Sep 15, 2016, from: http://www.millionsoul.iitb.ac.in/outreach_blogs.html

Howell, J. B., & Saye, J. W. (2016). Using lesson study to develop a shared professional teaching knowledge culture among 4th grade Social Studies teachers. *Journal of Social Studies Research*, *40*(1), 25–37. doi:10.1016/j.jssr.2015.03.001

Huber, U., & Gould, W. (2003). *The effect of orphanhood on primary school attendance reconsidered: the power of female-headed households in Tanzania*. HIV/AIDS Impact on Education Clearinghouse. Retrieved from http://iiep.tomoye.com/ev.php

Huizingh, E. K. R. E. (2011). Open innovation: State of the art and future perspectives. *Technovation*, *31*(1), 2–9. doi:10.1016/j.technovation.2010.10.002

Hunzicker, J. (2010). Effective professional development for teachers: A checklist. *Professional Development in Education, 37*(2), 177-179.

Hussein, K., & Nelson, J. (2016). *Sustainable livelihoods and livelihood diversification*. Academic Press.

Huxham, C., & Vangen, S. (2003, July). Researching organizational practice through action research: Case studies and Design Choices. *Organizational Research Methods, 6*(3), 383–403. doi:10.1177/1094428103254454

Hyland, N. E. (2009). Opening and closing communicative space with teachers investigating race and racism in their own practice. *Action Research, 7*(3), 335–354. doi:10.1177/1476750309336717

Illeris, K. (2003). Workplace learning and learning theory. *Journal of Workplace Learning, 15*(4), 167–178. doi:10.1108/13665620310474615

Initiatives of Change. (2013). *Friends of Moral Re-Armament (India), an educational public charitable trust established in 1961 and registered under the Bombay Public Trust Act, 1950*. Panchgani, Maharashtra, India. Retrieved Sep 15, 2016, from: http://in.iofc.org/

Integrated Strategic Planning Framework for Teacher Education and Development. (2011). *Together, taking responsibility for teacher education and development. Department of Basic education and Department of Higher Education and Training*. Pretoria: Government Press.

International Economic Development Council. (2011a). *Introduction to economic development*. Washington, DC: International Economic Development Council.

International Economic Development Council. (2011b). *Neighbourhood Development Strategies*. Washington, DC: International Economic Development Council.

Irby, D., Mawhinney, L., & Thomas, K. (2013). Re-examining Participatory Research in dropout prevention planning in urban communities. *Educational Action Research, 21*(2), 267–283. doi:10.1080/09650792.2013.789732

Isdale, K. (2016, August). *Smooth, staggered or stopped? Educational transitions in the South African Youth Panel Study*. Labour Market Intelligence Partnership-Human Sciences Research Council (LMIP-HSRC).

Israel, B.A., Schulz, A.J., Parker, E.A., & Becker, A.B. (1998). Review of community-based research: assessing partnership approaches to improve public health. *Annual Review of Public Health, 19*, 173-202.

Israel, B. A., Schulz, A. J., Parker, E. A., & Becker, A. B. (1998). Review of community-based research: Assessing partnership approaches to improve public health. *Annual Review of Public Health, 19*(1), 173–202. doi:10.1146/annurev.publhealth.19.1.173 PMID:9611617

ITEA. (2000). *Standards for technological literacy: Content for the study of technology*. Retrieved February 9, 2016 from: http://scimathmn.org/wp-content/uploads/2013/08/tech_standards.pdf

Jacobs, C. (2007). Towards a critical understanding of the teaching of discipline-specific academic literacies: Making the tacit explicit. *Journal of Education, 41*, 59–81.

Jacobs, J., Yamamura, E., Guerra, P. L., & Nelson, S. W. (2013). Emerging leaders for social justice: Negotiating the journey through action research. *Journal of School Leadership, 23*, 91–121.

Jaeger, P. T. (2013). Internet Justice: Reconceptualizing the Legal Rights of Persons with Disabilities to Promote Equal Access in the Age of Rapid Technological Change. *Review of Disability Studies, 9*(1), 39–59.

Janes, J. E. (2016). Democratic encounters? Epistemic privilege, power, and community-based participatory action research. *Action Research, 14*(1), 72–87. doi:10.1177/1476750315579129

Janosik, S. M., Carpenter, S., & Creamer, D. G. (2006). Beyond Professional Preparation Programs: The Role of Professional Associations in Ensuring a High Quality Workforce. *College Student Affairs Journal, 25*(2), 228–237.

Jayadev, A., Motiram, S., & Vakulabharanam, V. (2007). Patterns of wealth disparities in India during the liberalisation era. *Economic and Political Weekly*, 3853–3863.

Jensen, B., & Sonnemann, J. (2014). *Turning Around Schools: It Can Be Done*. Melbourne: Grattan Institute.

Jensen, K. L., Iipito, H. N., Onwordi, M. U., & Mukumbira, S. (2012). Toward an mPolicing solution for Namibia: leveraging emerging mobile platforms and crime mapping. *Proceedings of the South African Institute for Computer Scientists and Information Technologists Conference* (pp. 196 - 205). New York, NY: ACM. doi:10.1145/2389836.2389860

Jita, L. C., & Mokhele, M. L. (2013). Shifting research identities through communities of practice: Reflections on a faculty development programme for teacher education in South Africa. *Progressio. South African Journal for Open and Distance Learning Practice*, *35*(2), 126–142.

Johnsey, R. (1997). Improving Children's Performance in the Procedure of Design and Technology. *Journal for Design and Technology Education*, *2*(3), 201–207.

Johnson-Burel, D., Drame, E., & Frattura, E. (2014). Participatory research in support of quality public education in New Orleans. *Educational Action Research*, *22*(3), 288–305. doi:10.1080/09650792.2013.872576

Johnson, D. R. (2012). Campus racial climate perceptions and overall sense of belonging among racially diverse women in STEM majors. *Journal of College Student Development*, *53*(2), 336–346. doi:10.1353/csd.2012.0028

Johnson, M. (2012). *Business Consulting: What you Need to Know For IT Operations Management*. Emero Publishing.

Johnston, M. (2005). The lamp and the mirror: Action research and self studies in the social studies. In K. Barton (Ed.), *Research methods in social studies education: Contemporary issues and perspectives* (pp. 57–83). Greenwich, CT: Information Age Publishing.

Jones, M.G. & Brader-Araje, L. (2002). The Impact of Constructivism on Education: Language, Discourse, and Meaning. *American Communication Journal, 5*(3).

Jones, A. (2003). The development of a national curriculum in Technology for New Zealand. *International Journal of Technology and Design Education*, *13*(1), 83–99. doi:10.1023/A:1022355410425

Jones, A. (2007). Reflecting on the development of a new school subject: The development of Technology Education in New Zealand. *Waikato Journal of Education*, *13*, 273–294.

Jones, L., & Wells, K. (2007). Strategies for academic and clinician engagement in community-participatory partnered research. *Journal of the American Medical Association*, *297*(4), 407–410. doi:10.1001/jama.297.4.407 PMID:17244838

Juahir, H., Toriman, M. E., Azid, A., Abdullah, A. M., Aziz, N. A. A., & Kamarudin, M. K. A. (2014). Young Environmental Scientist (YES) program to enhance the environmental awareness among the Malaysian school students. In *INTED2014 Proceedings* (pp. 7122-7128). IATED.

Kalanda, K. (2005). *Factors influencing college students' attitudes toward Technology in Lesotho* (Unpublished MEd dissertation). UNISA, Pretoria, South Africa.

Kalanda, K., & De Villiers, M. R. (2013). E-Learning in the Science curriculum: A study in selected high schools in Lesotho. In J. Herrington (Ed.), *Proceedings of World Conference on Educational Media and Technology* (pp. 1664-1673). Victoria, Canada: Association for the Advancement of Computing in Education (AACE).

Kamal, R. (2014). The ethical conundrums of Community Engagement in South Arica. *Unisa e-news*. Available online: https://staff.unisa.ac.za/e-news/2014/07/09/the-ethical-conundrums-of-ce-in-south-africa

Kamuju, N. (2016). Soil Erosion and Sediment Yield Analysis Using Prototype & Enhanced SATEEC GIS System Models. *International Journal of Advanced Remote Sensing and GIS*, *5*(1), 1471–1482. doi:10.23953/cloud.ijarsg.39

Kankanhalli, A., Tanudidjaja, F., Sutanto, J., & Tan, B. (2003). The role of IT in successful knowledge management initiatives. *Communications of the ACM*, *46*(9), 69–73. doi:10.1145/903893.903896

Kapila, U. (2009). *Indian Economy since independence*. Academic Foundation.

Kapoor, D., & Jordan, S. (Eds.). (2009). *Education, Participatory Action Research, and social change: International perspectives*. New York, NY: Palgrave Macmillan. doi:10.1057/9780230100640

Kassam, A. (2002). Ethnotheory, ethnopraxis: Ethnodevelopment in the Oromia regional state of Ethiopia. In P. Sillitoe, A. Bicker, & J. Pottier (Eds.), *Participating in development: Approaches to indigenous knowledge* (pp. 64–81). New York, NY: Routledge. doi:10.4324/9780203428603_chapter_4

Keesstra, S. D., Bouma, J., Wallinga, J., Tittonell, P., Smith, P., Cerdà, A., & Fresco, L. O. et al. (2016). The significance of soils and soil science towards realization of the United Nations Sustainable Development Goals. *Soil*, *2*(2), 111–128. doi:10.5194/soil-2-111-2016

Keiny, S. (2004). *Ecological thinking. A new approach to educational change*. Boston: University Press of America.

Kelly, M. J. (2000). *The encounter between HIV/AIDS and education*. Harare, Zimbabwe: UNESCO, SubRegional Office for Southern Africa.

Kemmis, S., & McTaggart, R. (1988). *The action research planner* (3rd ed.). Victoria, Australia: Deakin University.

Kemmis, S., & McTaggart, R. (1988). *The Action Research Planner*. Victoria, Australia: Deakin University Press.

Kemmis, S., & McTaggart, R. (2003). Participatory action research. In N. K. Denzin & Y. S. Lincoln (Eds.), *Strategies of qualitative inquiry* (pp. 151–183). Thousand Oaks, CA: Sage.

Kemmis, S., & McTaggart, R. (2005). Participatory Action Research: Communicative action and the public sphere. In N. K. Denzin & Y. S. Lincoln (Eds.), *Strategies of qualitative inquire* (3rd ed.; pp. 271–330). Thousand Oaks, CA: Sage Publications.

Kesby, M. (2000). Participatory diagramming as a means to improve communication about sex in rural Zimbabwe: A pilot study. *Social Science & Medicine*, *50*(12), 1723–1741. doi:10.1016/S0277-9536(99)00413-X PMID:10798328

Kezar, A., & Rhoads, R. A. (2001). The dynamic tensions of service learning in higher education: A philosophical perspective. *The Journal of Higher Education*, *72*(2), 148–171. doi:10.2307/2649320

Kilpatrick, L., & Loughland, T. (2012). Critical literacy and interactivity year 1 in a classroom. *Scan: The Journal for Educators*, *31*(3).

Kimaro, H. C., & Titlestad, O. H. (2008). Challenges of user participation in the design of a computer based system: The possibility of participatory customisation in low income countries. *Journal of Health Informatics in Developing Countries*, *2*(1), 1–9.

Kindon, S., Pain, R., & Kesby, M. (Eds.). (2007). *Participatory Action Research approaches and methods: Connecting people, participation, and place*. New York, NY: Routlege.

Knoblich, G., Butterfill, S., & Sebanz, N. (2011). 3 Psychological research on joint action: Theory and data. *Psychology of Learning and Motivation-Advances in Research and Theory*, *54*, 59–101. doi:10.1016/B978-0-12-385527-5.00003-6

Koch, T., Selim, P., & Kralik, D. (2002). Enhancing lives through the development of a community-based, participatory action research programme. *Journal of Clinical Nursing*, *11*(1), 109–112. doi:10.1046/j.1365-2702.2002.00563.x PMID:11845746

Kohli, A. (2006). Politics of economic growth in India, 1980-2005: Part I: The 1980s. *Economic and Political Weekly*, 1251–1259.

Kosinksi, S. (2009). State of uncertainty: Citizenship, statelessness, and discrimination in the Dominican Republic. *Boston College International and Comparative Law Review*, *32*(2), 377–398.

Kothari, A., Mohanty, P., Yadav, R., & Sisodia, T. (2016). A Study of Social Media Usage in Rural Development Schemes of Government of India. *International Journal of Engineering Applied Sciences and Technology*, *2*(2), 1–5.

Kremer, M., Brannen, C., & Glennerster, R. (2013). The challenge of education and learning in the developing world. *Science*, *340*(6130), 297–300. doi:10.1126/science.1235350 PMID:23599477

Krishnamurthy, S., Joseph, S., & Bharathi, V. (2014b). Creating environment friendly projects in rural India–A synergy framework for sustainable renewable energy. *International Journal of Applied Engineering Research*, *9*(24), 26719–26738.

Krishnamurthy, S., Joseph, S., & Bharathi, V. (2014b). Creating Environment Friendly Projects in Rural India–A Synergy Framework for Sustainable Renewable Energy. *International Journal of Applied Engineering Research*, *9*(24), 26719–26738.

Krishnamurthy, S., Joseph, S., Bharathi, V., & Pradhan, V. (2014a). Inclusion of sustainability education in business schools – An Indian B-school case study of MBA-ITBM curriculum. *International Journal of Applied Engineering Research*, *9*(23), 22703–22725.

Krishnamurthy, S., Joseph, S., Bharathi, V., & Pradhan, V. (2014a). Inclusion of Sustainability Education in Business Schools–An Indian B-School Case Study of MBA-ITBM Curriculum. *International Journal of Applied Engineering Research*, *9*(23), 22703–22725.

Kumar, K. L. (2002). *Impact of new technology on teaching and learning in technology education*. Paper presented at the Design and Technology Association International Research Conference.

Kuwahara, J. L. (2013). Impacts of a place-based science curriculum on student place attachment in Hawaiian and Western cultural institutions at an urban high school in Hawaii. *International Journal of Science and Mathematics Education*, *11*(1), 191–212. doi:10.1007/s10763-012-9387-3

Kwaira, P., & Gumbo, M. T. (2014). Applying Design and Technology Education in addressing farmers' problems in the Makonde Rural District, Zimbabwe. *Journal of Human Ecology (Delhi, India)*, *48*(1), 103–113.

Lackritz, J. R. (2004). Exploring burnout among university faculty: Incidence, performance, and demographic issues. *Teaching and Teacher Education*, *20*(7), 713–729. doi:10.1016/j.tate.2004.07.002

Ladegård, G. (2011). Stress management through workplace coaching: The impact of learning experiences. *International Journal of Evidence Based Coaching and Mentoring*, *9*, 29–43.

Ladson-Billings, G. (1998). Teaching in dangerous times: Culturally relevant approaches to teacher assessment. *The Journal of Negro Education*, *67*(3), 255–267. doi:10.2307/2668194

Lahman, M., & Park, S. (2004). Understanding children from diverse cultures: Bridging perspectives of parents and teachers. *International Journal of Early Years Education*, *12*(2), 131–142. doi:10.1080/0966976042000225525

Lamoureux, S. T. (2013). *Five Trends in Leveraging Leadership Development to Drive a Competitive Advantage.* Bersin Research Bulletin.

Landerholm, E., Gehrie, C., & Hao, Y. (2004). Educating early childhood teachers for the global world. *Early Child Development and Care, 174*(7–8), 593–606. doi:10.1080/0300443042000187095

Landsberg, S., Ozment, S., Stickler, M., Henninger, N., Treweek, J., Venn, O., & Mock, G. (2011). *Ecosystem services review for impact assessment: Introduction and guide to scoping.* WRI Working Paper. World Resources Institute.

Lave, L., & Wenger, E. (2014). Jean Lave, Etienne Wenger and communities of practice, Conference: innovation in Youth Work – Creative practice in challenging times. YMCA George Williams College.

Lave, J., & Wenger, E. (1991). *Situated learning: Legitimate peripheral participation.* Cambridge, UK: Cambridge University Press. doi:10.1017/CBO9780511815355

Le Roux, J. J., Newby, T. S., & Sumner, P. D. (2007). Monitoring soil erosion in South Africa at a regional scale: Review and recommendations. *South African Journal of Science, 103*, 329–335.

Lee, K. (2015). *Culture of STEM Education.* Presentation at the SOWA 2015 Bridge conference. Retrieved from https://bridge2015.sched.org/event/4AGR/culture-of-stem-education

Leidenfrost, B., Strassnig, B., Schabmann, A., Spiel, C., & Carbon, C. C. (2011). Peer Mentoring Styles and Their Contribution to Academic Success Among Mentees: A Person-Oriented Study in Higher Education. *Mentoring & Tutoring: Partnership in Learning, 19*(3), 347–364. doi:10.1080/13611267.2011.597122

Leijdekkers, S., Marpaung, Y. M., Meesters, M., Naser, A. K., Penninx, M., van Rookhuijzen, M., & Willems, M. (2015). *Effective Interventions on littering behaviour of youngsters. What are the ingredients?.* MSc programme.

Leitch, R., & Day, C. (2000). Action research and reflective practice: Towards a holistic view. *Educational Action Research, 8*(1), 179–193. doi:10.1080/09650790000200108

Letsekha, T. (2013). Revisiting the debate on the Africanisation of higher. *The Independent Journal of Teaching and Learning, 8*, 5–18.

Lewin, K. (1958). *Group decision and social change.* New York: Holt, Rinehart and Winston.

Lewin, K. (1946). Action research and minority problems. *The Journal of Social Issues, 2*(4), 34–46. doi:10.1111/j.1540-4560.1946.tb02295.x

Li-Hua, R. (2009). Definitions of technology. In J. K. B. Olsen, S. A. Pedersen, & V. E. Hendricks (Eds.), *A comparison to the philosophy of technology* (pp. 18–22). Malden, MA: Wiley-Blackwell. doi:10.1002/9781444310795.ch2

Lin, L. (2015). Exploring collaborative learning: Theoretical and conceptual perspectives. In Investigating Chinese HE EFL Classrooms (pp. 11-28). Heidelberg, Germany: Springer.

Lindsey, E., Stajduhar, K., & McGuinness, L. (2001). Examining the process of community development. *Journal of Advanced Nursing, 33*(6), 828–835. doi:10.1046/j.1365-2648.2001.01722.x PMID:11298221

Liu, J., Dietz, T., Carpenter, S. R., Alberti, M., Folke, C., Moran, E., & Taylor, W. W. et al. (2007). Complexity of coupled human and natural systems. *Science, 317*(5844), 1513–1516. doi:10.1126/science.1144004 PMID:17872436

Lleras-Muney, A. (2005). The relationship between education and adult mortality in the United States. *The Review of Economic Studies, 72*(1), 189–221. doi:10.1111/0034-6527.00329

Locke, T., Alcorn, N., & ONeill, J. (2013). Ethical issues in collaborative action research. *Educational Action Research, 21*(1), 107–123. doi:10.1080/09650792.2013.763448

Lotz-Sisitka, H. (2004). Environmental education research and social change: Southern African perspectives. *Environmental Education Research, 10*(3), 291–295. doi:10.1080/1350462042000258143

Love, T. (2015). *An examination of preparation experiences that influence technology and engineering educators' teaching of science concepts.* Presentation at the 102nd Mississippi Valley Technology Teacher Education Conference, Nashville, TN.

Lyons, M. (1998). *The impact of HIV and AIDS on children, families and communities: risks and realities of childhood during the HTV epidemic.* United Nations Development Programme: HIV and development programme. Issues Paper 30. Retrieved from www.undp.org/hiv/ publications.html

Mabala, R., & Allen, K. (2002). Participatory action research on HIV/AIDS through a popular theatre approach in Tanzania. *Evaluation and Program Planning.*

MacColl, I., Cooper, R., Rittenbruch, M., & Viller, S. (2005). Watching Ourselves Watching: Ethical Issues in Ethnographic Action Research. *Proceedings of the Annual Conference of the Computer-Human Interaction Special Interest Group (CHISIG) of the Human Factors and Ergonomics Society of Australia (HFESA) (OZCHI)* (pp. 1 - 4). Canberra, Australia: ACM.

Macdonald, D. (2003). Curriculum change and the post-modern world: Is the school curriculum-reform movement an anachronism? *Journal of Curriculum Studies, 35*(2), 139–149. doi:10.1080/00220270210157605

Mack, L. (2010). *The philosophical underpinnings of educational research.* Academic Press.

MacLeod, A., Skinner, M. W., & Low, E. (2012). Supporting hospice volunteers and caregivers through Community-based Participatory Research. *Health & Social Care in the Community, 20*(2), 190–198. doi:10.1111/j.1365-2524.2011.01030.x PMID:21978371

Madhlangobe, L., & Gordon, S. P. (2012). Culturally responsive leadership in a diverse school: A case study of a high school leader. *NASSP Bulletin, 96*(3), 177–202. doi:10.1177/0192636512450909

Mahomed, H. (2004). *Challenges in Curriculum Transformation in South Africa.* Paper presented at the 5th Annual Speaking Conference, Gauteng, South Africa.

Mahoney, J. L., Larson, R., & Eccles, J. S. (Eds.). (2005). *Organized activities as developmental contexts: Extracurricular activities, after-school and community programs.* Mahwah, NJ: Lawrence Erlbaum Associates.

Makgato, M. (2012). Identifying constructivist methodologies and pedagogic content knowledge in the teaching and learning of technology. *Procedia: Social and Behavioral Sciences, 47*, 1398–1402. doi:10.1016/j.sbspro.2012.06.832

Maloney, M. M. (2012). Cultivating Community, Promoting Inclusivity: Collections as Fulcrum for Targeted Outreach. *New Library World, 113*(5–6), 281–289. doi:10.1108/03074801211226364

Maltbia, T. E., Marsick, V. J., & Ghosh, R. (2014). Executive and organizational coaching: A review of insights drawn from literature to inform hrd practice. *Advances in Developing Human Resources, 16*(2), 161–183. doi:10.1177/1523422313520474

Maluleke, H. M. (2015). *Curriculum policy implementation in the South African context, with reference to environmental education within the natural sciences* (D.Ed. thesis). University of South Africa.

Manfra, M. M. (2009). Action research: Exploring the theoretical divide between practical and critical approaches. *Journal of Curriculum and Instruction, 3*(1), 32–46.

Mansour, N., Heba, E.-D., Alshamrani, S., & Aldahmash, A. (2014). Rethinking the theory and practice of continuing professional development: Science teachers perspectives. *Research in Science Education, 44*(6), 949–973. doi:10.1007/s11165-014-9409-y

Mapotse, T. A. (2012). *The teaching practice of senior phase Technology Education teachers in selected schools of Limpopo Province: An Action Research study* (DEd thesis). University of South Africa.

Mapotse, T. A. (2012). *The teaching practice of Senior Phase Technology Education teachers in selected schools of Limpopo Province: An action research study* (Unpublished DEd thesis). UNISA, Pretoria, South Africa.

Mapotse, T. A. (2015). An emancipation framework for technology education teachers: an action research study. *International Journal of Technology and Design Education, 25*(2), 213–225. Available at http://link.springer.com/article/10.1007/s10798-014-9275-y/fulltext.html

Mapotse, T. A. (2015). *From etic action research practitioner to emic action learning facilitator: A community engagement case.* Paper Presented at ALARA World Congress: 9th Action Learning Action Research and 13th Participatory Action Research, Centurion, South Africa.

Mapotse, T. A. (2017). Development of a Technology Education Cascading Theory through community engagement site-based support. *International Journal of Technology and Design Education.* Retrieved from http://download.springer.com

Mapotse, T. A. (2015a). An Emancipation Framework for Technology Education teachers: An Action Research study. *International Journal of Technology and Design Education, 25*(2), 213–225. doi:10.1007/s10798-014-9275-y

Mapotse, T. A. (2015b). A reconnaissance study as a *sine qua non* factor for *status quo* change among senior phase Technology teachers. *African Education Review, 12*(4), 583–599.

Mapotse, T. A. (2016). *Programme in Technology Education. PTECAL-H Study Guide. Centre for Community Training and Development.* Pretoria: Unisa Press.

Mapotse, T. A. (2017a). Development of a Technology Education Cascading Theory through community engagement site-based support. *International Journal of Technology and Design Education.* doi:10.1007/s10798-017-9411-6

Mapotse, T. A. (2017b). *Open and Distance Learning practitioners sustain the teaching of Technology Education through Action Learning: Success stories of ODL institutions. In Open and Distance Learning for Sustainable Growth in Emerging Nations.* IGI Global.

Mapotse, T. A. (2017c). Technology Education teachers' professional development through Action Research. In P. J. Williams & D. Barlex (Eds.), *Contemporary Issues in Technology Education: helping teachers develop research-informed practice.* Singapore: Springer. doi:10.1007/978-981-10-2819-9_11

Maravić, M., Cvjetićanin, S., & Ivković, S. (2014). Level of environmental awareness of students in republic of Serbia. *World Journal of Education, 4*(3), 13–18. doi:10.5430/wje.v4n3p13

Marquardt, M. (2004). *Optimising the power of action learning: Solving problems and building leaders in real time.* London: Davies-Black.

Marquardt, M. J. (1999). *Action learning in action: Transforming problems and people for world class organizational learning.* Palo Alto, CA: Davies-Black Press.

Marquardt, M., Leonard, S., Freedman, A., & Hill, C. (2009). *Action Learning for Developing Leaders and Organizations through Action Learning.* Washington, DC: American Psychological Association Press.

Marshak, R. J., & Heracleous, L. (2005). A discursive approach to organization development. *Action Research*, *3*(1), 69–88. doi:10.1177/1476750305049966

Marsick, V. J. (2003). Invited Reaction: Informal Learning and the Transfer of Learning: How Managers Develop Proficiency. *Human Resource Development Quarterly*, *14*(4), 389–395. doi:10.1002/hrdq.1075

Marsick, V. J., & ONeil, J. (1999). The many faces of action learning. *Management Learning*, *30*(2), 159–177. doi:10.1177/1350507699302004

Martin, S., Brannigan, J., & Hall, A. (2005). Sustainability, systems thinking and professional practice. *Journal of Geography in Higher Education*, *29*(1), 79–89. doi:10.1080/03098260500030389

Masera, O., & López-Ridaura, S. (2000). *Sustentabilidad y Sistemas Campesinos. Cinco experiencias de evaluación en el México rural*. MundiPrensa-GIRA-UNAM.

Maslach, C., & Leiter, M. P. (2005). Reversing burnout. *Stanford Social Innovation*, 43-49.

Mason, J. (2002). *Qualitative researching* (2nd ed.). London: Sage Publications.

Massachusetts Science and Technology/Engineering Curriculum Framework. (2006). *Purpose and nature of science and technology/engineering*. Malden,MA: Massachusetts Department of Education.

Mastropieri, M. A., & Scruggs, T. E. (2013). Inclusive classroom: Strategies for effective differentiated instruction (5th ed.). Prentice Hall.

Mathe, R. (2014). *Effects of the Biographic factors and Religious Convictions on Littering to Enhance Waste Management in Pretoria City* (MSc dissertation). University of South Africa.

Matthews, P. (1999). Workplace learning: Developing a holistic model. *The Learning Organization*, *6*(1), 18–29. doi:10.1108/09696479910255684

Mauser, W., Klepper, G., Rice, M., Schmalzbauer, B. S., Hackmann, H., Leemans, R., & Moore, H. (2013). Transdisciplinary global change research: The co-creation of knowledge for sustainability. *Current Opinion in Environmental Sustainability*, *5*(3–4), 420–431. doi:10.1016/j.cosust.2013.07.001

Mbatha, N. M. (2003). *Impact of Environmental Education on Environmental awareness levels of grade 9 learners in Soweto* (MA dissertation). Rand Afrikaans University.

McEwen, K. E. (2015). *The primary school factors that shape the environmental Knowledge, attitudes, and behaviours of children* (MSc dissertation). University of Waikato.

McGill, I., & Beaty, L. (2001). *Action Learning: a guide for professional, management & educational development*. Psychology Press.

McIntyre, A. (2008). *Participatory Action Research*. Thousand Oaks, CA: Sage Publications, Inc.

McIsaac, E., Park, S., & Toupin, L. (2013). *Shaping the Future: Leadership in Ontario's Nonprofit Labour Force*. Toronto: ONN and The Mowat Centre.

McKay, V. (2014). The ethical conundrums of Community Engagement in South Arica. *Unisa e-news*. Available online: https://staff.unisa.ac.za/e-news/2014/07/09/the-ethical-conundrums-of-ce-in-south-africa

McKay, T. J. M. (2013). Embedding academic support within an academic discipline: A teaching model. *South African Journal of Higher Education*, *27*(3), 682–695.

McKay, T. J. M. (2016). Do tutors matter? Assessing the impact of tutors on first year academic performance at a South African university. *Journal of Student Affairs in Africa*, *4*(1), 53–64. doi:10.14426/jsaa.v4i1.144

McKay, T., & Simpson, Z. (2013). The Space Between: Pedagogic Collaboration between a Writing Centre and an Academic Department. *Perspectives in Education*, *31*(4), 27–43.

McLagan, P. (2015). Preface: Making sense out of complexity. *Advances in Developing Human Resources*, *17*(2), 139–144. doi:10.1177/1523422315572617

McLain, M., Barlex, D., Bell, D., & Hardy, A. (2015). Teacher perspectives on pedagogical modelling and explaining in design & technology: A Q Methodology study. In Plurality and Complementarity of Approaches in Design and Technology Education. Available at: https://hal.archives-ouvertes.fr/hal-01161553

McLain, M., Bell, D., & Pratt, A. (2013). *Show-how know-how (Part 1): Theory and practice for demonstrating in Design and Technology.* Welles Bourne, UK: Design and Technology Association.

McNiff, J. (1988). *Action research: Principles and practice.* London: Routledge. doi:10.4324/9780203305676

McNiff, J. (2013). *Action research: principles and practice* (3rd ed.). Oxon, UK: Routledge.

McTaggart, R. (1991). Principles for Participatory Action Research. *Adult Education Quarterly*, *41*(3), 168–187. doi:10.1177/0001848191041003003

McTaggart, R. (1994). Participatory action research: Issues in theory and practice. *Educational Action Research*, *2*(3), 313–337. doi:10.1080/0965079940020302

Meeussen, L., Delvaux, E., & Phalet, K. (2014). Becoming a group: Value convergence and emergent work group identities. *The British Journal of Social Psychology*, *53*(2), 235–248. doi:10.1111/bjso.12021 PMID:23316715

Mendelsohn, M., Mulholland, E., & Shamshiri, N. (2011). *Strengthening the Third Pillar of the Canadian Union.* Toronto: The Mowat Centre.

Merrell, J. (2000). Ambiguity: Exploring the complexity of roles and boundaries when working with volunteers in well woman clinics. *Social Science & Medicine*, *51*(1), 93–102. doi:10.1016/S0277-9536(99)00442-6 PMID:10817472

Merrill, C. (2017). *Action research and Technology Education.* Retrieved May 18, 2017 from: https://www.iteea.org/File.aspx?id=85497&v=6eec1be

Messiou, K., Ainscow, M., Echeita, G., Goldrick, S., Hope, M., Paes, I., & Vitorino, T. et al. (2016). Learning from differences: A strategy for teacher development in respect to student diversity. *School Effectiveness and School Improvement*, *27*(1), 45–61. doi:10.1080/09243453.2014.966726

Mhina, C. H. (2009). Research and agency: The case of rural women and land tenure in Tanzania. In D. Kapoor & S. Jordan (Eds.), *Education, Participatory Action Research, and social change: International perspectives* (pp. 155–168). New York, NY: Palgrave MacMillan. doi:10.1057/9780230100640_11

Miles, S., & Ainscow, M. (2011). *Responding To Diversity in Schools.* London: Routledge.

Miller, A., Christensen, E. M., Eather, N., Sproule, J., Annis-Brown, L., & Lubans, D. R. (2015). The PLUNGE randomized controlled trial: Evaluation of a games-based physical activity professional learning program in primary school physical education. *Preventive Medicine*, *74*, 1–8. doi:10.1016/j.ypmed.2015.02.002 PMID:25668220

Miller, W. R., & Rollnick, S. (2012). *Motivational interviewing: Helping people change.* Guilford press.

Minkler, M., & Wallerstein, N. (2003). *Community-Based Participatory Research for Health.* San Francisco, CA: Jossey-Bass.

Minkler, M., & Wallerstein, N. (2005). Improving health through community organization and community building: A health education perspective. In M. Minkler (Ed.), *Community organizing and community building for health* (2nd ed.; pp. 26–50). New Brunswick, NJ: Rutgers University Press.

Mirowsky, J., & Ross, C.E. (2004, Oct.). Education, social status, and health. *Canadian Journal of Sociology,* 242.

Mirtl, M., Orenstein, D. E., Wildenberg, M., Peterseil, J., & Frenzel, M. (2013). Development of LTSER platforms in LTER-Europe: challenges and experiences in implementing place-based long-term socio-ecological research in selected regions. In *Long Term Socio-Ecological Research* (pp. 409–442). Springer Netherlands. doi:10.1007/978-94-007-1177-8_17

Misa, T. J. (2009). History of technology. In J. K. B. Olsen, S. A. Pedersen, & V. E. Hendricks (Eds.), *A companion of the philosophy of technology* (pp. 7–17). Malden,MA: Wiley-Blackwell. doi:10.1002/9781444310795.ch1

Mitra, S. K., & Singh, V. B. (1999). *Democracy and social change in India: a cross-sectional analysis of the national electorate.* Sage Publications Pvt. Ltd.

Mlitwa, N. W., & Koranteng, K. (2013). Integration of ICT onto Curricula in the Western Cape Schools: The Activity Theory Perspective. *The Journal of Community Informatics*, *9*(4). Retrieved from http://ci-journal.net/index.php/ciej/article/view/957/1055

Moalosi, R. (2007). *The impact of socio-cultural factors upon human-centred design in Botswana* (Unpublished PhD thesis). Queensland University of Technology, Queensland, Australia.

Moalosi, R., Popovic, V., Hudson, A., & Kumar, K. L. (2005). *Integration of culture within Botswana product design.* Paper presented at International Design Congress, Yunlin, Taiwan.

Moen, F. (2011). Executive coaching and the effect on causal attribution. *International Journal of Coaching in Organizations*, *8*(2), 90–105.

Moini, M., Fackler-Lowrie, N., & Jones, L. (2005). Community Engagement: Moving From Community Involvement to Community Engagement – A Paradigm Shift. Santa Monica, CA: PHP Consulting.

Mokhele, M. L. (2011). *Teachers' perspectives on continuing professional development: A case study of the Mpumalanga Secondary Science Initiative (MSSI) project* (DEd thesis). University of South Africa.

Moll, L. (2000). *Exploring biliteracy and beyond: Developing a case example of educational sovereignty.* Retrieved May 4, 2009 from http://www.gened.arizona.edu/tmiller/comlit/moll.htm

Mollinga, P. (2013). Boundary concepts for interdisciplinary analysis of water resources management in South Asia. *Crop Science, 50*(S1).

Morisky, D., Snehendu, B.K., Chaudrhry, A.S., Chen, K.R., Shaheen, M., & Chickering, K. (2002). Condom use among female sex workers in China: Role of gatekeepers practices in Pakistan. *Pakistan Journal of Nutrition, 1,* 137–142.

Morrow, W. (2007). *Learning to teach in South Africa.* Cape Town: HSRC Press.

Msezane, S. B. (2014). *An exploration of the impact of environmental education innovation on students in Sustaining land resources: A case of Mkhondo village* (MEd dissertation). University of South Africa.

Mumford, E. (1994). New treatments or old remedies: Is business process reengineering really socio-technical design? *The Journal of Strategic Information Systems*, *3*(4), 313–326. doi:10.1016/0963-8687(94)90036-1

Murphy, J. W. (2014). *Community-based interventions: Philosophy and action*. New York, NY: Springer. doi:10.1007/978-1-4899-8020-5

Musick, M. A., & Wilson, J. (2008). *Volunteers: A social profile*. Bloomington, IN: Indiana University Press.

Nair, A., Dahiya, M., Gupta, N., Yadav, R., & Mehta, R. (2016). *Educating Consumers about Digital Wallets* (Vol. 3). IJCSN International Journal of Computer Science and Network.

Nath, V. (1973). Female infanticide and the Lewa Kanbis of Gujarat in the nineteenth century. *The Indian Economic and Social History Review*, *10*(4), 386–404. doi:10.1177/001946467301000404

National Staff Development Committee. (1996). *Action learning in vocational education and training*. Chadstone, Australia: NSDC Press.

Nelson, N., & Wright, S. (1995). Power and participatory development: Theory and practice. Intermediate Technology Publications.

Netting, F. E., OConnor, M. K., Thomas, M. L., & Yancey, G. (2005). Mixing and phasing of roles among volunteers, staff, and participants in faith-based programs. *Nonprofit and Voluntary Sector Quarterly*, *34*(2), 179–205. doi:10.1177/0899764005275204

Newell, S., Robertson, M., Scarborough, H., & Swan, J. (2009). *Managing knowledge work and innovation*. Hampshire, UK: Palgrave Macmillan. doi:10.1007/978-0-230-36641-1

Nguyen, D. J., & Larson, J. B. (2015). Dont forget about the body: Exploring the curricular possibilities of embodied pedagogy. *Innovative Higher Education*, *40*(4), 331–344. doi:10.1007/s10755-015-9319-6

Nguyen, N. C., Graham, D., Ross, H., Maani, K., & Bosch, O. (2012). Educating systems thinking for sustainability: Experience with a developing country. *Systems Research and Behavioral Science*, *29*(1), 14–29. doi:10.1002/sres.1097

Nielsen, P., & Sæbø, J. (2016). Three Strategies for Functional Architecting: Cases from the Health Systems of Developing Countries. *Information Technology for Development*, *22*(1), 134–151. doi:10.1080/02681102.2015.1026304

Niemeijer, D., & de Groot, R. S. (2008). A conceptual framework for selecting environmental indicator sets. *Ecological Indicators*, *8*(1), 14–25. doi:10.1016/j.ecolind.2006.11.012

Nilsson, P. (2008). Recognising the needs – Student teachers' learning to teach. *NORDINA*, *4*(1), 92–106.

NITI Aayog. (2015). *National Institution for Transforming India*. Retrieved from niti.gov.in

Noe, R. A., Clarke, A. D. M., & Klein, H. J. (2014). Learning in the twenty-first century workplace. *Annual Review of Organizational Psychology Organizational Behavior*, *1*(1), 245–275. doi:10.1146/annurev-orgpsych-031413-091321

Nolen, A. L., & Vander Putten, J. (2007). Action research in education: Addressing gaps in ethical principles and practices. *Educational Researcher*, *36*(7), 401–440. doi:10.3102/0013189X07309629

Norris, P. E., O'Rourke, M., Mayer, A. S., & Halvorsen, K. E. (2016). Managing the wicked problem of transdisciplinary team formation in socio-ecological systems. *Landscape and Urban Planning*, 1–8. doi:10.1016/j.landurbplan.2016.01.008

Novak, A. D., & Bartelheim, F. J. (2012). General Education Students' Changing Perceptions of Students with Special Needs. *Current Issues in Education*, *15*(2), 1–10.

O'Hara, S., Beaty, L., Lawson, J., & Bourner, T. (1997). Action learning comes of age-part 2: action learning for whom?. *Education+ Training*, *39*(3), 91-95.10.1108/00400919710164080

O'Neil, J., & Marsick, V. J. (2007). *Understanding action learning*. AMACOM Div American Mgmt Assn.

Oates, B. J. (2006). *Researching Information Systems and Computing*. London: Sage.

Obikeze, D. S. (2011). *Indigenous knowledge systems and the transformation of the academy in Africa: The CULPIP model*. Unpublished paper.

Ogunbure, A. A. (2011). The possibilities of technological development in Africa: An evaluation of the role of culture. *The Journal of Pan African Studies*, *4*(3), 86–100.

ONeil, J., & Marsick, V. J. (2014). Action Learning Coaching. *Advances in Developing Human Resources*, *16*(2), 202–221. doi:10.1177/1523422313520202

Ontario Ministry of Education. (2014). *Achieving Excellence: Ensuring Equity*. Queen's Printer for Ontario. Retrieved from http://www.edu.gov.on.ca/eng/about/ equity.html

Opfer, V. D., & Pedder, D. (2011). Conceptualizing teacher professional learning. *Review of Educational Research*, *81*(3), 376–407. doi:10.3102/0034654311413609

Orenstein, D. E., & Groner, E. (2014). In the eye of the stakeholder: Changes in perceptions of ecosystem services across an international border. *Ecosystem Services*, *8*, 185–196. doi:10.1016/j.ecoser.2014.04.004

Orenstein, D. E., & Shach-Pinsly, D. (2015). *Searching for Spatial Sustainability: One Goal, Multiple Paths?. Annual Conference of the Associations of the European Schools of Planning (AESOP)*, Prague, Czech Republic.

Orock, S. O. (2017). *The Effects of Littering on Tourism in Limbe, Southwest Region of Cameroon*. BSc course.

Ortega, C., & Ortega, R. (1995). Integrated elementary Technology Education. *Technology Teacher*, *54*(5), 11–15.

Overton, H., Wrench, A., & Garrett, R. (2017). Pedagogies for inclusion of junior primary students with disabilities in PE. *Physical Education and Sport Pedagogy*, *22*(4), 414–426. doi:10.1080/17408989.2016.1176134

Owens, K. (2015). Changing the teaching of mathematics for improved Indigenous education in a rural Australian city. *Journal of Mathematics Teacher Education*, *18*(1), 53–78. doi:10.1007/s10857-014-9271-x

Page, S. E. (2007). *The Difference: How the Power of Diversity Creates Better Groups, Firms, Schools and Societies*. Princeton University Press.

Pandor, N. (2008). *Pandor slams SA's failing education system*. University of Witwatersrand. Available online: http://www.mg.co.za/article/2008-08-05-pandor-slams-sas-failing-education-system

Pappamihiel, N. E. (2004). Hugs and smiles: Demonstrating caring in a multicultural early childhood classroom. *Early Child Development and Care*, *174*(6), 539–548. doi:10.1080/0300443042000187059

Park, P. (1999). People, knowledge, and change in participatory research. *Management Learning*, *30*(2), 141–157. doi:10.1177/1350507699302003

Patterson, C. (2008). *Country Report Local Economic Development in South Africa' Prepared for the GTZ strengthening local governance programme's LED component in South Africa*. Retrieved September 10, 2016, from http://www2.giz.de/wbf/4tDx9kw63gma/Country_Report_LED_in_South_Africa.pdf

Pavlova, M. (2005). Social change: How should Technology Education respond? *International Journal of Technology and Design Education*, *15*(3), 199–215. doi:10.1007/s10798-004-5867-2

Pea, R. D. (2004). The social and technological dimensions of scaffolding and related theoretical concepts for learning, education, and human activity. *Journal of the Learning Sciences*, *13*(3), 423–451. doi:10.1207/s15327809jls1303_6

Pedler, M. (1997). Interpreting action learning. *Management learning: Integrating perspectives in theory and practice*, 248-264.

Pedler, M. (2003). *Action learning in practice* (3rd ed.). Aldershot, UK: Gower.

Pedler, M., & Abbott, C. (2008). Am I doing it right? Facilitating action learning for service improvement. *Leadership in Health Services*, *21*(Iss: 3), 185–199. doi:10.1108/17511870810893010

Pedler, M., Burgoyne, J., & Brook, C. (2005). What has action learning learned to become? *Action Learning*, *2*(1), 49–68. doi:10.1080/14767330500041251

Pelech, J., & Pieper, G. W. (2010). *The comprehensive handbook of constructivist teaching: From theory to practice*. Charlotte, NC: Information Age Pub.

Pemsel, A. C. (1891). Commissioners for Special Purposes of Income Tax v. Academic Press.

Petrovic, O., Kittl, C., & Teksten, R. D. (2001). *Developing business models for ebusiness*. Available at SSRN 1658505.

Pickett, K., & Vanderbloemen, L. (2015). *Mind the gap: tackling social and educational inequality*. Cambridge Primary Review Trust.

Pickett, T. A., Kolasa, J., & Jones, C. G. (1994). *Ecological understanding. The nature of theory and theory of nature*. San Diego, CA: Academic Press.

Pilkington, C., & Gelderblom, H. (2010). Using the Karplus learning cycle to teach learners with ADHD introductory computer programming. *African Journal of Research in MST Education*, *14*(1), 73–84.

Pineteh, E. A. (2014). The Academic Writing Challenges of Undergraduate Students: A South African Case Study. *International Journal of Higher Education*, *3*(1), 12–22.

Plantin, J., Lagoze, C., Edwards, P., & Sandvig, C. (2016). Infrastructure studies meet platform studies in the age of Google and Facebook. *New Media & Society*. Available at: http://nms.sagepub.com/content/early/2016/08/02/1461444816661553.abstract

Plummer, R., & Armitage, D. (2007). A resilience-based framework for evaluating adaptive co-management: Linking ecology, economics and society in a complex world. *Ecological Economics*, *61*(1), 62–74. doi:10.1016/j.ecolecon.2006.09.025

Pollock, N., Williams, R., & DAdderio, L. (2007). Global Software and its Provenance: Generification Work in the Production of Organizational Software Packages. *Social Studies of Science*, *37*(2), 254–280. doi:10.1177/0306312706066022

Ponte, P., Beijard, D., & Ax, J. (2004). Don't wait till the cows come home: Action research and initial teacher education in three different countries. *Teachers and Teaching*, *10*(6), 591–621. doi:10.1080/1354060042000304792

Portelli, J. P. (1993). Exposing the Hidden Curriculum. *Journal of Curriculum Studies*, *25*(4), 343–358. doi:10.1080/0022027930250404

Potgieter, C. (2004). The impact of the implementation of technology education on in-service teacher education in South Africa. *International Journal of Technology and Design Education*, *14*(3), 205–218. doi:10.1007/s10798-004-2270-y

Poverty, E. (2015). *Millennium development goals*. United Nations. Retrieved August 20, 2016, from: http://www.un.org/millenniumgoals/

Pradhan, V. S., Sharma, R., & Krishnamurthy, S. (2016), (in Press). Analyses of factors affecting employee perception on organizational Corporate Social Responsibility, SCIT Journal 2016, Volume XVI, August 2016 (ISSN 0974-5076)

Pradhan, V. S., Sharma, R., & Krishnamurthy, S. (2016). Analyses of factors affecting employee perception on organizational Corporate Social Responsibility, SCIT Journal 2016, Volume XVI, August 2016 (ISSN 0974-5076)

Prahalad, C. K., & Ramaswamy, V. (2004). Co-creation experiences: The next practice in value creation. *Journal of Interactive Marketing*, *18*(3), 5–14. doi:10.1002/dir.20015

Pretty, J. N. (1995). Participatory learning for sustainable agriculture. *World Development*, *23*(8), 1247–1263. doi:10.1016/0305-750X(95)00046-F

Price, J. N. (2001). Action research, pedagogy and change: The transformative potential of action research in pre-service teacher education. *Journal of Curriculum Studies*, *33*(1), 43–74. doi:10.1080/00220270118039

Prince, M. J., & Felder, R. M. (2006). Inductive teaching and learning methods: Definitions, comparisons, and research bases. *The Journal of Engineering Education*, *95*(2), 123–138. doi:10.1002/j.2168-9830.2006.tb00884.x

Principles of Community Engagement. (2011). *Clinical and Translation Science Awards Consortium; Community Engagement Key Function Committee Task; Force on the Principle of Community Engagement* (2nd ed.). National Institute of Health.

Pritchett, L., & Beatty, A. (2012). *The Negative Consequences of Overambitious Curricula in Developing Countries.* CGD Working Paper 293. Washington, DC: Centre for Global Development.

Pudi, T. I. (2005). Educator roles for Technology Education teacher- educator. *Africa Education Review*, *34*(1), 147–167. doi:10.1080/18146620508566297

Pudi, T. I. (2006). *Technology education (Study Guide).* Pretoria, South Africa: UNISA Press.

Pudi, T. I. (Ed.). (2007). *Understanding Technology Education from a South African perspective.* Pretoria: Van Schailk.

Pule, S. (2017). Mixed Media Modelling of Technological Concepts in Electricity. In P. J. Williams & D. Barlex (Eds.), *Contemporary Issues in Technology Education: helping teachers develop research-informed practice*. Singapore: Springer. doi:10.1007/978-981-10-2819-9_6

Putnam, R. W. (2014). Theories of Action. In D. Coghlan & M. Brydon Miller (Eds.), *The Sage Encyclopedia of Action Research* (pp. 778–781). London, UK: Sage Publications Ltd. Retrieved from http://actionsmithnetwork.net/wp-content/uploads/2015/09/Theories-of-Action-Putnam-2014.pdf

Pyles, L. (2015). Participation and other ethical considerations in Participatory Action Research in post-earthquake rural Haiti. *International Social Work*, *58*(5), 628–645. doi:10.1177/0020872815581912

Qablan, A., Mansour, N., Alshamrani, S., Aldahmash, A., & Sabbah, S. (2015). Ensuring effective impact of continuing professional development: Saudi science teachers' perspective. *Eurasia Journal of Mathematics, Science & Technology Education*, *11*(3), 619–631.

Quasthoff, U., & Wild, E. (2014). Learning in context from an interdisciplinary perspective. *Learning. Cultures and Social Interaction*, *3*(2), 69–76. doi:10.1016/j.lcsi.2013.12.004

Queensland Department of Emergency Services. (2001). *Charter for community engagement, Community Engagement Unit, Strategic and Executive Services. The State of Victoria, 1996-2017.* Author.

Quick, H. E., Holtzman, D. J., & Chaney, K. R. (2009). Professional development and instructional practice: Conceptions and evidence of effectiveness. *Journal of Education for Students Placed at Risk*, *14*(1), 45–71. doi:10.1080/10824660802715429

Quinn, J. (2005). Building effective practices and policies for out-of school time. In J. L. Mahoney, R. Larson, & J. S. Eccles (Eds.), *Organized activities as developmental contexts: Extracurricular activities, after-school and community programs* (pp. 479–495). Mahwah, NJ: Lawrence Erlbaum Associates.

Quirke, P. (n.d.). *Action learning & research.* Retrieved from http://philseflsupport.com/actionlearning.htm

Rabie, B. (2011). *Improving the systematic evaluation of local economic development results in South African local government* (Unpublished PhD thesis). Stellenbosch University, Stellenbosch.

Rai, R. K. (2012). A Participatory Action Research training initiative to improve police effectiveness. *Action Research, 10*(3), 225–243. doi:10.1177/1476750312439901

Rajan, R. G. (2011). *Fault lines: How hidden fractures still threaten the world economy.* Princeton University Press. doi:10.1515/9781400839803

Rakotomamonjy, S. N., Jones, J. P. G., Razafimanahaka, J. H., Ramamonjisoa, B., & Williams, S. J. (2015). The effects of environmental education on childrens and parents knowledge and attitudes towards lemurs in rural Madagascar. *Animal Conservation, 18*(2), 157–166. doi:10.1111/acv.12153

Rao, P., & Patil, Y. (2015). Integrating Energy and Environment in Postgraduate Management Education: A Case Study from Symbiosis Institute of International Business, Pune, India. In Implementing Campus Greening Initiatives (pp. 265-276). Springer International Publishing.

Rao, P., & Patil, Y. (2015). Integrating energy and environment in postgraduate management education: A case study from Symbiosis institute of international business, Pune, India.. In *Implementing Campus Greening Initiatives* (pp. 265–276). Springer International Publishing. doi:10.1007/978-3-319-11961-8_20

Rao, P., Patil, Y., & Gupte, R. (2013). Education for sustainable development: trends in Indian business schools and universities in a post liberalization era. In *Sustainability Assessment Tools in Higher Education Institutions* (pp. 417–432). Springer International Publishing. doi:10.1007/978-3-319-02375-5_24

Raudenbush, L., & Marquardt, M. (2008). Growing leaders at the US Department of Agriculture: A case study of leadership development using action learning. *Public Administration Quarterly*, 147–173.

Raymaker, D. M. (2016). Intersections of Critical Systems Thinking and Community-Based Participatory Research: A learning organization example with the autistic communities. *Systemic Practice and Action Research, 29*(5), 405–423. doi:10.1007/s11213-016-9376-5 PMID:27833398

Ray, R., Fisher, D. R., & Fisher-Maltese, C. (2016). School gardens in the city: Does environmental equity help close the achievement gap?. *Du Bois Review: Social Science Research on Race, 13*(2), 379–395. doi:10.1017/S1742058X16000229

RBI. (2016). *Withdrawal of Legal Tender Status for ₹ 500 and ₹ 1000 Notes: RBI Notice* (Revised). Reserve Bank of India.

Ready, D., Conger, J., Hill, L., & Stecker, E. (2010). The anatomy of a high potential. *Business Strategy Review, 21*(3), 52–55. doi:10.1111/j.1467-8616.2010.00685.x

Reason, P., & Bradbury, H. (2001). Introduction: Inquiry and participation in search of a world worthy of human aspiration. In P. Reason & H. Bradbury (Eds.), *Handbook of action research* (pp. 1–45). London: Sage. doi:10.1016/B978-075065398-5/50002-9

Reason, P., & Bradbury, H. (Eds.). (2001). *The SAGE handbook of action research: Participative inquiry and practice* (1st ed.). London: SAGE.

Reeve, J., Bolt, E., & Cai, Y. (1999). Autonomy-supportive teachers: How they teach and motivate students. *Journal of Educational Psychology*, *91*(3), 537–548. doi:10.1037/0022-0663.91.3.537

Regional Bureau for Education in Latin America and the Caribbean. United Nations Educational, Scientific and Cultural Organization. (2011). *Regional report on Education for All in Latin America and the Caribbean.* Retrieved May 1, 2015, from http://unesdoc.unesco.org/ images/0019/001916/191666e.pdf

Reinelt, C. (2006). A Framework for Leadership Investment and Evaluation. In *Investing in Leadership* (vol. 1). Grantmaker's Framework for Understanding Nonprofit Leadership Development.

Republic of South Africa. (1996). *Constitution of the Republic of South Africa Act no. 108 of 1996.* Pretoria: Government Gazette.

Republic of South Africa. (2000). *Local Government: Municipal Systems Act no. 32 of 2000.* Pretoria: Government Gazette.

Republic of South Africa. (2011a). *The National Development Plan.* Pretoria: The Presidency- National Planning Commission.

Republic of South Africa. (2011b). *The New Growth Path: Framework.* Pretoria: Economic Development Department.

Revans, R. W. (2011). ABC of action learning. Farnham, UK: Gower.

Revans, R. W. (1972). Action learning: A management development program. *Personnel Review*, *1*(4), 36–44. doi:10.1108/eb055215

Revans, R. W. (1982). *Origins and Growth of Action Learning*. Bromley: Chartwell Bratt.

Revans, R. W. (2011). *ABC of action learning*. Farnham, UK: Gower.

Reynolds, M. (2009). Environmental ethics. In M. Reynolds, C. Blackmore, & M. J. Smith (Eds.), *The Environmental Responsibility Reader* (pp. 40–51). London: Zed Books and the Open University.

Rhedding-Jones, J. (2002). An undoing of documents and other texts: Toward a critical multi-culturalism in early childhood education. *Contemporary Issues in Early Childhood*, *3*(1), 90–116. doi:10.2304/ciec.2002.3.1.10

Richardson, V. (1997). *Constructivist teacher education: Building new understandings*. Bristol, PA: Psychology Press.

Rich, R., & Chavis, D. (1990). Benefits, costs, incentive management and participation in voluntary organizations: A means to understanding and promoting empowerment. *American Journal of Community Psychology*, *18*(1), 117–149. doi:10.1007/BF00922691

Rippin, A., Booth, C., Bowie, S., & Jordan, J. (2002). A complex case: Using the case study method to explore uncertainty and ambiguity in undergraduate business education. *Teaching in Higher Education*, *7*(4), 429–441. RetrievedSep152016 doi:10.1080/135625102760553928

Risley, C. L., & Drake, L. J. (2007). *Impact of HIV/AIDS on education in the Greater Mekong sub-region.* Workshop Presentation. Siem Reap.

Roberts, J. (2006). Limits to communities of practice. *Journal of Management Studies*, *43*(3), 623–639. doi:10.1111/j.1467-6486.2006.00618.x

Robinsons, M. (2013). A call for teachers. In *The Teacher newspaper. October 2013.* M & G Media Ltd.

Rogers Stanton, C. (2014). Crossing methodological borders: Decolonizing Community-Based Participatory Research. *Qualitative Inquiry*, *20*(5), 573–583. doi:10.1177/1077800413505541

Rogers, R. G., Hummer, R., & Nam, C. B. (2000). *Living and dying in the USA: behavioral, health, and social differentials of adult mortality.* San Diego, CA: Academic Press.

Romme, G., & Dillen, R. (1997). Mapping the landscape of organizational learning. *European Management Journal, 15*(1), 68–78. doi:10.1016/S0263-2373(96)00075-8

Rose, S., Spinks, N., & Canhoto, A. I. (2015). *Management research: Applying the principles.* New York: Routledge.

Rosing, H. (2015). Tracking Culture: The Meanings of Community Engagement Data Collection in Higher Education. *Metropolitan Universities, 26*(2), 147–164.

Rosing, H., Reed, S., Ferrari, J. R., & Bothne, N. J. (2010). Understanding student complaints in the service learning pedagogy. *American Journal of Community Psychology, 46*(3-4), 472–481. doi:10.1007/s10464-010-9338-5 PMID:20859675

Ross, C. E., & Wu, C. (1995). The links between education and health. *American Sociological Review, 60*(5), 719–745. doi:10.2307/2096319

Royal Bank of Canada. (2012). *RBC Diversity Blueprint 2012-2015.* Retrieved from http://www.rbc.com/ diversity/pdf/rbc-diversity-blueprint.pdf

Russell, T., & Munby, H. (1992). *Teachers and Teaching: from classrooms to reflection.* London: Falmer Press.

Saft, E. W., & Pianta, R. C. (2001). Teachers perceptions of their relationships with students: Effects of child age, gender, and ethnicity of teachers and children. *School Psychology Quarterly, 16*(2), 125–141. doi:10.1521/scpq.16.2.125.18698

Sagor, R. (2000). *Guiding school improvement with action research.* Alexandria, VA: Association for Supervision and Curriculum Development.

Saint-Paul, G. (2003). *Information sharing and cumulative innovation in business networks. CEPR Discussion Paper Series (No. 4116).* London: Centre for Economic Policy Research. Available at https://ssrn.com/abstract=480643

Saldaña, J. (2012). *The coding manual for qualitative researchers* (2nd ed.). Thousand Oaks, CA: SAGE Publications.

Salomon, G., & Perkins, D. N. (1998). Individual and social aspects of learning. *Review of Research in Education, 23*(1), 1–24. doi:10.3102/0091732X023001001

Salzmann, O., Ionescu-Somers, A., & Steger, U. (2005). The business case for corporate sustainability: Literature review and research options. *European Management Journal, 23*(1), 27–36. doi:10.1016/j.emj.2004.12.007

Samuels, D. (2014). *The Culturally Inclusive Educator: Preparing for a Multicultural World.* Teachers College Press, Columbia University.

Sandelowski, M. (1986). The problem of rigor in qualitative research. *ANS. Advances in Nursing Science, 8*(3), 27–37. doi:10.1097/00012272-198604000-00005 PMID:3083765

Sandelowski, M. (1991). Telling stories: Narrative approaches in qualitative research. *Journal of Nursing Scholarship, 23*(3), 161–166. doi:10.1111/j.1547-5069.1991.tb00662.x PMID:1916857

Sankaran, S., & Brown, S. (2012). Coaching collaborative creativity and innovation: An action-based method for sustainable innovation learning and development in business organization. In O. Zuber-Skerritt (Ed.), Action research for sustainable development in a turbulent world (pp. 127–150). Bingley: Emerald Publishing.

Santos, D. (2016). Re-signifying participatory action research (PAR) in higher education: What does P stand for in PAR? *Educational Action Research, 24*(4), 635–646. doi:10.1080/09650792.2015.1103658

Sawal, N., Venkateswaran, J., Solanki, C. S., & Narayanan, N. C. (2015, December). Million Solar Urja Lamp Programme: A supply chain experience. In *Industrial Engineering and Engineering Management (IEEM), 2015 IEEE International Conference on* (pp. 342-346). IEEE. doi:10.1109/IEEM.2015.7385665

Schwartz, R., Weinberg, J., Hagenbunch, D., & Scott, A. (2008). *The Voice of Nonprofit Talent: Perceptions of Diversity in the Workplace*. Retrieved on February 10, 2014 from: http://www.cgcareers.org/diversityreport.pdf

Sekhar, K. (2002). Three decades of the National Service Scheme The social work component and development issues. *International Social Work*, *45*(1), 99–114. doi:10.1177/0020872802045001322

Senanayake, S. G. J. N. (2006). Indigenous knowledge as a key to sustainable development. *The Journal of Agricultural Science*, *2*(1), 87–94. doi:10.4038/jas.v2i1.8117

Senge, P. M., Schamer, C. O., Jaworski, J., & Flowers, B. S. (2005). *Presence: Exploring profound change in people, organizations, and society*. New York, NY: Doubleday.

Shackelford, J. L., & Maxwell, M. (2012). Sense of community in graduate online education: Contribution of learner to learner interaction. *International Review of Research in Open and Distance Learning*, *13*(14), 228–250. doi:10.19173/irrodl.v13i4.1339

Sharma, G., & Sturges, P. (2007). Making use of ICT to facilitate poor peoples access to public services: An action research programme. *Information Development*, *23*(1), 15–23. doi:10.1177/0266666907075623

Shaw, C. F. (1930). Is "Pedology" Soil Science?. Proceedings of: American Soil Survey Association.

Shephard, K. (2008). Higher education for sustainability: Seeking affective learning outcomes. *International Journal of Sustainability in Higher Education*, *9*(1), 87–98. doi:10.1108/14676370810842201

Sherman, S., & Freas, A. (2004). The wild west of executive coaching. *Harvard Business Review*, *82*(11), 82–93. PMID:15559448

Shipley, R., & Utz, S. (2012). Making it count: A review of the value and techniques for public consultation. *Journal of Planning Literature*, *27*, 22–42.

Shulman, L. S. (1986). Those who understand knowledge growth in teaching. *Educational Researcher*, *15*(2), 4–14. doi:10.3102/0013189X015002004

Sibanyoni, J. J., Tshabalala, P. A., & Tabit, F. T. (2017). Food safety knowledge and awareness of food handlers in school feeding programmes in Mpumalanga, South Africa. *Food Control*, *73*, 1397–1406. doi:10.1016/j.foodcont.2016.11.001

Silver, H. F., Strong, R. W., & Perini, M. J. (2000). So each may learn: Integrating learning styles and multiple intelligences. Association for Supervision and Curriculum Development.

Simalumba, P. M. (2011). *The implementation of environmental learning in grades 8-10 geography in the Caprivi region. Namibia* (MEd dissertation). University of South Africa.

Simpson, Z., & McKay, T. J. M. (2013). Assessment Rubrics: Artefacts That Speak In Tongues? *Lingua*, *29*(1), 15–32. doi:10.5785/29-1-539

Simsekli, Y. (2015). An implementation to raise environmental awareness of elementary education students. *Procedia: Social and Behavioral Sciences*, *191*, 222–226. doi:10.1016/j.sbspro.2015.04.449

Singh, S. J., Haberl, H., Chertow, M., Mirtl, M., & Schmid, M. (Eds.). (2012). *Long term socio-ecological research: studies in society-nature interactions across spatial and temporal scales* (Vol. 2). Springer Science & Business Media.

Sipos, Y., Battisti, B., & Grimm, K. (2008). Achieving transformative sustainability learning: Engaging head, hands and heart. *International Journal of Sustainability in Higher Education*, *9*(1), 68–86. doi:10.1108/14676370810842193

Skinner, J. (Ed.). (2012). *The interview: An ethnographic approach*. New York, NY: Bloomsbury Academic.

Skoglund, A. (2006). Do not forget about your volunteers: A qualitative analysis of factors influencing volunteer turnover. *Health & Social Work*, *31*(3), 217–220. doi:10.1093/hsw/31.3.217 PMID:16955660

Slåtten, T., Svensson, G., & Sværi, S. (2011). Service quality and turnover intentions as perceived by employees: Antecedents and consequences. *Personnel Review*, *40*(2), 205–221. doi:10.1108/00483481111106084

Smit, B., & Romm, N. (2013). *Focus group field guide, 500 schools' project: making Schools Better. Unisa, College of Education*. Pretoria: Unisa Press.

Smith, C. L., Lopes, V. L., & Carrejo, F. M. (2011). Recasting paradigm shift: "True" sustainability and complex systems. *Human Ecology Review*, *18*(1), 67–74.

Smith, G. (2015). The Impact of a Professional Development Programme on Primary Teachers Classroom Practice and Pupils Attitudes to Science. *Research in Science Education*, *45*(2), 215–239. doi:10.1007/s11165-014-9420-3

Smith, K., & Sela, O. (2005). Action research as a bridge between pre-service teacher education and in-service professional development for students and teacher educators. *European Journal of Teacher Education*, *28*(3), 293–310. doi:10.1080/02619760500269418

Smith, P. A., & Sharma, M. (2002). Rationalizing the promotion of non-rational behaviors in organizations. *The Learning Organization*, *9*(5), 197–201. doi:10.1108/09696470210442132

Smith, P., & ONeil, J. (2003). A Review of Action Learning Literature 19942000 Part 2. *Journal of Workplace Learning*, *15*(4), 154–166. doi:10.1108/13665620310474606

Smith, W. A., Mustaffa, J. B., Jones, C. M., Curry, T. J., & Allen, W. R. (2016). You make me wanna holler and throw up both my hands! Campus culture, Black misandric microaggressions, and racial battle fatigue. *International Journal of Qualitative Studies in Education*, *29*(9), 1189–1209. doi:10.1080/09518398.2016.1214296

Solanki, C. S., Narayanan, N. C., Venkateswaran, J., Joshi, L., Arora, N., & Rajagopalan, S. (2016). *Concurrent Evaluation Report of Million SoUL Program in India*. IIT Bombay. Retrieved from http://www.millionsoul.iitb.ac.in/Infographics/Four%20States%20Final.pdf

Solis, R. (2015). *Teachers' experiences of learning through a reflective inquiry process focused on the relationship between teaching beliefs and behaviors* (Doctoral dissertation).

Somekh, B. (2006). *Action research: A methodology for change and development*. Berkshire, UK: Open University Press.

Somekh, B. (2006). *Action Research: A methodology for change and development*. New York, NY: Open University Press.

Sommer, R. (2009). Dissemination in Action Research. *Action Research*, *7*(2), 227–236. doi:10.1177/1476750308097028

Standal, O. F., & Rugseth, G. (2014). Practicum in adapted physical activity: A Dewey-inspired action research project. *Adapted Physical Activity Quarterly; APAQ*, *31*(3), 219–239. doi:10.1123/apaq.2013-0105 PMID:25028475

Stanley, A. M., Snell, A., & Edgar, S. (2013). Collaboration as effective music professional development: Success stories from the field. *Journal of Music Teacher Education*.

Stark, J. L. (2014). The potential of Deweyan-inspired action research. *Education and Culture*, *30*(2), 87–101. doi:10.1353/eac.2014.0013

Stark, S. (2006). Using action learning for professional development. *Educational Action Research*, *14*(01), 23–43. doi:10.1080/09650790600585244

Stars in Education. (2011). The Teacher and the Teaching Career. In *Education Handbook: Connecting to make a difference*. Stellenbosch: Argo.

Statistics South Africa. (2011c). Retrieved September 5, 2016, from www.statssa.gov.za/?page_id=4286&id=171

Steier, F. (1991). Introduction: Research as self-reflexivity, self-reflexivity as social process. In F. Steier (Ed.), *Research and reflexivity* (pp. 1–11). Newbury Park, CA: SAGE Publications, Inc. doi:10.1017/CBO9780511553547.001

Steiner, G., & Posch, A. (2006). Higher education for sustainability by means of transdisciplinary case studies: An innovative approach for solving complex, real-world problems. *Journal of Cleaner Production*, *14*(9), 877–890. doi:10.1016/j.jclepro.2005.11.054

Stevens, A. (2006). Technology teacher education in South Africa. In International Handbook of Technology Education: Reviewing the past twenty years. Rotterdam: Sense.

Stevens, D. (2017). Teaching Bioethics: the intersection of values and the application that advances in Technology-Make-Possible. In P. J. Williams & D. Barlex (Eds.), *Contemporary Issues in Technology Education: helping teachers develop research-informed practice*. Singapore: Springer. doi:10.1007/978-981-10-2819-9_9

Stewart, R., & Bhagwanjee, A. (1999). Promoting group empowerment and self-reliance through participatory research: A case study of people with physical disability. *Disability and Rehabilitation*, *21*(7), 338–345. doi:10.1080/096382899297585 PMID:10471164

Steyn, J. (2011). Paradigm Shift Required for ICT4D. In J. Steyn & G. Johanson (Eds.), *ICTs and Sustainable Solutions for the Digital Divide: Theory and Perspectives* (pp. 19–44). Hershey, PA: IGI Global; doi:10.4018/978-1-61520-799-2.ch002

Stokols, D. (2006). Toward a science of transdisciplinary action research. *American Journal of Community Psychology*, *38*(1-2), 63–77. doi:10.1007/s10464-006-9060-5 PMID:16791514

Struyven, K., Dochy, F., & Janssens, S. (2003). Students' perceptions about new modes of assessment in higher education: A review. In *Optimising new modes of assessment: In search of qualities and standards* (pp. 171–223). Springer. doi:10.1007/0-306-48125-1_8

Studer, S., & von Schnurbein, G. (2013). Organizational factors affecting volunteers: A literature review on volunteer coordination. *Voluntas*, *24*(2), 403–440. doi:10.1007/s11266-012-9268-y

Subban, P. (2006). Differentiated instruction: A research basis. *International Education Journal*, *7*(7), 935–947.

Suchman, L. (1988). Designing with the user: book review of Computers and democracy: A Scandinavian challenge. ACM Transactions on Information Systems, 6(2), 173–183.

Sudha, S. S. I. R., & Rajan, S. I. (1999). Female demographic disadvantage in India 1981–1991: Sex selective abortions and female infanticide. *Development and Change*, *30*(3), 585–618.

Surrette, T. N., & Johnson, C. C. (2015). Assessing the Ability of an Online Environment to Facilitate the Critical Features of Teacher Professional Development. *School Science and Mathematics*, *115*(6), 260–270. doi:10.1111/ssm.12132

Sutherland, K. A. (2009). Nurturing undergraduate tutors role in the university teaching community. Mentoring & Tutoring. *Partnership in Learning*, *17*(2), 147–164. doi:10.1080/13611260902860091

Sutinen, E., & Tedre, M. (2010). ICT4D: A Computer Science Perspective. In T. Elomaa, H. Mannila, & P. Orponen (Eds.), *Algorithms and Applications* (Vol. 6060, pp. 221–231). Berlin: Springer-Verlag. doi:10.1007/978-3-642-12476-1_16

Swanepoel, H. (1997). *Community Development. Putting plans into action* (3rd ed.). Cape Town: Juta.

Swan, K. (2003). A constructivist model for thinking about learning online. In J. Bourne & J. C. Moore (Eds.), *Elements of quality online education, practice and direction* (pp. 13–30). Needham, MA: Sloan Center for Online Education.

Swantz, M. L. (2008). Participatory action research as practice. In P. Reason, & H. Bradbury (Eds.), The SAGE Handbook of Action Research: Participative Inquiry and Practice. Sage Publications. doi:10.4135/9781848607934.n8

Symbiosis International University. (2012), Social Activities, http://siu.edu.in/social-initiatives.php, retrieved Aug 2016.

Symbiosis International University. (2015), Vision and Mission, http://siu.edu.in/symbiosis-international-university.php#visi, retrieved Aug 2016.

Tai, F. M. (2014). Exploring multiple intelligences. *The Journal of Human Resource and Adult Learning*, *10*(1), 11.

Tainter, J. A. (2006). Social complexity and sustainability. *Ecological Complexity*, *3*(2), 91–103. doi:10.1016/j.ecocom.2005.07.004

Takavarasha, S., Hapanyengwi, G., Chimanikire, D., & Kabanda, G. (2013, October-December). An IT Project Management Framework for Assessing the Dynamism of Culture under Globalization: Evidence from Zimbabwe. *International Journal of Information Technology Project Management*, *4*(4), 66–81. doi:10.4018/ijitpm.2013100104

Tangkuampien, J. (2013, February). *A Community-based Approach to New Medium Integration in South African Education: A Combination of ICT4D Process Approach and Ethnographic Action Research Techniques* (PhD Thesis). Cape Town, South Africa: University of Cape Town. Retrieved January 28, 2017, from http://open.uct.ac.za/bitstream/handle/11427/10671/thesis_sci_2013_phdthesis_ttangkuampien_jakkaphan.pdf?sequence=1

Tan, K. (2013). *A Framework for Assessment for Learning: Implications for Feedback Practices within and beyond the Gap*. ISRN Education. doi:10.1155/2013/640609

Tavares, C. (2016). Human rights education and the research process; action research as a tool for reflection and change. *Educational Action Research*, *24*(4), 617–634. doi:10.1080/09650792.2015.1124045

Taylor, J. A. (2011). *Introduction to criminal justice for high school students: A curriculum plan* (Unpublished masters thesis). California State University, Sacramento, CA.

Taylor, T., Mallinson, C., & Bloch, K. (2008). Looking for a few good women: Volunteerism as an interaction between two organizations. *Nonprofit and Voluntary Sector Quarterly*, *37*(3), 389–410. doi:10.1177/0899764007310420

Teacher Quality Enhancement Center. (2009). *Co-teaching strategies & examples*. Retrieved May 1, 2015, from http://www.tacoma.uw.edu/sites/default /files/global/ documents/education/co-teaching_definitions_x_examples.pdf

The Teacher. (2013). Understanding the school's location. Principal report by Phumzile Nokuthula Langa. In *The Teacher, Your Guide to Education, October 2013*. M & G Media Ltd.

Thomas, I. (2009). Critical thinking, transformative learning, sustainable education, and problem-based learning in universities. *Journal of Transformative Education*, *7*(3), 245–264. doi:10.1177/1541344610385753

Timperley, H. S. (2005). Distributed leadership: Developing theory from practice. *Journal of Curriculum Studies*, *37*(4), 395–420. doi:10.1080/00220270500038545

Tooley, W. R. (2000). *Political rationality & government mechanisms: Maori education policy in the new millennium* (MA in Education dissertation). Auckland: University of Auckland.

Topping, K. J. (1996). The effectiveness of peer tutoring in further and higher education: A typology and review of the literature. *Higher Education*, *32*(3), 321–345. doi:10.1007/BF00138870

Towell, M. (2013). *How can coaching improve effectiveness of performance management conversations?* (Unpublished doctoral dissertation). Oxford Brookes University, Oxford, UK.

Townsend, E. (1997). Occupation: Potential for personal and social transformation. *Journal of Occupational Science*, *4*(1), 18–26. doi:10.1080/14427591.1997.9686417

Trantopoulos, K., Schlopfer, M., & Helbing, D. (2011). Toward sustainability of complex urban systems through techno-social reality mining. *Environmental Science and Technology-Columbus*, *45*(15), 6231–6232. doi:10.1021/es2020988 PMID:21744879

Turner, R. K., Georgiou, S., & Fisher, B. (2008). *Valuing ecosystem services: The case of multi-functional wetlands.* London, UK: Earthscan. Retrieved from http://books.google.co.il/books?id=ZacCAWBKHW8C

Tushman, M. L., OReilly, C. A., Fenollosa, A., Kleinbaum, A. M., & McGrath, D. (2007). Relevance and rigor: Executive education as a lever in shaping practice and research. *Academy of Management Learning & Education*, *6*(3), 345–362. doi:10.5465/AMLE.2007.26361625

U.S. Department of State. (2013). *Dominican Republic: Country reports on human rights practices-2012.* Retrieved May 1, 2015, from http://www.state.gov/j/drl/ rls/hrrpt/humanrightsreport/index.htm#wrapper

UNESCO (United Nations Educational, Scientific and Cultural Organisation). (2012). Education for All Global Monitoring Report: Education Counts. France: UNESCO.

UNESCO Institute for Statistics. (2012). *Opportunities lost: The impact of grade repetition and early school leaving.* Retrieved September 1, 2015, from Library/Documents/global-education-digest-opportunities-lost-impact-grade-repetition- early-school-leaving-2012-en.pdf

UNESCO. (1996). *Learning, the Treasure Within: Report to UNESCO of the International Commission on Education for the Twenty-first Century: Highlights.* UNESCO Publication.

UNESCO. (1997). *Educating for a Sustainable Future.* Retrieved from http://www.unesco.org/education/tlsf/TLSF/theme_a/mod01/uncom01t05s01.htm#edu

UNESCO. (1997). United Nations Educational Scientific and Cultural Organization. In *Educating for a Sustainable Future.* Retrieved Oct 3, 2016, from http://www.unesco.org/education/tlsf/TLSF/theme_a/mod01/uncom01t05s01.htm#edu

UNESCO. (2015). *Education for all 2000-2015: achievements and challenges.* Paris: UNESCO.

UNICEF. (2012). *Dominican Republic: Country programme document 2012-2016.* Retrieved May 1, 2015, from http://www.unicef.org/about/execboard/files/ Dominican_Republic_final_aproved_2012-2016_English_10Feb2012.pdf

Unisa Senate Community Engagement Policy. (2013). *Community Engagement Budgets 2014.* Pretoria: Unisa.

United Nations Educational, Scientific and Cultural Organization (UNESCO). (2016). *Education for people and planet: Creating sustainable futures for all. Global education monitoring report.* UNESCO.

United Nations Human Rights Council. (2008). *Report of the Special Rapporteur on contemporary forms of racism, racial discrimination, xenophobia and related intolerance, Doudou Diène, and the independent expert on minority issues, Gay McDougall: addendum: Mission to Dominican Republic.* Retrieved May 1, 2015, from http://www.unhcr.org/refworld/docid/47e23b252.html

United Nations Human Rights Council. (2010). Report *submitted by the Special Rapporteur on the human rights of migrants, addendum: Communications sent to governments and replies.* Retrieved May 1, 2015 from http://www.unhcr.org/ refworld/docid/ 4c29bb2b2.html

United Nations. (1992). *The Rio declaration on environment and development.* Rio de Janeiro: UN.

Vaartjes, V. (2005). Integrating action learning practices into executive coaching to enhance business results. *International Journal of Evidence Based Coaching and Mentoring*, *3*(1), 1–17.

Vaismoradi, M., Turunen, H., & Bondas, T. (2013). Content analysis and thematic analysis: Implications for conducting a qualitative descriptive study. *Nursing & Health Sciences*, *15*(3), 398–405. doi:10.1111/nhs.12048 PMID:23480423

Van Beinum, H. (1999). On the design of the ACRES program. In D. J. Greenwood (Ed.), *Action Research: From practice to writing in an international action research development program* (pp. 3–24). Philadelphia, PA: John Benjamins Publishing Co. doi:10.1075/dowi.8.03bei

Van Biljon, J., & Lotriet, H. (2014, October). ICT for Development in Southern Africa: Guest editorial. *South African Computer Journal*, *54*, v–vi. doi:10.18489/sacj.v54i0.282

Van der Riet, M., Hough, A., & Killian, B. (2005). Mapping HIV/AIDS as a barrier to education: A reflection on the methodological and ethical challenges to child participation. *Journal of Education*, *35*, 75–98.

Van Der Vliet, V. (1996). *The politics of AIDS.* London: Bowerdean Publishing Company Ltd.

Van Driel, J. H., Verloop, N., & De Vos, W. (1998). Developing science teachers pedagogical content knowledge. *Journal of Research in Science Teaching*, *35*(6), 673–695. doi:10.1002/(SICI)1098-2736(199808)35:6<673::AID-TEA5>3.0.CO;2-J

Van Zyl, I., & Vannini, S. (2013). Participatory re-action: reflecting on a Design-Based Research approach in ICT4D. In J. Steyn, & A. G. Van der Vyver (Ed.), *Public and private access to ICTs in developing regions: Proceedings of the 7th International Development Informatics Conference* (pp. 206 - 223). Bangkok: Development Informatics Association.

Veal, W. R., & Makinster, J. G. (1999). Pedagogical content knowledge taxonomies. *Electronic Journal of Science Education, 3*(4). Available at: http://unr.edu/homepage/crowther/ejse/ejsev3n4.html

Vecina, M. L., Chacón, F., Sueiro, M., & Barrón, A. (2012). Volunteer engagement: Does engagement predict the degree of satisfaction among new volunteers and the commitment of those who have been active longer? *Applied Psychology*, *61*(1), 130–148. doi:10.1111/j.1464-0597.2011.00460.x

Venter, M. (2014). From lip-service to service delivery in Local Economic Development: Guidelines to set up an agency for action. *Journal of Economic and Financial Sciences*, *7*(3), 721–736.

Verma, S. K. (2016). Environmental awareness: A need of nature. *Journal of Global Resources*, *2*, 182–185.

Viswanathan, M., Ammerman, A., & Eng, E. (2004). *Community-based participatory research: Assessing the evidence.* Rockville, MD: Agency for Healthcare Research and Quality.

Vitousek, P. M., Mooney, H. A., Lubchenco, J., & Melillo, J. M. (1997). Human Domination of Earths Ecosystems. *Science, 277*(5325), 494–499. doi:10.1126/science.277.5325.494

Volz-Peacock, M., Carson, B., & Marquardt, M. (2016). Action Learning and Leadership Development. *Advances in Developing Human Resources, 18*(3), 318–333. doi:10.1177/1523422316645884

Von Hippel, E. (1986). Lead users: An important Source of Novel Product Concepts. *Management Science, 32*(7), 791–805. doi:10.1287/mnsc.32.7.791

Vygotsky, L. (1978). Interaction between learning and development. In *Mind and Society: The development of higher psychological processes* (pp. 79–91). Cambridge, MA: Harvard University Press.

Waddell, S. (2007). Realising Global Change: Developing the Tools; Building the Infrastructure. *Journal of Corporate Citizenship, 2007*(26), 69–84. doi:10.9774/GLEAF.4700.2007.su.00011

Waddell, S., McLachlan, M., Meszoely, G., & Waddock, S. (2015). Large scale change action research. In H. Bradbury (Ed.), *Action research handbook* (pp. 538–548). Sage Publications Ltd. doi:10.4135/9781473921290.n55

Waddill, D., & Marquardt, M. (2003). Action Learning and the Five Schools of Adult Learning. *Human Resource Development Review, 2*(4), 406–429.

Wakatsuki, T., & Rasyidin, A. (1992). Rates of weathering and soil formation. *Geoderma, 52*(3–4), 251–263. doi:10.1016/0016-7061(92)90040-E

Wallerstein, N., & Duran, B. (2003). The conceptual, historical, and practice roots of Community Based Participatory Research and related participatory traditions. In M. Minkler & N. Wallerstein (Eds.), *Community-Based Participatory Research for health* (pp. 27–52). San Francisco, CA: Jossey-Bass.

Walsham, G., & Sahay, S. (2006). Research on information systems in developing countries: Current landscape and future prospects. *Information Technology for Development, 12*(1), 7–24. doi:10.1002/itdj.20020

Warhurst, J., Crawford, K., Ireland, J., Neale, D., Pickering, J., Rathmell, C., Watson, G., & Ewing, R. (2010). Improving literacy pedagogy and outcomes through teaching imaginatively. *Practically Primary, 15*(3).

Warikoo, N. K., & de Novais, J. (2015). Colour-blindness and diversity: Race frames and their consequences for White undergraduates at elite US universities. *Ethnic and Racial Studies, 38*(6), 860–876. doi:10.1080/01419870.2014.964281

Warner, S. A., & Morford, L. L. (2004). The status of design in Technology teacher education in United States. *Journal of Technology Education, 15*(2), 33–45. doi:10.21061/jte.v15i2.a.3

Wasylyshyn, K. M. (2003). Executive coaching: An outcome study. *Consulting Psychology Journal: Practice and Research, 55*(2), 94–106. doi:10.1037/1061-4087.55.2.94

Waters-Adams, S. (2006). *Action research in education.* Faculty of Education, University of Plymouth.

Watkins, K.E., Marsick, V.J., & O'Neil, J. (2011). *Action Learning Toolkit.* Academic Press.

Watkins, S. C., & Swidler, A. (2007). *Conversations into texts: a method for studying public culture.* Paper originally presented as "hearsay ethnography" at the annual meeting of the Population Association of America, Philadelphia, PA.

Watson, H., & Chambers, D. W. (1989). *Singing the land, signing the land.* Geelong: Deakin University Press.

Weber, M. (1968). *Economy and society: An outline of interpretive sociology* (G. Roth & C. Wittich, Eds., E. Fischoff, Trans.). New York, NY: Bedminister Press. (Original work published 1921)

Weber, R. (2009). Research on ICT for development: Some reflections on rhetoric, rigor, reality, and relevance. In J. Steyn, P. Plantinga, B. Van der Vyver, & J. Meyer (Eds.), *Proceedings of the 3rd IDIA Conference* (pp. 2–27). Berg-en-Dal, Kruger National Park: School of Information Technology, Monash University (South Africa). Retrieved from http://arrow.monash.edu.au/hdl/1959.1/514447

Wegner, E. (2000). Communities of Practice and Social Learning Systems. *Organization*, *7*(2), 225–246. doi:10.1177/135050840072002

Wenger, E. (2007). *Communities of practices: A brief introduction.* Available online: http://www.ewenger.com/theory/

Wenger, E. (1998). *Communities of practices: Learning, Meaning, and Identity.* Cambridge University Press. doi:10.1017/CBO9780511803932

Wheatley, L., McInch, A., Fleming, S., & Lord, R. (2015). Feeding back to feed forward: Formative assessment as a platform for effective learning. *Kentucky Journal of Higher Education Policy and Practice, 3*(2).

White, D., & Maddox, J. (2012). Organization development in Guatemala: Rapid engagement with long term impact. *Organization Development Journal, 30*(4), 32–43.

Whitehead, J., & McNiff, J. (2006). *Action Research Living Theory.* London: Sage. doi:10.4135/9781849208536

Whitmore, J. (2003). *Coaching for Performance.* London: Nicholas Brealey Publishing.

Whitworth, B. A., & Chiu, J. L. (2015). Professional development and teacher change: The missing leadership link. *Journal of Science Teacher Education, 26*(2), 121–137. doi:10.1007/s10972-014-9411-2

Wikipedia. (2016). *Community Engagement.* Available online: www.wikipedia.org

Wilkie, K. J., & Clarke, D. (2015). Pathways to professional growth: Investigating upper primary school teachers' perspectives on learning to teach algebra. *Australian Journal of Teacher Education, 40*(4), 6.

Wilkinson, R. G., & Pickett, K. (2009). *The spirit level: Why more equal societies almost always do better* (Vol. 6). London: Allen Lane.

Williams, P. J. (2011). Engineering: Good for Technology Education?. In M. J. De Vries (Ed.), Positioning Technology Education in the curriculum (pp. 87-100). Rotterdam, The Netherlands: Sense.

Williams, P. J., & Gumbo, M. T. (2011). New Zealand Technology pedagogical content knowledge. In Pupils' Attitude Towards Technology (PATT) 25 and Centre for Research in Primary Technology (CRIPT) 8 on "Perspective on Learning in Design & Technology Education". Goldsmiths, University of London.

Williams, P. J., & Gumbo, M. T. (2011). New Zealand Technology pedagogical content knowledge. In Pupils' Attitude Towards Technology (PATT) 25 and Centre for Research in Primary Technology (CRIPT) 8. Goldsmiths, University of London.

Williams, D. A. (2013). *Strategic Diversity Leadership: Activating Change and Transformation in Higher Education.* Sterling, VA: Stylus Publishing, LLC.

Williams, J., Eames, C., Hume, A., & Lockley, J. (2012). Promoting pedagogical content knowledge development for early career secondary teachers in science and technology using content representations. *Research in Science & Technological Education, 30*(3), 327–343. doi:10.1080/02635143.2012.740005

Williamson, M. (2009). Connecting the other three billion. *Engineering & Technology, 4*(4), 70–73. doi:10.1049/et.2009.0417

Wilson, J. (2012). Volunteerism research: A review essay. *Nonprofit and Voluntary Sector Quarterly, 41*(2), 176–212. doi:10.1177/0899764011434558

Winch, A., & Ingram, H. (2002). Re-defining the focus of workplace learning. *International Journal of Contemporary Hospitality Management, 14*(7), 361–367. doi:10.1108/09596110210440666

Wischmeier, W. H., & Smith, D. D. (1978). Predicting rainfall erosion losses—a guide to conservation planning. USDA, ARS, Agricultural Hand-Book #537.

Wittgenstein, L. (1958). *The blue and brown books.* New York, NY: Harper and Row.

Wood, D., Bruner, J. S., & Ross, G. (1976). The Role of Tutoring in problem solving. *Journal of Child Psychology and Psychiatry, and Allied Disciplines, 17*(2), 89–100. doi:10.1111/j.1469-7610.1976.tb00381.x PMID:932126

Wood, G. D. A. (2012). What Is Sustainability Studies?. *American Literary History, 24*(1), 1–15. doi:10.1093/alh/ajr044

Wooding, B. (2008). Contesting Dominican discrimination and statelessness. *Peace Review, 20*(3), 366–375. doi:10.1080/10402650802330279

World Bank. (2007). *World Bank Participation source book.* Available from http://www.worldbank.org/wbi/sourcebook/sbintro.pdf

World Health Organization/UNAIDS. (2000). *AIDS epidemic update.* Geneva, Switzerland: World Health Organization.

Wu, Y. C. J., Shen, J. P., & Kuo, T. (2015). An overview of management education for sustainability in Asia. *International Journal of Sustainability in Higher Education, 16*(3), 341–353. doi:10.1108/IJSHE-10-2013-0136

Yaduvanshi, N. R., Myana, R., & Krishnamurthy, S. (2016). Circular Economy for Sustainable Development in India. *Indian Journal of Science and Technology, 9*(46). doi:10.17485/ijst/2016/v9i46/107325

Yager, R. E. (2013). Successes and continuing challenges: Implementing the changes in professional development for teachers in the U.S. *International Journal of Education in Mathematics, Science and Technology, 1*(3), 184–189.

Yeravdekar, V. R., & Tiwari, G. (2012). The higher education system in India and its impact on the economy. doi:10.2139/ssrn.2139894

Yeravdekar, V. R., & Tiwari, G. (2014). Chinas Lead in Higher Education: Much to Learn for India. *Procedia: Social and Behavioral Sciences, 157*, 369–372. doi:10.1016/j.sbspro.2014.11.041

Yiu Yuen, P., & Cheong Cheng, Y. (2000). Leadership for teachers action learning. *International Journal of Educational Management, 14*(5), 198–209. doi:10.1108/09513540010373144

Yoders, S. (2014). Constructivism Theory and Use from 21st Century Perspective. *Journal of Applied Learning Technology, 4*(3).

Yorks, L., Lamm, S., & ONeil, J. (1999). Transfer of Learning from Action Learning Programs to the Organizational Setting. *Advances in Developing Human Resources, 1*(2), 56–74. doi:10.1177/152342239900100205

Zuber-Skeritt, O. (2002). The concept of action learning. *The Learning Organization, 9*(3), 114–124. doi:10.1108/09696470210428831

Zuber-Skerrit, O. (2015). Participatory Action Learning and Action Research (PALAR) for Community Engagement: A Theoretical Framework. Educational Research for Social Change, 4(1), 5-25.

Zuber-Skerritt, O. (1996). Emancipatory action research for organization change and management development. In O. Zuber-Skerritt (Ed.), New Directions in Action Research (pp. 83-105). London: Falmer.

Zuber-Skerritt, O. (1993). Improving Learning and Teaching Through Action Learning and Action Research. *Higher Education Research & Development, 12*(1), 45–58. doi:10.1080/0729436930120105

Zuber-Skerritt, O. (2001). *Action learning, action research and process management: theory, practice, praxis*. Brisbane: Action Research Unit, Faculty of Education, Griffith University.

Zuber-Skerritt, O. et al. (2009). *Action Learning and Action Research Songlines through Interviews*. Rotterdam, The Netherlands: Sense Publishers.

Zuber-Skerritt, O. (2015). *Action Research*. Palgrave Macmillan.

Zuber-Skerritt, O. (Ed.). (2012). *Action research for sustainable development in a turbulent world*. Bingley, UK: Emerald Group Publishing.

Zuber-Skerritt, O. D. (2009). *Action learning and action research: Songlines through interviews*. Sense Publishers.

About the Contributors

Tomé Mapotse was a Technical Education teacher; Departmental Head of Mathematics, Science and Technology Education; School Principal for Leaners with Special Needs; Technology Education (TE): teacher; lecturer and senior lecturer. Currently, he is an Associate Professor at University of South Africa (Unisa) where he has completed my doctoral study in TE within Action Research (AR) design. He is passionate about Community Engagement with TE teachers and Open Distance Learning online pedagogy.

***

Noa Avriel-Avni is a researcher in the field of environmental education and ecology, and leading the process of establishing the LTSER Negev Highlands platform. Avriel-Avni Specializes in the research Participatory Action Research.

Neeta Baporikar is currently Director/Professor (Business Management) at Harold Pupkewitz Graduate School of Business (HP-GSB), Namibia University of Science and Technology, Namibia. Prior to this she was Head-Scientific Research, with Ministry of Higher Education CAS-Salalah, Sultanate of Oman, Professor (Strategic Management and Entrepreneurship) at IIIT Pune and BITS India. With more than a decade of experience in industry, consultancy and training, she made a lateral switch to research and academics in 1995. Dr. Baporikar holds D.Sc. (Management Studies) USA, PhD in Management, University of Pune INDIA with MBA (Distinction) and Law (Hons.) degrees. Apart from this, she is also an External Reviewer, Oman Academic Accreditation Authority, Accredited Management Teacher, Qualified Trainer, Doctoral Guide and Board Member of Academics and Advisory Committee in accredited B-Schools. Reviewer for international journals, she has to her credit 5 conferred doctorates, several refereed research papers, and authored books in the area of Entrepreneurship, Strategy, Management and Higher Education.

Angela Carter received her BA in English from Georgia State University, then embarked on a 16 year career as a manager in the shopping center industry. She received her MEd in Workforce Development from the University of Arkansas in 2012, and an EdD in Adult Education in 2015 from University of Georgia to culminate a career change. In 2015 she was hired as a limited term Assistant Clinical Professor by the Learning, Leadership, and Organization Development program at University of Georgia. She writes and presents widely on issues concerning human resource and organization development, leadership, coaching, and action learning.

Elisabeth Fruijtier has received her MSc. in Global Health from Trinity College Dublin in 2013. In 2014 she started her PhD fellowship at the University of Oslo at the Department of Informatics within the Global Infrastructure group. The focus of her research is on the collaborative development of an Open Source software solution called DHIS2 (District Health Information System 2) as part of the Health Information Systems Programme (HISP). HISP aims at strengthening health information systems in low resource settings.

Georgina Gomba is a Lecturer in the Department of Adult Education and Youth Development, UNISA. Areas of specialization include Adult Education, Performance management and staff development.

Leila Goosen was appointed as full professor in the Department of Science and Technology Education in the School of Teacher Education, College for Education, at the University of South Africa (UNISA) from December 2015. Prof. Goosen currently supervises ten Masters and Doctoral students, and has successfully completed supervision of 62 students at postgraduate level. She was also a grant holder under the Women in Research Support Programme (WiR-SP) of UNISA as the leader of a project entitled e-Schools' Community Engagement. Her research interests also include Massive Open Online Courses and Information and Communication Technology (ICT) for Development (ICT4D). Previously, she was an Associate Professor in the School of Computing of the College for Science, Engineering and Technology (CSET), also at UNISA. There, Prof. Goosen was the module leader and designer of the CSET Signature module, which has had in excess of 90 000 registered students over the six semesters since its first roll-out in 2013, with an average of more than 15 000 students per semester. Before coming to UNISA, she was a Deputy Director at the South African (SA) National Department of Basic Education. In this capacity, she was required to develop ICT in Education strategies for implementation, and guidelines for the coordination and support to implement ICT in Education in SA provinces. She also promoted, coordinated, managed, monitored and evaluated ICT in Education policies and strategies, and drove the research agenda of ICT in Education.

Elli Groner is a desert ecologist. He has been working on long term ecosystem processes as part of the international Long Term Ecological Research (LTER) network. He works on biodiversity of insects, especially beetles, but also on climate change and Acacia tree populations. He worked on desert ecosystem services and has recently extended that work to the notion of Ecological Integrity and is now developing indices for desert ecological integrity.

Mishack Gumbo received his PhD in Technology Education in from Vista University in 2003. He worked as a Lecturer in the Faculty of Education at Vista University in 1996 to 2004 and University of Pretoria in 2005 to 2008. He is currently a Full Professor in the Department of Science and Technology Education in the College of Education, University of South Africa. His research interest areas include Indigenous Knowledge Systems, Technology Teachers' Pedagogical Content Knowledge and Distance Education and E-Learning. He writes and presents papers in these areas and heads research and community engagement projects.

Jennifer Holzer is a PhD Student.

Saravan Krishnamurthy is a graduate in Industrial Management. He has taught Business Research, Strategic Management, Operations Management, Globalization and Consumer Behaviour. He has published papers in peer reviewed journals, and has keen interests in research on Social responsibility, Sustainability and Renewable Energy. He is currently pursuing PhD at Symbiosis International University, Pune.

Machaba MM is a senior lecturer in the Department of Early Childhood Education at University of South Africa. She joined UNISA after being a teacher for several years. She has many years of experience in open distance education. Her responsibilities include teacher training (tuition), postgraduate supervision, research, community engagement and international collaborative research projects. Her area of interest is Foundation Phase Mathematics, Inclusive Education, Teaching Practice and Qualitative research. She has presented papers locally and abroad.

Meahabo Magano is a full professor in the Department of Psychology of Education at University of South Africa (Unisa) specialising in Psychology of Education, with a passion for teacher education. Once a teacher, head of department (HoD) and a deputy principal.

Tsebo Mashiloane is an Acton Research practitioner in Environmental Education.

Elias Mathipa holds his PhD from University of Pretoria. Once a teacher, HoD and a school Principal. He also has an experience as the former colleges of education Rector. He is a full Professor within Unisa in the College of Education under the Department of Language Education, Art & Culture.

Sizakele Matlabe is an Action Research practitioner belonging to Educational Foundations department.

Tracey McKay is a senior lecturer at Unisa who is currently registered for a PhD under Prof Chris Rogerson at the University of Johannesburg. Her research interests are school access, higher education and adventure tourism. She has won 4 teaching awards, and has multiple peer reviewed publications.

Bronwyn Mehorter is a currently studying a PhD at the University of Wollongong. The focus of her PhD is on the effective professional development and learning of primary teachers within Australian primary schools. Her areas of interest are in: Action Learning as a form of effective professional development for teachers and professional learning and professional development for school development and improvement. She stems from a background in primary education and is a qualified and accredited primary teacher.

Milton Milaras is a writer and geographer with research interests in mine closure, soil science, pedagogy and action research.

Soane Mohapi is a Senior lecturer in the Department of Early Childhood and manager of the Centre for Continuing Education and Training at Unisa's College of Education. She is a graduate of the University of Pretoria with a DPhil in education management and policy studies.

Monkie Muriel Moseki is a Lecturer in Psychology of Education, and registered Educational Psychologist.

Daniel Orenstein is a socio-ecologist whose current research focus considers the way people perceive and use their natural environment.

Karie Peralta is Assistant Professor of Sociology in the Department of Sociology and Anthropology at The University of Toledo in Toledo, Ohio, United States. Her areas of interest are social and community development, community-based research, race and ethnic relations, and sociology of education.

Vishal Pradhan is a graduate in Statistics and an Accredited Management Teacher (AMT), AIMA, New Delhi. He has taught Research Methodology and Business Statistics. His research interests include Psychometric Analysis and Time Series Modelling, SAS programming and SAS Analytics, Business Forecasting and Demand Planning. He had taught in India program of Carleton University (Sprott School of Business). He is currently pursuing PhD at Symbiosis International University, Pune.

Moshe Shachak is an Ecologist, head of the LTER network in Israel.

Lloyd Tlale is an associate professor in the Department of Psychology of Education at UNISA. His responsibilities include teacher training, postgraduate supervision, research, community engagement and international collaborative research projects. He completed his DEd (Inclusive Education) with UNISA. His areas of interests are Gifted Child Education, Children in conflict with the law, Orphans and Vulnerable Children, Social justice and processes of qualitative research. He has presented conference papers on these subjects locally and abroad and published articles on topics such as Children in conflict with the law in an ecosystemic context, the role of the facilitator in focus group research, and reflective research practice.

Marius Venter is the Director of the Centre for Local Economic Development (CENLED) at the University of Johannesburg. Previously I was the Chief Executive Officer of the Overstrand Local Economic Development Agency. I am also the Director for the PASCAL Observatory for the Africa office. My interests are in professionalising the careers of persons working in the economic development sphere, entrepreneurship and learning cities.

Index

T

U

V

W

www.ingramcontent.com/pod-product-compliance
Lightning Source LLC
LaVergne TN
LVHW062135300926
849161LV00010B/2538